3 Degrees More

Klaus Wiegandt

Editor

3 Degrees More

The Impending Hot Season and How Nature
Can Help Us Prevent It

 Springer

Editor
Klaus Wiegandt
Foundation Forum für Verantwortung
Seeheim-Jugenheim, Hessen, Germany

ISBN 978-3-031-58143-4 ISBN 978-3-031-58144-1 (eBook)
https://doi.org/10.1007/978-3-031-58144-1

Translation from the German language edition: "3 Grad Mehr: Ein Blick in die drohende Heißzeit und wie uns die Natur helfen kann, sie zu verhindern" by Klaus Wiegandt, © Oekom Verlag 2022. Published by Oekom Verlag. All Rights Reserved.

This Springer imprint is published by the registered company Springer Nature Switzerland AG
The registered company address is: Gewerbestrasse 11, 6330 Cham, Switzerland

Paper in this product is recyclable.

*To our grandchildren Livia, Tim,
and Theo—
Representing four billion young people*

Preface to the English Edition

In the global public climate debate of the past 20 years, the effects of an increase in the average global surface temperature by 1.5 °C have been described and discussed at length: glaciers around the world will melt, parts of Bangladesh will be permanently flooded, various islands in the Pacific will disappear, the Arctic will become ice-free in summer, and Greenland will lose its ice cover within 150–200 years. These effects are catastrophic for millions of people, yet not a serious threat to humanity.

Such well-known scenarios are, however, far too innocuous because we are now heading for 3 °C of global warming by the end of the century. Such a "three-degree world" involves not a mere doubling of the adverse effects of a "1.5-degree world." The effects of global warming are not linear but worsen dramatically as temperatures rise. A three-degree world would have catastrophic impacts on humans and other living beings all around the world.

Because this fact has found virtually no entry into media reporting, public debates, or even expert planning and policy making, I brought together, in mid-2021, a group of scientists to produce a climate book focused on the looming three-degree scenario.

This examination of Earth's impending hot era was meant to show what the second half of this century may have in store for 10 billion people, including some 2.6 billion of our children and grandchildren now under 20 years of age. To this end, the authors outlined different earth and social systems in a world that is three degrees hotter. But they also made clear that we still have it in our power to prevent such a scenario, to limit global warming to 2 °C at least, if we substantially strengthen our efforts initiated with the 2015 Paris Agreement.

Shortly after our book *3 Grad mehr* appeared in August 2022, a news item from the world of climate change science received much attention: prominent climatologists around lead author Luke Kemp of Cambridge University had published an article in the scientific journal *PNAS*, entitled "Climate Endgame." In this paper, they accused the Intergovernmental Panel on Climate Change (IPCC) of having paid scant attention for years to the devastating effects of global warming by three degrees or more: "Facing a future of accelerating climate change while blind to worst-case scenarios is naive risk management at best and fatally foolish at worst." The authors therefore called on the IPCC to study the "three-degree scenario" and to publish a special report on it.

This dramatic call was well aligned with our book, in which scientists conclude that we must never allow a "three-degree scenario" because of its devastating effects. Only if the public becomes aware of what such a world would really look like can democratic processes be mobilized to limit global warming at least to 2 °C. But this goal can be achieved only if we act NOW!

Seeheim-Jugenheim, Hessen, Germany   Klaus Wiegandt

Acknowledgement

For the English edition, I would like to thank the publisher's coordinators for this book, Aaaron Schiller, Ragavendar Mohan and Aravajy Meenahkumary, for their helpful and easy cooperation. Special thanks go to Professor Dr Thomas Pogge, who was willing to proofread the AI-assisted translation and to ensure that its language and content matched the original. Many thanks also to the staff of my foundation, above all Anne Marschner and Petra Lauermann, who helped overcome the challenges relating to Open Access regulations.

Cover: Midjourney, Werner Marschall
Infographics: Esther Gonstalla, www.gonstalla.com
Translation: AI-assisted (DeepL)
Proofread by: Thomas Pogge

Contents

About the Authors

Jutta Allmendinger, PhD, is president of the Science Center Berlin for Social Research (WZB) and professor at the Humboldt University as well as honorary professor at the Free University of Berlin. Previously, she was a professor at the Ludwig-Maximilians-University Munich and Director of the Institute for Labor Market and Occupational Research of the Federal Employment Agency in Nuremberg. In her research and as author (most recently published by Ullstein: *Es geht nur gemeinsam!*), she deals, among other things, with gender and educational equity.

Dina Ionesco is co-director of the new master's program in Migration, Climate Change and Environment (MAMCE) at Webster University in Geneva and works as manager at the Secretariat of the United Nations Framework Convention on Climate Change (UNFCCC). Previously, she headed the Migration Division, Environment and Climate Change at the United Nations Agency for Migration (IOM).

Hans Joosten is a biologist and professor emeritus of peatland science and paleo-ecology at the University of Greifswald. He is co-founder of the Greifswald Moor Centre, which acts as an expert forum, think tank, and interface between science, politics, and Practice understands. For his peatland research and his commitment to highlighting the importance of the peatlands to a broad public, he was awarded the renowned German Environmental Award.

Bernhard Kegel studied biology and chemistry in Berlin and since 1996 has been a full-time active as an author. As one of the few German writers, Kegel is known both for his novels as well as through his popular non-fiction books. His most recent publications are *The Nature of the Future and Extinct Animals* at Dumont.

Stefan Klotz is head of the thematic area "Ecosystems of the Future" and the Department of Biocenosis Research of the Helmholtz Centre for Environmental Research—UFZ in Halle (Saale). In addition to numerous international publications, the plant ecologist is lead author for the biodiversity section of the German climate report.

Friderike Kuik studied physics in Marburg and Potsdam and received her PhD in Meteorology from the University of Berlin. Meanwhile, she conducted research in Potsdam at the Institute for Advanced Sustainability Studies (IASS). Since 2017, she has been a staff member of the European Central Bank and currently serving on the Board of Directors for Economics.[1]

Reinhard Mosandl was full professor of the Department of Silviculture from 1997 to 2018 at the Technical University of Munich. His research interests are primarily in the management of forest ecosystems in temperate latitudes and in the tropics. His main focus has been on the field of Restoration Ecology with projects in Ecuador, Ethiopia and China, among others. His special concern is the connection of forest science and forest practice.

Edgar Peiter is responsible at Martin Luther University Halle-Wittenberg for the Crop Sciences program and chairman of the German Society for Plant Nutrition. His current research focuses on plant mechanisms of stress responses as well as special aspects of plant nutrition.

Stefan Rahmstorf is one of the most renowned climate researchers worldwide. He was one of the lead authors of the Fourth Assessment Report of the Intergovernmental Panel on Climate Change (IPCC) and is regarded as one of the leading oceanographers. Rahmstorf is the author or co-author of some 135 scientific publications in international journals, including *Nature* and *Science*. He is a sought-after and multi-award-winning climate communicator and is one of the top five climate scientists with the most Twitter followers worldwide.

Hans Joachim Schellnhuber is director emeritus of the Potsdam Institute for Climate Impact Research (PIK), which he founded in 1992. He is a visiting professor at the Tsinghua-University (China) and elected member of numerous scholars' associations such as the Pontifical Academy of Sciences. Since 2019, Schellnhuber has been working intensively on the transformation of the built environment and the potential of wood buildings as carbon sinks. He is the founder of Bauhaus Erde and a member of the New European Bauhaus High-Level Roundtable of the European Commission.

Hans Peter Schmidt is managing director of the Ithaka Institute, an international network for carbon strategies and climate farming. The institute is developing methods to make carbon permanently in materials and agricultural soils. Schmidt

---

[1] The views expressed herein are those of the author and do not necessarily correspond to those of the European Central Bank.

conducts research in the areas of carbon and plant nutrient cycles, ecology and environmental engineering.

Wolfgang Schroeder is head of the department "Political System of the Federal Republic of Germany – Statehood in Transition" at the University of Kassel. From 2009 to 2014 he was state secretary in the Ministry of Labor, Social Affairs, Women and Family of the State of Brandenburg. Since 2016, he was appointed fellow at the Social Science Research Center Berlin (WZB), and in 2022 he became a member of the Council of the Working Environment of the Federal Ministry of Labor and Social Affairs (BMAS).

Stefan Schwarzer is a physical geographer and permaculture designer. Over 20 years he worked for the United Nations Environment Programme (UNEP) in Geneva. Schwarzer is organizer of the symposium and the webinar series "Constructive Agriculture" as well as the network "Climate Landscapes."

Ralf Seppelt is a mathematician and holds a PhD in Geoecology. He teaches at the Martin-Luther University Halle-Wittenberg and is department head at the Helmholtz Center for Environmental Research in Leipzig. His field of research is landscape ecology and resource economics. In addition to numerous scientific publications, he is co-author of the Global Report of the Intergovernmental Platform on Biodiversity and Ecosystem Services (IPBES).

Mariam Traore Chazalnoël is a senior policy officer with specific expertise on migration, environment and climate change at the United Nations Agency for Migration (IOM). Since 2013 she concentrates on global policy issues related to climate change and migration and works to integrate these issues into the global agendas of climate change and making migration policy visible.

Martin Volk is geographer and geoecologist at the University of Halle-Wittenberg and deputy department head and working group leader at the Helmholtz Center for Environmental Research (UFZ) in Leipzig. His field of research is the assessment of the influence of land use on the landscape water and material balance as well as biodiversity.

Leonie Wenz is a working group leader and deputy head of department at the Potsdam Institute for Climate Impact Research. Her research findings on the socio-economic impacts of climate change have been published in *Nature*, *Science Advances* and *PNAS*, among others. As mathematician with a doctorate in climate physics, she completed a postdoctoral fellowship at the UC Berkeley in environmental economics and is a member of the Young Academy of the Leopoldina.

Klaus Wiegandt is founder and board member of Forum for Responsibility. His foundation promotes education, science and research, especially in the field of sustainable development. Klaus Wiegandt was CEO of an international trading company.

Susanne Winter studied forestry in Munich (TU) and Göttingen. For her doctoral thesis, habilitation and as head of the professorship for regional culture and nature conservation, she worked alternately in Dresden (TU) and Munich. For many years, her research interest was focused on the question of how forest management can be improved, in order to enhance the biodiversity and climate protection performance of our forests while simultaneously improve wood utilization. She has been applying this knowledge at WWF Germany since 2016, where she heads the Forest Program as Forest Policy Director and is internationally involved in the forest conservation and the restoration of forest landscapes committed.

Part I
Hot Season Ahead: What a 3-Degree Warmer World Looks Like

Climate and Weather at 3 Degrees More

Earth as We Don't (Want to) Know It

Stefan Rahmstorf

What does 3 °C of global warming mean for us? So far, according to the Intergovernmental Panel on Climate Change (IPCC, 2021), we have reached 1.1 degrees of warming, relative to the late nineteenth century (which is generally used as the baseline in this article because it is also the baseline for the Paris target of 1.5 degrees). We are already seeing many negative effects. Three degrees of warming would be nearly 3 times more. However, the effects would be considerably worse than just 3 times the current impacts, as we will see in this chapter.

A useful perspective on a warming of 3 degrees is provided by the Earth's history. According to current knowledge, one must go back about three million years, to the Pliocene, to find a similarly high global temperature. This already indicates that large parts of today's biosphere are not adapted by evolution to such a warm Earth. Many species would not survive it. In the Pliocene, our ancestors, the australopitheci, still lived partly on trees.

The course of global temperature over the last 20,000 years since the peak of the last ice age can now be reconstructed quite accurately thanks to numerous sediment and ice cores (Fig. 1). The graph shows three important things:

1. Today's temperatures already exceed the range prevailing in the Holocene and thus the entire history of human civilisation since humans developed agriculture and became sedentary.
2. Modern global warming is about ten times faster than the natural warming from the Ice Age into the Holocene, making adaptation massively more difficult.

S. Rahmstorf (✉)
Potsdam Institute for Climate Impact Research, Potsdam, Germany
e-mail: rahmstorf@pik-potsdam.de

© The Author(s) 2024
K. Wiegandt (ed.), *3 Degrees More*,
https://doi.org/10.1007/978-3-031-58144-1_1

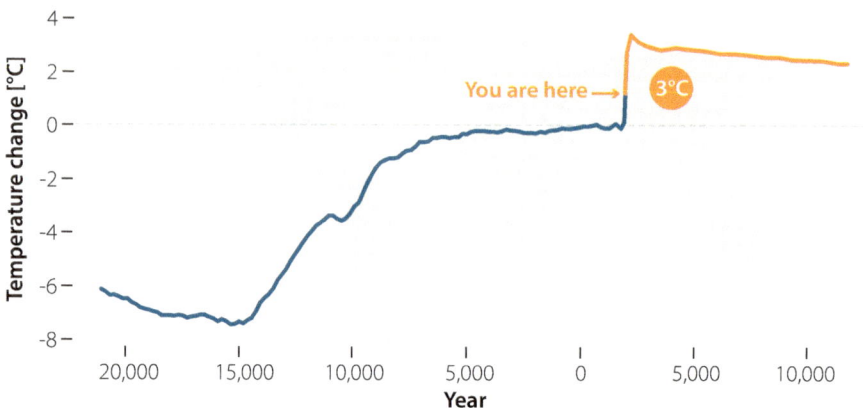

Fig. 1 Evolution of the global temperature since the last ice age (about 20,000 years before our era) and for the next 10,000 years under a scenario with 3 degrees of global warming. (Data from Dessler, 2021)

3. Modern warming will last for tens of thousands of years—unless gigantic amounts of carbon dioxide can be actively removed from the atmosphere.

Model simulations at the Potsdam Institute for Climate Impact Research (PIK), which correctly reproduce the ice age cycles of the last three million years (driven by the known Milankovitch cycles of the Earth's orbit), show that we have probably already added enough CO_2 to the atmosphere to prevent the next ice age, which would otherwise be due in 50,000 years. If we heat up the Earth by as much as 3 °C, the natural ice age cycles of the next half million years will probably not occur. A few generations of humans are changing our planet Earth massively and for long geological periods.

The US has already reached around 1.7 °C of warming (Fig. 2). Because the US is a land area, this is not surprising, since many land areas are warming about twice as fast as the global mean, 70% of which draws on ocean temperatures. The average warming of all land areas in 2020 was 2.0 °C. With 3 degrees of global warming, we can therefore expect around 6 degrees of warming on land.

A 6-degree rise in the annual average temperature—that's a lot. This would make New York roughly as warm as Los Angeles is today. And while some might dream of California weather conditions, this completely new climate will not please farmers or the local flora and fauna at all. It is likely to lead to widespread drought problems, wild fires and forest dieback.

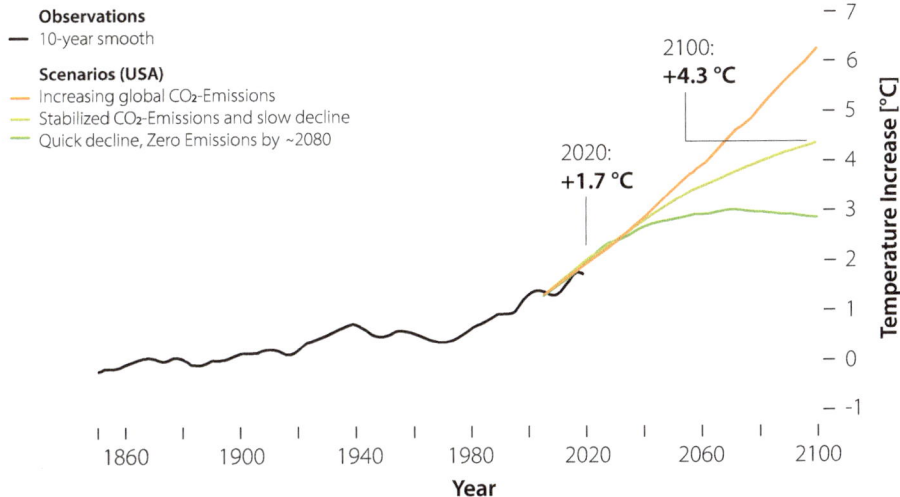

Fig. 2 Temperature trend in the US according to data from the Berkeley Earth Surface Temperature Project. The scenario with 3 degrees of global warming lies between the light green and orange colored future scenarios. (Data from Berkeley Earth)

Extreme Heat

Even more important than the average temperatures are the extremes. Where New Yorkers used to groan under a heat wave of 40 °C, it will be an unbearable 46 degrees. Or even more with drying out of soils, which can intensify the heat.

in Europe, the summer of 2003 was called "summer of the century" and claimed around 70,000 heat-related deaths (Robine et al., 2008, pp. 171–175). The peak in excess mortality in France (where the center of the heat was) was significantly higher than the spikes during the Covid 19 pandemic. The city of Paris had to set up refrigerated tents for the many dead in August 2003 because the morgues were overflowing. For summer 2022 the heat-related mortality in Europe has been estimated as exceeding 61,000.

The human body's cooling system works by sweating, i.e., by evaporating water on skin surface. How well this works depends on temperature and humidity: the more humid the air, the lower its ability to absorb additional water vapor. The relevant measure of heat stress is the theoretical limit of cooling: the lowest temperature that can be reached by direct evaporative cooling. It is also called wet bulb temperature because it can be measured with a ventilated thermometer wrapped in wet cloth.

According to new research, the human body's stress limit is at a sustained wet bulb temperature of 35 °C, but even temperatures below 30 degrees can be dangerous, because we must keep our body temperature at around 37 degrees and must be able to dissipate the heat generated by our metabolism and physical exertion. In the heat wave of 2003, wet bulb temperatures of 28 °C occurred in Europe.

At a humidity of 70%, the wet bulb temperature of 35 degrees, which is lethal after a few hours even for healthy people, is reached at an air temperature of 40 °C. Today, this wet bulb temperature is exceeded, anywhere on Earth, only rarely and for brief periods, mainly in the Persian Gulf or on the Mexican coast. According to a recent study (Raymond et al., 2020), the frequency of dangerous values has already more than doubled since 1979 and, in the Persian Gulf, monthly values of sea water temperatures exceeded the 35-degree limit for the first time in 2017—the moisture-saturated breeze from the sea can be deadly at such temperatures. In Qatar, since May 2021, workers are no longer allowed to work outdoors between 10 am and 3.30 pm during summer.

With a global warming of 3 degrees—which, as I said, is equivalent to 6 or more degrees Celsius on many land areas—the areas that are deadly hot during heat waves will expand massively, making it increasingly dangerous to stay outdoors and thus impairing work in agriculture or on building sites, for example.

Extreme Precipitation and Droughts

Temperatures still behave approximately linearly—that is, they increase in proportion to our cumulative emissions of carbon dioxide. Unfortunately, this is not true for many effects of warming. Many physical effects increase more than proportionally. This is true, for example, of the atmosphere's ability to hold water vapor, which increases exponentially with temperature. This is captured in the Clausius-Clapeyron equation, a basic law of physics describing the saturation vapor pressure of water vapor that has been known since the nineteenth century.

The same increase also applies to the vapor deficit of the atmosphere. This vapor deficit is the amount of water vapor that the air can still absorb at a given relative humidity. This is relevant because as the earth heats, the relative humidity remains approximately constant, and the vapor deficit therefore increases exponentially. It is this vapor deficit of the air that causes soils and vegetation to dry out on hot days, withering crops and increasing the risk of forest fires.

The data show that extreme precipitation has already significantly increased—as climate models have predicted for three decades. This applies to the Earth overall and has now been shown also for many regions (IPCC, 2021). Because of the stronger natural fluctuations on a regional scale and the smaller number of cases of extremes, the signal becomes statistically detectable later for smaller regions. The current report of the Intergovernmental Panel on Climate Change also includes Central and Northern Europe among the regions where an increase can already be detected. In 2020, a study by ETH Zurich also showed a statistically significant increase in extreme rainfall events is observed in these countries (Zeder & Fischer, 2020).

Overall, precipitation increases worldwide with warming because the evaporation rate from the oceans increases by about 2–3% per degree.

However, almost the entire increase falls from the sky in heavy rainfall events, for which the amount of water vapor in saturated air masses is important, which, according to the Clausius-Clapeyron equation mentioned above, increases by 7% per degree Celsius of warming—i.e., faster than the water supply through evaporation. As a result, heavy rainfall increases, days with little precipitation become rarer, and periods without precipitation become longer. Overall, therefore, both heavy rainfall events and periods of drought increase.

The destruction that extreme precipitation can cause has been demonstrated by a devastating series of flooding events in 2023. In September 2023 alone, massive floods occurred in Libya, Greece, Spain and in Hong Kong and New York City (Ombadi, 2023).

There are also large regional and seasonal differences in precipitation. Certain regions such as the Mediterranean, the Midwest of the USA, South Africa and Australia are increasingly drying out. What matters for agriculture and natural ecosystems, is agricultural drought, i.e., loss of soil moisture and drying out of vegetation. Drought understood in this way increases even if precipitation remains unchanged, because in a warmer climate water loss through evaporation increases. The current IPCC report also reports an increase in drought caused by anthropogenic warming for the majority of the world's land areas.

In addition to the simple fact that warm air can absorb more water vapor, there are also changes in atmospheric dynamics. Current research indicates that the persistence, i.e., duration, of certain weather conditions has increased in large parts of Europe in recent decades (Hoffmann et al., 2021). Thus, a few hot days turn into a health-threatening heat wave, or a dry phase into a prolonged drought. This increasing persistence is attributed to a slowing of the general westerly wind circulation, including the jet stream in summer, which is probably related to the strong warming of the Arctic land areas (Coumou et al., 2015). According to a recent study (Voossen, 2021), the Arctic has actually warmed four times as much as the rest of the globe over the last 40 years which flattens the temperature gradient from the tropics to the Arctic which drives the mid-latitude westerly winds. In addition, there are occasional large and persistent variations in the jet stream which reach around the entire northern hemisphere, causing simultaneous extremes there (Kornhuber et al., 2017). A nightmare scenario for some climate researchers is a drought with crop failures simultaneously hitting the large granaries of the northern hemisphere: Western North America, Russia, Western Europe, and Ukraine (Kornhuber et al., 2019).

Already in the drought and fire disaster in the summer of 2010, Russia stopped exporting grain because of crop failures, which led to massive price increases for North African buyers and thereby contributed to the "Arab Spring", which was partly ignited by high bread prices. Similarly, the revolt in Syria, which began in March 2011, followed the worst drought there in more than a century of weather records (Kelley et al., 2015). Conflict-ridden, weak states can be destabilised by extreme events and crop failures, with implications for global politics.

Tropical Cyclones

Tropical cyclones are a significant hazard in the tropical and subtropical regions of the world. For example, in September 2017, Category 5 Hurricane Maria destroyed large parts of the island of Puerto Rico and caused the loss of more than 3000 lives.

Global warming invests tropical cyclones with additional energy—because these storms draw their destructive power from the heat energy stored in the upper ocean. This is why they only form in regions with water temperatures above 26.5 °C; in more temperate latitudes, the seawater has heretofore been too cold. Climate researchers have therefore been predicting for decades that tropical storms would become stronger. For a long time, however, an increase could not be demonstrated from the observational data. Not because the data did not show an increase (they did), but because it was unclear how reliably the older data captured the strength of tropical storms and whether some tropical storms far from land areas might have been missed before the satellite era.

More recently, a real climatic increase in tropical storm intensity can be seen from the data (Kossin et al., 2020). The current IPCC report states for the first time that the proportion of especially strong tropical storms (categories 3–5) has increased, for which anthropogenic climate change is the main cause. Anyone familiar with earlier IPCC reports' extremely cautious and restrained statements on this question will understand the significance of this conclusion. In addition, there is evidence that tropical storms can intensify more rapidly, travel more slowly (affecting areas below the storm for longer) and move to higher latitudes—in Europe, for example, approaching the coast of Portugal.

It has long been undisputed that extreme precipitation, which is often the main reason for the devastation caused by tropical storms, has increased due to warming, which again is due to the Clausius-Clapeyron equation and, more particularly, to the increase in evaporation of warmer seawater under the storm.

Hurricane Harvey hit Houston in August 2017 and became the costliest tropical storm in US history ($125 billion in damage), on a par with Hurricane Katrina in New Orleans in 2005. Harvey brought the heaviest rainfall ever recorded in the USA: 1539 mm of rainfall in 4 days at its peak. For comparison: the 3-day precipitation total in the German Ahr valley was 115 mm during the flood disaster in July 2021.

It is also indisputable that rising sea levels due to the warming of Earth aggravate storm surges caused by tropical or other storms. It is often the last additional decimeters that cause the greatest damage, when water reaches areas where no one had previously expected a storm surge. Like Hurricane Sandy in 2012, whose storm surge flooded tunnels of the New York subway. Or Typhoon Haiyan, whose storm surge in 2013 leveled the city of Tacloban in the Philippines and claimed over 6300 lives.

Sea Level and Ice Sheets

The perspective of Earth's history also helps when it comes to sea level. In the Pliocene, three million years ago, sea levels were between 5 and 25 m higher than today because there was much less ice on the continents. Conversely, at the peak of the last ice age 20,000 years ago, sea levels were 120 m lower than today. The current continental ice masses, especially in Antarctica and Greenland, are so large that they can provide enough water for 65 m of global sea level rise.

Our Australopitheci ancestors in the Pliocene are unlikely to have been bothered by the higher sea level. But our planet's current coastlines are home to more than 130 cities larger than a million inhabitants, plus other infrastructure such as ports, airports, and some 200 nuclear power plants with seawater cooling (such as Sizewell B on the British North Sea coast). Even 1 m of sea rise would be a disaster. So far, the rise since the late nineteenth century has been around 20 cm, which is already causing problems on some coasts—and this not merely during storm surges. Even normal tidal cycles occasionally cause streets in cities on the U.S. East Coast to be inundated in what is called "nuisance flooding"—not a disaster, but a growing nuisance.

In the case of sea level rise, it appears thus far that the speed of the rise is increasing roughly in proportion to the temperature rise. This means that, with 3 degrees of warming, sea levels should rise roughly three times faster than today. This is partly because the warmer it gets, the faster the continental ice masses melt. The rise in sea level is already accelerating—this is visible not merely in the long data series of harbor tide gauges, but meanwhile even in the satellite measurements that have been available only since 1993 (Fig. 3).

However, there are also more complex effects here that add to this simple logic. Ice does not merely melt on the surface, it can also slide into the sea, or more precisely: it flows like a viscous slow river. If meltwater gets under the ice, ground friction is reduced and the ice flows faster, thus accelerating sea level rise. In Antarctica, moreover, floating ice shelves located in front of the outlet glaciers are gradually disappearing because they are melted from below by warmer sea water. These ice shelves also impede the flow of continental ice, which will therefore accelerate once these shelves disappear.

And it gets even more complicated: continental ice sheets have tipping points. A tipping point is a point at which further development into a fundamentally different state becomes an unstoppable self-sustained process, driven by reinforcing feedback effects. Greenland's ice sheet has such a tipping point, after which it will melt completely. The reinforcing feedback consists in the fact that, as the approximately 3000-m-thick ice sheet melts from the top, its surface reaches ever lower altitudes and thus warmer layers of air. After a certain point, the ice is destined to melt completely, even without further global warming. The end result is that global sea levels will rise by 7 m due to the loss of Greenland ice. This tipping point is probably somewhere between 1 and 3 degrees of global warming (Robinson et al., 2012).

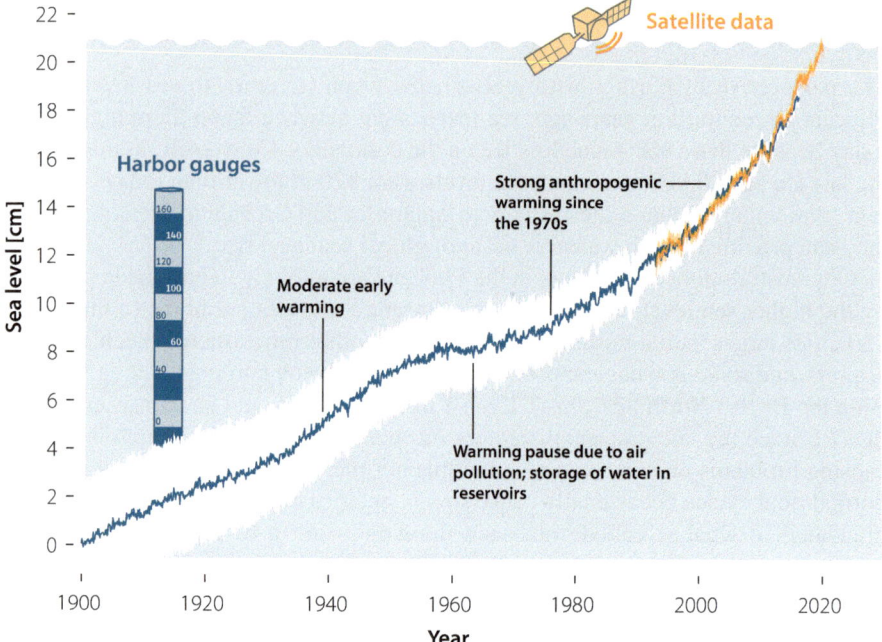

Fig. 3 Development of global sea level, measured by harbor gauges (blue) as well as by satellites (orange). Over the last 60 years, this rise has accelerated continuously. (Data from Dangendorf et al., 2019)

The situation of the West Antarctic Ice Sheet is similar. Here we are talking about another 3 m of sea leave rise, which is due to a different feedback effect: marine ice sheet instability, which can cause continental ice to slide unstoppably. There are studies suggesting that this tipping point has already been passed so that the loss of this ice sheet has already been triggered (Joughin et al., 2014).

The current IPCC report expects a sea level rise of 70 cm (compared to the late nineteenth century) before the end of this century with a warming of 3 °C. According to the report, the 1-m mark will be reached between 2100 and 2150. However, there are considerable one-sided risks to sea-level rise—that is, it could get much worse if, especially in the Antarctic, large ice masses are destabilized. The IPCC writes that if emissions are high, more than 2 m by 2100 and even 5 m by 2150 cannot be ruled out—a global catastrophe of unimaginable proportions.

This risk assessment is new to the IPCC. In its fourth report of 2007, it had still anticipated a 1990–2100 sea level rise in the range of 26–59 cm for the highest emission scenario (envisioning up to 5.2 degrees of warming), which corresponds to about 41–74 cm relative to the late nineteenth century. In that report, the IPCC remarked that ice sliding might possibly add another 10–20 cm, thus expressing assurance that, even with extreme warming, the sea level rise by 2100 would remain below 1 m.

Several colleagues, myself included, believed at the time that the IPCC was significantly underestimating sea level risks—not least because the measured rise to date was already some 50% faster than in the IPCC's model scenarios. In addition, the IPCC assumed that Antarctica would contribute practically nothing to the future rise, again in contrast to the ice loss already shown by satellite data. However, anyone in climate research who takes a more pessimistic view of things than the traditionally very cautious IPCC must deal with being accused of "alarmism" in some media—even if their assessment is correct and later shared by the IPCC.

The current IPCC report further warns that sea levels will continue to rise for millennia after global temperatures will have stabilized and that this rise is—with "very high confidence"—irreversible in the foreseeable future. In its landmark ruling on climate protection in 2021, the German Federal Constitutional Court emphasized intergenerational justice. When sea levels rise, countless generations after us will have to suffer the effects of our decisions today.

What is at stake here is not merely the sea level rise in this century, which we may be able to adapt to. Rather, sea levels will continue to rise for millennia, at 3 degrees of warming by about a meter per century, eroding the Earth's coastal zones, washing away beaches, threatening all infrastructure with ever-increasing storm surge risks, and making permanent coastal cities as we know them today almost impossible to sustain.

The Tipping Points of the Climate System

Regarding ice sheets, we have already mentioned two tipping points of the climate system at which further development becomes an unstoppable self-sustained process and thus gets out of control. There are more such tipping points, because ultimately all that is needed is a reinforcing feedback, a simple non-linearity, as occurs in many physical systems. For example, a kayak will right itself if you tilt it a little to one side—it stabilizes itself in a horizontal position and resists attempts to tip it. But only up to a certain point—beyond which it continues to turn by itself and then stabilizes in a new position: upside down. This critical point is literally the tipping point.

Greenland also has two stable equilibria under today's climatic conditions: with the ice sheet as we know it today and without it. The ice is self-stabilising because once it is there, the surface of the 3000-m-thick layer of ice is at such a high altitude that the air is too cold for it to melt. This is called ice-elevation feedback. If, on the other hand, the ice were gone, Greenland's surface—close to sea level and hence not cold enough to form a new ice sheet—would remain permanently ice-free. The Greenland ice sheet was formed in a colder climate during one of the earlier ice ages.

Such tipping points exist not only in physics, but also for ecological systems, which are self-stabilizing, but can also "tip over" when a stress limit is exceeded. The human body regulates its own temperature—up to a critical heat limit, above which the self-cooling system is overloaded, organs increasingly fail and the person

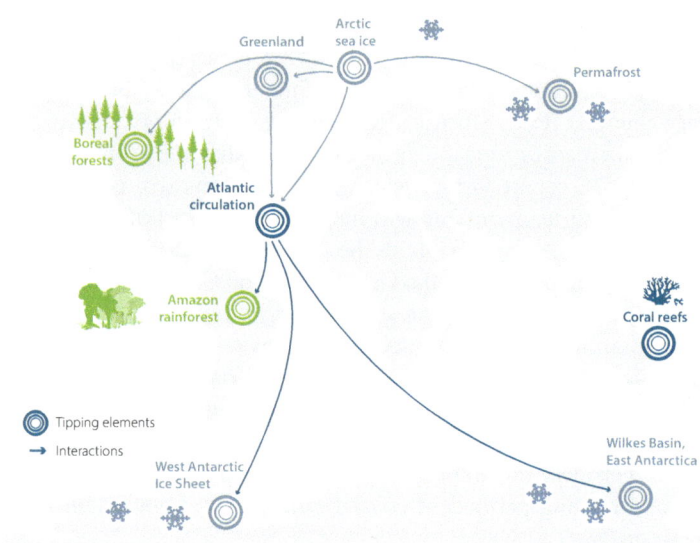

Fig. 4 Some of the most important tipping elements of the climate system. The arrows indicate interactions whereby the subsystems could cause one another to tip. (Data from Lenton et al., 2019)

finally dies. So, we too have our personal tipping point. This also applies to societies—the fall of the Berlin Wall was a tipping point of the GDR state. The term tipping point does not describe a value judgment, but merely a certain type of dynamic; the change that is triggered can of course also be desirable, that is in the eye of the beholder.

An overview of the most important tipping points of the climate system is provided in Fig. 4. All these tipping points are in danger of being exceeded at 3 degrees of global warming. For some, such as the Greenland ice sheet and the West Antarctic ice sheet, this is very likely, and for the Arctic summer sea ice cover and the Earth's coral reefs it even certain. The IPCC concludes that at 2 degrees of warming, almost all coral reefs will die; if we limit this to 1.5 degrees, we could still save 10–30% of the corals. Our planet has already been in a global coral die-off since 2015 (Hughes et al., 2018).

The Atlantic overturning circulation (sometimes called the Gulf Stream system) is a large circulation of the Atlantic Ocean in which warm surface water flows from the South Atlantic across the equator to the far north of the Atlantic, where it cools and releases heat into the air. The whole thing works like a central heating system for the North Atlantic region all the way to Europe. This current is endangered above all by freshwater input due to increased precipitation and ice melt. Freshwater is lighter than salt water and thus impedes the sinking of the water into the depths and thus the driving force of the Atlantic overturning circulation (Rahmstorf, 2024).

Models suggest a weakening of the current due to global warming, but its extent is uncertain, ranging from very small to 50% this century. There is serious evidence that many models systematically overestimate the stability of the Gulf Stream system. A striking cooling of the waters in the subpolar North Atlantic since the middle of the twentieth century indicates a weakening of 15% so far (Caesar et al., 2018). Several studies published since 2021 already found "early warning signs" that we are approaching the tipping point of the Atlantic overturning circulation, possibly even already in the next few decades (e.g. Boers, 2021). If this is confirmed, it will be extremely worrying.

The effects of a breakdown of the flow would be massive and partly unforeseeable, ranging from extreme weather in Europe to the collapse of important ecosystems in the North Atlantic to increased sea level rise on the US coast (up to an additional 1 m).

The rainforests of the Amazon region are already directly affected by climate change. Satellite data and on-site measurements have shown that increasing droughts are transforming the Amazon forest from a carbon sink into a carbon source (Brienen et al., 2015). Already today, parts of the tree populations are not able to cope with the new climate conditions and are dying. At least as important is the expansion of agriculture and the associated deforestation, which exacerbate the effects of climate change. As a result, the Amazon forest is losing resilience as deforestation continues. The tipping point leading to widespread loss of this unique ecosystem will be reached at lower global warming the more deforestation occurs. Today's forest loss is already estimated at 20%.

With increasing warming, the existence of coniferous forests in the north, which are adapted to cold climatic conditions, may also be increasingly threatened, among other things by fire and insect infestation. In recent years, there have already been extensive forest fires in Canada (e.g., Fort McMurray 2016 and the record-breaking fire season of 2023) and Russia (2010 in the European part) and even within the Arctic Circle (2017 in Greenland, 2018 in Sweden). In the transition zone of the northern forest belt to the steppe, tree stand regeneration may be threatened by increasing drought and heat stress.

A current focus of research is the risk of a cascade of tipping points that trigger one another like dominoes. For example, ice melt in the Arctic Ocean and on Greenland could dilute the North Atlantic water with fresh water to such an extent that the Atlantic overturning circulation would cease. This in turn would shift the tropical rainfall belts and could destabilize parts of the Amazon forest and the monsoons. And as if this were not enough, it could drive the Antarctic ice sheets beyond their tipping point. A quantitative assessment of these risks is still not possible.

The IPCC also attributes a strongly growing importance to tipping points. While the term "tipping point" was mentioned only 27 times in the 5th IPCC report, it was already mentioned 97 times in the 6th report. In its 2023 Synthesis Report of the latter, the IPCC concludes that "risks associated with large-scale singular events or tipping points, such as ice sheet instability or ecosystem loss from tropical forests, transition to high risk between 1.5 °C and 2.5 °C" global warming.

Self-Amplification of Global Warming

There is much public discussion about whether, beyond subsystems, global warming as a whole could, beyond a critical point, become an unstoppable self-amplified process. Mostly, the release of methane from permafrost is mentioned here as a reinforcing feedback. In 2018, a study on this topic appeared in the scientific journal *Proceedings of the National Academy of Sciences*, which made a big media splash as the "hothouse Earth paper" (Steffen et al., 2018).

The study investigated the extent to which feedback effects in the carbon cycle, which have not yet been taken into account in climate models, could exacerbate global warming. Not only methane release from permafrost was estimated, but also CO_2 release from dying or burning forests and a decreasing CO_2 uptake by the oceans.

The permafrost region is a globally significant carbon reservoir, which contains 1300–1600 billion tons of carbon and thus probably 50% of the total carbon stored in the soil worldwide. The permafrost areas have already warmed by up to 4 °C between 1990 and 2016. When the permafrost thaws, the soil carbon is decomposed by microbes. This could reduce the carbon stored in the permafrost by 15% by 2100.

Estimates of the carbon stored in living and dead plant material in the Amazon region (above and below ground) range from 80 to 120 billion tons. If this stored carbon were to be completely released in an extreme case, this would correspond to the amount of fossil CO_2 emissions that are currently released into the atmosphere in 8–12 years.

The result of the calculations (which unfortunately was somewhat neglected in many media reports) was that a warming of 2 degrees could become a warming of up to 2.5 degrees—if, as mentioned earlier, the carbon cycle changes and feedbacks are triggered.

This is by no means harmless and significantly exacerbates the climate crisis—but it does not mean that a global tipping point towards runaway warming has been passed. Fortunately, this risk is still considered very low, even if it cannot be completely ruled out. The methane problem should in any case be taken seriously, but is probably less dramatic within this century. In the long term, however, it is quite serious because thawing permafrost will create an uncontrollable source of greenhouse gas emissions for many centuries to come, which is likely to lead to further warming even after direct anthropogenic emissions will have been reduced to zero.

Conclusion

Without immediate, decisive climate protection measures, my children currently attending high school could already experience a 3-degree warmer Earth. No one can say exactly what this world would look like—it would be too far outside the entire experience of human history. But almost certainly this earth would be full of horrors for the people who would have to experience it. Weather chaos with deadly heat waves, devastating monster storms, and persistent widespread droughts that could trigger worldwide hunger crises. Rising sea levels that devastate our coasts. Collapsing ecosystems, devastating species extinctions, burning and withering forests, acidified oceans. Failed states, huge numbers of people on the run.

This sounds dark and dystopian and I find it hard to write it while thinking of my children. But it is likely. Most of this was predicted long ago and is now already being observed in its beginnings, which are by no means harmless for those affected. We must soberly face the fact that the conditions in a 3-degree world will most likely not "only" be three times worse than in a 1-degree world, considering the non-linear effects and the tipping points. I am not sure whether the more or less civilized coexistence of humans we enjoy now will endure under such conditions. Personally, I consider a 3-degree world to be an existential threat to human civilization.

What gives hope is that this 3-degree world is not an inevitable fate. It is still possible to limit warming to near the 1.5-degree mark—which was unanimously agreed by all countries in Paris in 2015 and to which almost all politicians in my country pay lip service. Global climate policy is certainly making progress: With the measures announced at the climate summit in Glasgow, the limit of 2 degrees is within reach, if these measures are not only promised but consistently implemented. And the International Energy Agency IEA has presented in September 2003 a feasible pathway to limiting global warming to 1.5 °C, kept open by the phenomenal global growth of renewable energies (International Energy Agency, 2023). Limiting the temperature to 2 degrees is not enough. In order to meet the 1.5 degree target, the world must finally switch into serious crisis mode, as the young people of Fridays for Future quite rightly demand. Climate protection must be given the highest priority.

References

Berkeley Earth. https://berkeleyearth.org/policy-insights/

Boers, N. (2021). Observation-based early-warning signals for a collapse of the Atlantic meridional overturning circulation. *Nature Climate Change, 11*(8), 680–688.

Brienen, R. J., et al. (2015). Long-term decline of the Amazon carbon sink. *Nature, 519*(7543), 344–348.

Caesar, L., et al. (2018). Observed fingerprint of a weakening Atlantic Ocean overturning circulation. *Nature, 556*(7700), 191–196.

Coumou, D., Lehmann, J., & Beckmann, J. (2015). The weakening summer circulation in the Northern Hemisphere midlatitudes. *Science, 348*(6232), 324–327.

Dangendorf, S., et al. (2019). Persistent acceleration in global sea-level rise since the 1960s. *Nature Climate Change, 9*(9), 705–710.

Dessler, A., on the basis of data from: Osman, M. B, et al. (2021). Globally resolved surface temperatures since the last glacial maximum. *Nature, 599*(7884), 239–244.

Hoffmann, P., et al. (2021). Atmosphere similarity patterns in boreal summer show an increase of persistent weather conditions connected to hydro-climatic risks. *Scientific Reports, 11*(1), 22893.

Hughes, T., et al. (2018). Spatial and temporal patterns of mass bleaching of corals in the Anthropocene. *Science, 359*, 80–83.

International Energy Agency. (2023). *Net zero roadmap: A global pathway to keep the 1.5 °C goal in reach.* https://www.iea.org/reports/net-zero-roadmap-a-global-pathway-to-keep-the-15-0c-goal-in-reach

IPCC. (2021). *Climate change 2021 the physical science basis.* Contribution of Working Group I to the Sixth Assessment Report of the Intergovernmental Panel on Climate Change Report.

Joughin, I., Smith, B. E., & Medley, B. (2014). Marine ice sheet collapse potentially under way for the Thwaites Glacier Basin, West Antarctica. *Science, 344*(6185), 735–738.

Kelley, C. P., et al. (2015). Climate change in the Fertile Crescent and implications of the recent Syrian drought. *Proceedings of the National Academy of Sciences of the United States of America, 112*(11), 3241–3246.

Kornhuber, K., et al. (2017). Summertime planetary wave resonance in the northern and southern hemispheres. *Journal of Climate, 30*(16), 6133–6150.

Kornhuber, K., et al. (2019). Amplified Rossby waves enhance risk of concurrent heatwaves in major breadbasket regions. *Nature Climate Change, 10*, 48–53.

Kossin, J. P., et al. (2020). Global increase in major tropical cyclone exceedance probability over the past four decades. *Proceedings of the National Academy of Sciences of the United States of America, 117*(22), 11975–11980.

Lenton, T. M., et al. (2019). Climate tipping points—Too risky to bet against. *Nature, 575*, 592–595.

Ombadi, M. (2023, September 19). As extreme downpours trigger flooding around the world, scientists take a closer look a global warming's role. *The Conversation.* https://theconversation.com/as-extreme-downpours-trigger-flooding-around-the-world-scientists-take-a-closer-look-a-global-warmings-role-213724

Rahmstorf, S. (2024). Is the Atlantic overturning circulation approaching a tipping point? Oceanography. https://doi.org/10.5670/oceanog.2024.501

Raymond, C., Matthews, T., & Horton, R. M. (2020). The emergence of heat and humidity too severe for human tolerance. *Science Advances, 6*(19), eaaw1838.

Robine, J. M., et al. (2008). Death toll ex-ceeded 70,000 in Europe during the summer of 2003. *Comptes Rendus Biologies, 331*(2), 171–175.

Robinson, A., Calov, R., & Ganopolski, A. (2012). Multistability and critical thresholds of the Greenland ice sheet. *Nature Climate Change, 2*(6), 429–432.

Steffen, W., et al. (2018). Trajectories of the earth system in the Anthropocene. *Proceedings of the National Academy of Sciences of the United States of America, 115*(33), 8252–8259.

Voossen, P. (2021). *The Arctic is warming four times faster than the rest of the world.* Science.

Zeder, J., & Fischer, E. M. (2020). Observed extreme precipitation trends and scaling in Central Europe. *Weather and Climate Extremes, 29*(11–12), 100266.

Biodiversity at the Tipping Point?

The Impact on Fauna and Flora

Bernhard Kegel

The creatures that inhabit this planet on land and sea are closely and in many ways connected to its atmosphere and the climate that prevails in it. Many hundreds of millions of years ago, marine unicellular organisms, the cyanobacteria, ensured that the oxygen produced as a by-product of their photosynthesis accumulated in the atmosphere over a long period of time, thus creating the conditions for the development of more complex organisms. By breathing this oxygen and using it to produce energy, living organisms are also significant producers of carbon dioxide. Methane is also to a large extent a product of living organisms, because it is produced, among other things, during the microbial decomposition of organic matter. Thus, the most important greenhouse gases are not least a product of biological processes. In view of the major problems caused by an anthropogenic increase in greenhouse gas concentrations, it is often forgotten that it is precisely thanks to this greenhouse effect caused by CO_2, water vapor, methane and other gases that tolerable temperatures prevail on Earth at all.

If it were not for this effect, our planet would be hurtling through space as a cold ball of rock or ice and would probably be inhabited by cold-resistant microbes at best.

The climate determines the distribution of living organisms on the planet, on the continents, above all through the distribution of water. If the climate changes, this has a direct impact on the composition, nature, temporal organization and spatial distribution of the biotic communities that exist on Earth. These processes are studied by a still young discipline within biology.

B. Kegel (✉)
Berlin, Germany
e-mail: info@bernhardkegel.de

"Climate Change Biology", defines Lee Hannah in his book of the same name, "is the study of the effects of climatic change on natural systems" (Hannah, 2011, p. 3). Of course, the main focus is on the effects of the current climate change, which is almost certainly caused by humans. However, in order to understand and correctly classify the changes that will affect us and the organisms of the Earth, Climate Change Biology also looks far back into the Earth's historical past, examines the changes in the world of organisms today and attempts to model future developments with the help of the most modern computer techniques (Kegel, 2021).

The prominent palaeobotanist John W. Williams has described the options that plants and animals have in a changing climate as "move, adapt, persist, or die," (Williams & Burke, 2019, p. 129) four options that will help us in what follows to classify the variety of responses that *Climate Change Biology* is investigating in nature.

Shifts in the Distribution Areas: "Move"

For years, biologists all over the world have been observing that the distribution ranges of plant and animal species have been on the move. These so-called *range shifts* accompanying rising or falling temperatures are also known from earlier times and are well documented in the fossil record (McInerney & Wing, 2011). In a warming world, living creatures move polewards in order to be able to continue living in their usual temperature range, i.e., northwards in the northern hemisphere and southwards in the southern hemisphere. The resulting shifts are already substantial. In Great Britain, of almost 330 animal species studied, 275 have migrated northwards at a rate of 14–25 km per decade. These include representatives of a wide range of animal groups, from mammals, birds, and fish to spiders, butterflies, dragonflies, and millipedes. Within a few decades, the poleward boundaries of their ranges have shifted up to 60 km to the north. The more the temperatures within the old distribution boundaries have risen, the more pronounced the changes have been (Hickling et al., 2006; Chen et al., 2011).

Of course, these are not migration or migration-related issues. In contrast to the migration of animals between summer and winter habitats, as practiced by many animal species, it is more a case of slow movement in one direction. Living creatures always try to gain a foothold outside their traditional distribution areas, but then it was too cold (or too warm) for them for a long time, their eggs did not develop, the young froze to death or they were displaced by climatically better adapted competitors—there are many reasons why expansion attempts fail. If they succeeded in single warm years, cold years pushed the species back into its old boundaries. Viewed in fast motion over longer periods of time, their distribution areas would have pulsated, as it were, without ultimately shifting. Now, with permanently increased temperatures, living beings can survive beyond their old boundaries. They reproduce and future generations can venture even farther—until they come up against new boundaries.

Studies from all over the world show that the distribution patterns of species are shifting in the mountains as well. Many animals and plants now live at higher altitudes than they did 20 or 30 years ago. On the slopes of the Antisana volcano in Ecuador, the plant species identified by Alexander von Humboldt and Aimé Bonpland at the beginning of the nineteenth century and located in their altitude profile are now found up to 266 m further uphill (Fig. 1) (Moret et al., 2019). This corresponds to a shift of 10–12 m of elevation gain per decade. New studies from Switzerland show that this value has been exceeded manifold by plant and animal species in the Alps during the last 50 years. The development has apparently accelerated considerably. Since 1970, average temperatures in the Swiss Alps have risen by 0.36 °C per decade, while at the same time the upper edge of the occurrence of various animal species has moved upwards by 47 to a maximum of 91 m per decade. For plants, it is 17–40 m. However, as isotherms have shifted by up to 71 m, these considerable shifts are not enough for most plant and animal species to keep pace with rising temperatures (Fig. 2) (Vitasse et al., 2021).

In the oceans, the distances bridged are even greater, averaging 72 km per decade, because the temperature gradient in the water is shallower. To continue living in the same ambient temperature, a fish at the same latitude has to move much farther

Fig. 1 Alexander von Humboldt's famous "Tableau Physique" (1807) schematically shows the altitude profile of the Andes. The plant species entered there grow up to around 250 m higher today. (Moret et al., 2019)

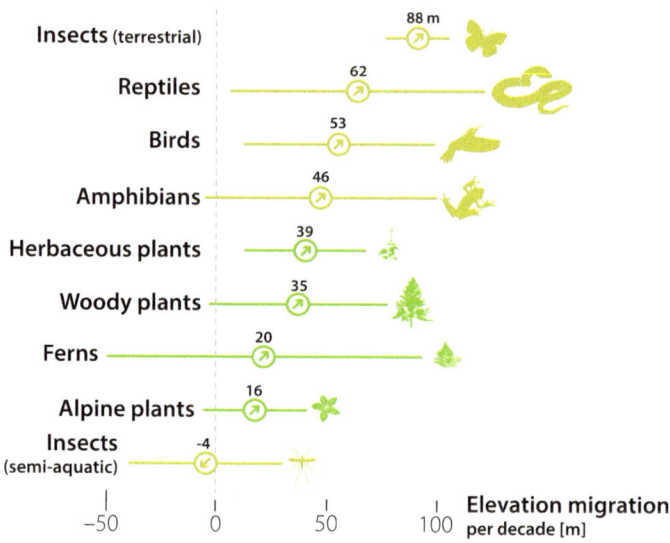

Fig. 2 The altitudinal migration of species is an excellent indicator of climate change. For the Swiss Alps, shifts of up to 90 m have been documented. (Data from Vitasse et al., 2021)

polewards than a mammal on the nearby coast (Parmesan, 2019). No wonder, then, that European scientists and fishermen all the way up to the south-west coast of Norway are increasingly catching fish species that were once native only to the waters off the coast of Portugal or in the Bay of Biscay. The North Sea is now home to several squid species that for a long time had been found there only sporadically (van der Kooij et al., 2016).

Hundreds of studies show that these shifts are happening in the same way all over the world, a uniformity that surprised even the experts (Newman et al., 2011).

In many cases, plant and animal species have increased the area they inhabit because (in the northern hemisphere) the southern edge of their range has not shifted as fast. This was the conclusion of a study of 80 British breeding bird species, for example. In terms of the size of their range, they have tended to benefit from rising temperatures in recent years (Massimino et al., 2015). As new species migrate into the existing communities, biodiversity in the temperate climate zones could even increase in the medium term. However, it is doubtful whether this will continue in the future if warming continues. There will be winners and losers among the plant and animal species at various stages of this long-lasting, perhaps centuries-long process of change. And today's beneficiaries are not necessarily tomorrow's winners.

It is also noticeable that about half the investigated species have not shifted their range: they persisted. Some find thermal refuges within their old distribution limits where they can survive. For others, there has been no need to move to cooler regions because the climatic changes were still tolerable or even beneficial for them. For example, a longer growing season due to warming enables some bird species to raise several generations per year. The same applies to bark beetles, which attack drought-weakened forests in Europe and North America and could thus become an even bigger problem.

However, the persistence of plants and animals in their old distribution limits means in many cases that these species cannot escape the rising temperatures. In mountainous regions, the peak regions are reached at some point, and in shallow shelf seas like the North Sea, there are no cooler depths. On land, humans have changed nature so massively that *range shifts* have been made difficult or almost impossible. What palaeobotanists have been able to document, for example, in the Bighorn Basin, a plateau in the U.S. state of Wyoming, would hardly be possible in today's man-made world.

The Bighorn Basin yields numerous fossils, including some from 56 million years ago, which is of great interest to palaeobiologists in connection with current climate change. With the greenhouse gas content of the atmosphere rising sharply, the temperature at that time rose by about 6–8 degrees within a few thousand years, albeit from a much higher level than today. In the Bighorn Basin, there were drastic changes in vegetation, among other things (Wing et al., 2005; McInerney & Wing, 2011). Before the temperature rise, birch, elm, walnut, laurel, and cypress trees grew there in a river landscape. Almost nothing of this remained during this so-called PETM, the Paleocene/Eocene temperature maximum. Only two plant species persisted, 27 disappeared and were replaced by 46 species from the tropics and subtropics. This new flora, adapted to heat and drought, dominated the Bighorn Basin for several tens of thousands of years, until temperatures slowly dropped back to their baseline. Afterwards, it was almost the same sight as before the temperature maximum: 22 plant species that used to be native here returned, only five were new immigrants. It was almost as if the dramatic climate change during the PETM had never happened.

This example impressively shows how living organisms react to climate change, so long as conditions are not turned upside down by catastrophes such as the impact of an asteroid or large-scale volcanic activity. They evade rising or falling temperatures, survive the unfavorable conditions in refuges and then, after the changes have subsided and if circumstances permit, return to their original settlement area.

In the present, however, the situation is fundamentally different—circumstances today prevent and hinder what has proven itself over Earth's history in climatic crisis situations. 50–70% of the mainland has been altered to a greater or lesser extent by humans and no longer supports natural vegetation (Barnosky et al., 2012). Huge, often pesticide-treated monocultures, roads, canals, and progressive urbanization have created insurmountable obstacles for many creatures to spread.

If temperatures continue to rise, species that cannot either *move*, or adapt on site will therefore sooner or later run into difficulties.

"A General Redistribution of Life on Earth"

Even if the observed *range-shifts* do not keep pace with climate change and many species do not or not yet shift their ranges, the range-shifts of the other half of the animal and plant world will reach a magnitude that scientists are concerned about: "the largest climate-driven redistribution of species since the last glacial maximum" 24,500–18,000 years ago. This process, as Australian marine ecologist Gretta Pecl and more than 40 scientists from around the world explain, "is a substantial challenge for human society" because range-shifts can amplify climate change in a positive feedback loop because altered vegetation also changes its reflectance and evaporation behavior (Pecl et al., 2017, p. eaai9214). Range shifts are affecting organisms such as the tiger mosquito *(Aedes albopictus)*, a vector of dengue fever and other viral diseases, which has already reached southern Germany. In Africa and South America in particular, millions of additional people will be threatened by malaria, as the vector, the Anopheles mosquito, spreads into the previously malaria-free highlands with the rising temperatures. Food production is also affected, not only by the relocation of cultivation areas, as in the case of coffee, but also by the migration of fish shoals in the oceans, the loss of pollinators, and the spread of pests on land.

Neobiota

The loss, decline or migration of native animal and plant species as well as the immigration of warmth-loving species will change the existing biotic communities and are already doing so today. But these two will be joined by a third group, the so-called neobiota. These are thousands of species of organisms that have been deliberately brought into the country by humans or introduced with the movement of goods and travel. The spectrum ranges from flatworms to hippos, from grasses to sequoias.

In Germany, about 400 neophyte species, mainly from Asia, North America, or the Mediterranean region are already considered established, i.e., they grow spontaneously without horticultural help from humans, reproduce and have already gone through several multiplication cycles. Most of them are rare and many are still limited in their occurrence to heat islands such as large cities or river valleys.

However, model studies (Kleinbauer et al., 2010) show that a large proportion of these alien plants have by no means exhausted their potential. They will spread beyond their current range, especially if warming continues. As the globalized movement of goods continues to distribute alien species, species will be added all over the world that we do not even know about now. A human-induced redistribution of species has already been in full swing for centuries, even without climate change (Kegel, 2013) (Fig. 3).

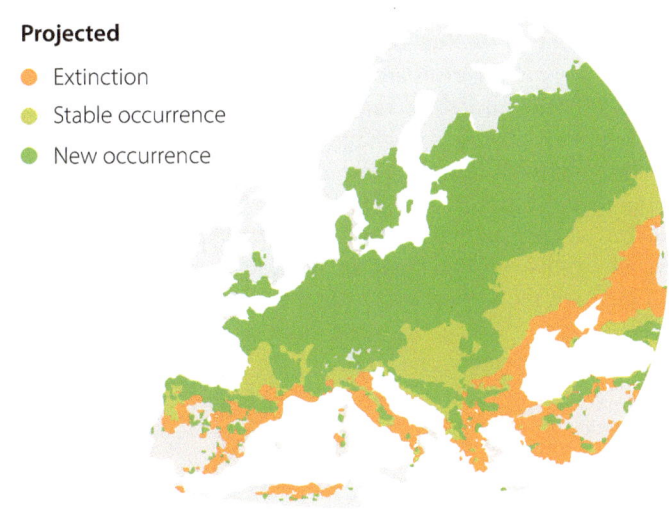

Fig. 3 Range shift of *Ambrosia artemisiifolia*: the highly allergenic mugwort ambrosia is increasingly spreading northwards. (Data from Cunze et al., 2013)

In general, aggressively spreading neobiota, so-called invasive species, are considered to be one of the most important reasons for the worldwide crisis of biodiversity, even if the problems caused by them are very unevenly distributed globally (Simberloff, 2013). Central Europe has so far come off rather lightly in this context, unlike New Zealand or Australia, for example, which have lost many native species to introduced rats, foxes and cats. On the other hand, there are more and more voices that see hope in foreign animal and plant species that could help fill the gaps in the native species population caused by climate change and take over their ecological functions (Pearce, 2015).

New Communities

The communities of the future will be made up of representatives of these three groups of species, the native species that tolerate the changed climate, the heat-loving immigrants and the neobiota. In many cases, especially in the tropics and subtropics, where temperatures are expected that *Homo sapiens* has never experienced before in its history, these will be so-called novel or *no-analog communities*, i.e., animal and plant communities that are completely new in their composition and for which there are no equivalents anywhere in the world today (Williams & Jackson, 2007).

This, of course, makes it difficult to predict how these new communities might behave and develop, how stable and resilient they will be, and to what extent they might replace or maintain the functions of current ecosystems. In the absence of empirical data, ecological surprises are almost certain. There are many examples of animal and plant species revealing very different characteristics and preferences from those in their native habitats when released on other continents. As recently as the middle of the twentieth century, no one would have thought it possible that tropical bird species could become breeding birds in temperate Central Europe. The collared parakeet has proved the opposite, and other parrot species could follow. South American nandus feel at home in Mecklenburg-Vorpommern, and Asian water buffalo graze in the Oderbruch. Apparently, at least some animal and plant species show only one of several possible faces in the native species structure of competitors, predators, partners, and rivals.

A Brief Look at the History of the Earth

The formation of such no-analog communities is also known from earlier epochs of Earth history, as transitional phenomena in times of climate change. A look into the geological past is helpful anyway when dealing with the possible effects of climate change on fauna and flora. It shows not what will or could happen under certain assumptions, but what actually happened. However, you will look in vain for an event that is similar in every respect to what is happening today.

Of particular importance is the aforementioned PETM, the temperature maximum at the Paleocene-Eocene boundary 56 million years ago, which was discovered in 1991 on the basis of isotope anomalies in Antarctic deep-sea cores (Kennett & Scott, 1991). It was triggered by large amounts of carbon (as CO_2 and/or CH_4) entering the atmosphere relatively quickly—in Earth-historical terms—and thus belongs to a series of rather rare phases in Earth history called *hyperthermals,* because global temperatures rose sharply during these times as a result of high greenhouse gas concentrations. During the PETM, the temperature increase was at least 6–8 °C, a value that would make large parts of the Earth uninhabitable for humans. Where this carbon came from in the case of the PETM, whether from thawing permafrost regions in Antarctica, volcanic activity, destabilized methane hydrate deposits on the ocean floor or even a meteorite impact, cannot be answered today.

There is also uncertainty about the speed of the release. It probably occurred at a rate of 1–2 Pg (10^{15} g) per year over several millennia, a rate not much below the average annual emissions of the last 150 years—though annual emissions have recently reached many times this amount. The total amount that entered the atmosphere at that time is estimated at 10,000 Pg. This is roughly equivalent to the total supply of fossil fuels believed to exist on Earth today.

Because warming is always slow in comparison to the input, it took 60,000 years until temperatures reached their maximum during the PETM (A. Sluijs cited by Dunne, 2017). Ocean acidification was also slow. Nevertheless, according to palaeobotanists Francesca McInerney and Scott Wing, the temperature rise had considerable biological consequences (McInerney & Wing, 2011). The oceans experienced oxygen depletion and massive algal blooms, and in the deep sea, which experienced a 5-degree temperature jump, nearly half of the foraminifera species, a widespread group of single-celled organisms with chambered calcareous shells, became extinct.

Climate-driven migratory movements began, the beginnings of which we are also experiencing today, and massive changes in fauna and flora occurred on the continents. In North America and Europe, the first representatives of cloven-hoofed and uncloven-hoofed animals appeared relatively suddenly, presumably migrating from Asia. The first modern apes emerged in America and Asia. Since the so-called meridional temperature gradient, the temperature difference between the equator and the poles, was only about 6 degrees during the PETM—today it is 22 degrees—and the continental masses in the northern hemisphere were even closer together, it was also possible for more sensitive species to move from one continent to the other (Sluijs et al., 2007).

As drastic as these changes were, leading scientists such as John Williams and Richard Zeebe fear that they could be surpassed by the current climate change, because the release of climate-altering carbon compounds now takes place in a much shorter time span than at the beginning of the PETM. Williams and his colleagues conclude that the expectation that the future will bring effects like those of the PETM "should be considered conservative." It is probably going to be worse. Richard Zeebe, a prominent oceanographer working in Hawaii, also sees catastrophic developments coming for the oceans, which will would be "completely without precedent" (Williams & Burke, 2019; R. Zeebe cited by Dunne, 2017).

In terms of Earth history, we are still in the Ice Age, or more precisely, in an interglacial, an intermediate warm period that we call the Holocene. We know from drilling in the ice sheet of Greenland, which began in the 1990s, that after the last retreat of the glaciers there were some abrupt temperature jumps that had a global impact on flora and fauna. They coincide with rising CO_2 concentrations in the atmosphere but were also triggered by changes in air mass circulation and disturbances in global ocean currents. The cause of the latter was, for example, the input of huge amounts of sweet glacial melt water 8200 years ago, which led to the so-called Misox oscillation, a drop in temperature of up to 5 degrees within a few years (Williams & Burke, 2019).

Investigations of post-glacial climate changes have the great advantage that they can be dated precisely to decades or even years with the help of ice cores and lake sediment deposits. In addition, numerous fossils are found in layers of lake sediments, especially pollen from plants growing in the lake environment. From this, palaeobotanists can infer the composition of the vegetation and follow its change through the increase or decrease of individual species.

A worrying result of these studies concerns the speed with which the plant communities in the vicinity of the lakes reacted to the temperature changes (Fig. 4).

Whether at Lake Gerzen in Switzerland or Meerfelder Maar in the Eifel, the researchers observe "near-immediate effects." "Forest response time", John Williams

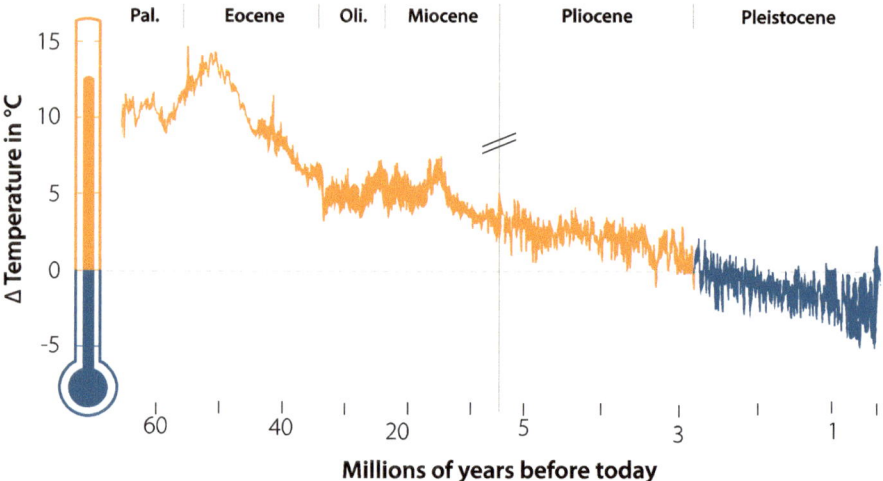

Fig. 4 The PETM at the Paleocene-Eocene boundary 56 million years ago stands out as a needle-pointed jag in the paleotemperature curve just before the Eocene temperature peak was reached. It lasted 200,000 years. (Data from All_palaeotemps_G2.svg)

and Kevin Burke summarize the state of knowledge, "were consistently less than 20–40 years, and often had no detectable time lag" (Williams & Burke, 2019, p. 139).

The temperature drop during the Misox Oscillation 8200 years ago also led to a "pronounced and immediate response of terrestrial vegetation" in Central Europe (Tinner & Lotter, 2001, p. 551). This can be seen, for example, in the stratigraphy of the sediments of the Swiss Soppensee in the canton of Lucerne. It took only about one tree generation, which corresponds to about one human generation, to collapse the predominant hazel stand there, from which 40% of all pollen grains originated before the temperature drop, and to replace it with a completely different forest of pines, birches and lime trees. Scientists speak of a regime shift in the case of such profound, more or less abrupt, but lasting changes in ecosystems.

Forests

It remains to be seen whether the damage from recent drought years, which became known as "forest dieback 2.0," is already the prelude to such a regime shift. The tree death caused by the drought and bark beetle infestation destroyed a forest area the size of Saarland throughout Germany (app. 2.500 km^2). The damage was mainly to spruce, but pine and important deciduous tree species were also affected. In the past 25 years, mortality in European forests has shown a "worrying upward trend." Especially endangered are conifers on "productive sites" (George et al., 2021).

Regime shifts are already occurring in Arctic habitats. The thawing of permafrost is reshaping the landscape in vast areas, shrubs are encroaching on the tundra, and the disappearance of ice is threatening marine communities (Kegel, 2021). Because their ice cover opens earlier and light can penetrate the water for longer, the character of many Arctic lakes has changed (Smol et al., 2005). Regime shifts are also looming in the boreal coniferous and tropical rainforests, which have been suffering from increasing drought for years. The latter even threaten to lose their important role as carbon sinks. The carbon uptake measured in the 1990s has never been reached again since then and is declining. Since 2010, this has also been the case for the long-stable African rainforests. Fire clearance and the resulting soot particles exacerbate the situation (Hubau et al., 2020).

Stressed by drought, shorter and milder winters, and the resulting bark beetle invasions, the largest contiguous forest area on Earth, the boreal coniferous forest, is also getting into trouble. In North America, the beetles killed 30 billion trees and the timber industry which, through its disastrous forest management, made problems of this magnitude possible felled another 30 billion (Hannah, 2011; Nikiforuk, 2011).

Bottom Up

Although climate change can alter entire ecosystems and drive them into regime shifts, it affects individuals first, every single plant and animal. The response of organisms to rising temperatures is "largely, if not entirely," a bottom-up process (Newman et al., 2011, p. 73). How tolerant an individual is to fluctuating environmental parameters and at what thresholds this tolerance ends depends primarily on its genetic make-up, possibly also on environmental experiences it has had. It is the individual animal that moves beyond the old distribution limits and tries to survive there. When many do this, the range-shifts discussed occur. On the way bottom-up, "up" through the food chains and communities, these effects will "combine, amplify, weaken and generally interact" (Newman et al., 2011, p. 73).

Since practically all life processes, especially biochemical reactions in the cells, are temperature-dependent, living organisms are directly affected by rising temperatures. Numerous studies already show that many animal individuals even change their body proportions by way of adapting to a warmer environment. These so-called *shape-shifts* follow two rules postulated over a hundred years ago, Allen's rule and Bergmann's rule, according to which the size of body appendages such as extremities, tails, and ears decreases towards the poles, i.e., with falling temperatures, while body size increases. Both have to do with the exchange of heat between the body and the environment. So, in a warmer world, body size should decrease and limb size should increase.

A recently published study of 77 bird species in the Brazilian rainforest shows that their body proportions actually change in step with rising temperatures. Since the 1980s, the animals have lost body mass and developed larger wings (Jirinec et al., 2021). Australian parrots have seen their beak area increase by 4–10% since 1871 (Campbell et al., 2015). American bison were 37% larger 40,000 years ago than they are today. The average annual temperature increased by 6 degrees during this period. If this trend were to continue at the same rate, bison would lose another 46% by the end of this century with a 4 degree warming compared to today's average mass of 665 kg, and would weigh only 357 kg on average (Martin et al., 2018). In the aforementioned Bighorn Basin, fossil finds prove that the prehistoric horses of the genus Sifrhippus living there shrank by 30% in the course of warming during the PETM, only to gain 76% in body mass again during the recovery phase when temperatures fell. Insects and worms even lost almost half of their body mass (Smith et al., 2009; Secord et al., 2012).

The biological effects of a warming world show themselves especially in polar habitats and the tropics because plants and animals living there are used to only small temperature fluctuations and do not tolerate larger deviations. The eggs of the Arctic cod, for example, begin to die at temperatures above 3 degrees. The marine species of the Antarctic have lived since time immemorial in a range between -1.9 degrees, the freezing point of salt water, and $+1.8$ degrees, the highest water temperature ever measured there. Many have to surrender at temperatures as low as 3 degrees, and even after a long acclimatization period, 6 degrees is the absolute maximum. Crocodile icefish have no haemoglobin, so they get into trouble when oxygen content drops with rising temperatures (Somero, 2010, 2012).

Tropical organisms generally live closer to the maximum temperatures they can tolerate than animals and plants in temperate zones. Therefore, a small increase of a few degrees can also be fatal for them. The distance between these maximum tolerable temperatures and the highest temperatures to which living organisms are actually exposed in their environment is called the "thermal safety margin". Its size depends largely on whether there are refuges in the habitat where organisms can avoid dangerously high temperatures. This can be a shady spot under trees or a body of water, but often a crevice in the rock or a stone under which one can hide is enough. If such refuges are missing or no longer offer protection, the thermal safety distance can drop to zero. This can lead to life-threatening hyperthermia, and the animals usually die of heart failure (Pinsky et al., 2019).

For mussels, which live in water at 21 degrees, the lethal limit is around 28 degrees. No wonder, then, that the animals have expanded their range on the US East Coast by 350 kilometers to the north. At the southern edge of their range, water temperatures have risen so high that hardly any mussels can survive (Somero, 2010, 2012).

However, living creatures are not completely defenseless against rising temperatures. Mussels, too, have such defensive means, which are quite old in phylogenetic terms. Organisms have always been confronted with dangerously high temperatures and have developed astonishingly effective repair mechanisms against them. Modern methods of molecular biology show that damage to protein molecules occurs in mussels even far below critical temperature values, to which the cells react immediately by activating the genes of so-called heat protection proteins (Hsp).

They help the damaged protein molecules to regain their correct three-dimensional structure, without which they cannot fulfil their function. If temperatures continue to rise, proteolytic enzymes are produced. So many proteins are now damaged that these enzymes can only break them down and dispose of them. If it remains warm, the cell cycle is stopped as a last resort. Cells can then no longer divide and the organism stops growing.

Mass Mortality Events

What happens beyond this threshold can now be experienced with sad regularity in Australia. For flying foxes, large fruit-eating bats, the critical temperature is 42 °C, a value that has been significantly exceeded several times during extreme heat waves in recent years. The flying foxes, which now live in large numbers in the Australian metropolises because their natural habitat had to make way for planta-tions, fell dead by the thousands from the treetops, where they spend the days hang-ing close together on the branches. The high point was 4 February 2014, when hot winds from the outback drove temperatures to 43 degrees and higher. In south-east Queensland, 45,000 animals died in a single day. The tropical black flying foxes were hit hardest. Half of the animals living in the area perished. Justin Welbergen, who has been researching these mass mortality events for years, dates the first docu-mented event of this kind to 1791, when it hit Sydney (Welbergen et al., 2008). So these die-offs have always existed. It is difficult to say whether they have increased, because in the past they usually took place in remote areas and remained undetected.

Welbergen and his group thus suffer from a problem that affects many areas of *climate change biology:* the lack of old data. Scientific studies that meet today's standards could not be carried out 50 let alone a hundred years ago, or only in rare and exceptional cases. Today's long-term studies, however, usually only focus on the last few decades, or at most 50 or 60 years. Data documenting the state of biotic communities a hundred years ago or even before the beginning of industrialization, such as the botanical surveys of Humboldt and Bonpland at the Antisana, are rare strokes of luck.

Mass mortality events (MMEs) such as that of Australian fruit bats can occur in all habitats and affect all types of animals, but they are especially common in aquatic habitats. However, due to a lack of old data, research cannot prove beyond doubt that mass mortality events have become more frequent. Before 1940, they were documented only sporadically. Since then, they have clearly increased in birds, fish, and marine invertebrates. However, it cannot be ruled out that we are just looking more closely today and registering MMEs that would have remained undetected in the past (Fey et al., 2015).

Marine *Heatwaves*

The data situation for heat waves in the oceans is somewhat better, although here too experts complain that "the scientific understanding of marine heat waves is still in its infancy" (Holbrook et al., 2019, p. 2). The conditions under which marine heat waves actually occur and the factors that determine their duration and intensity are still insufficiently understood. Since the beginning of the twentieth century, their frequency has increased by 36% and their duration by 17%. The North-East Pacific stood out with exceptionally long heat waves (Oliver et al., 2018, 2019).

An especially dramatic event of this kind took place there between 2013 and 2016 (Fig. 5). This heat wave was named "The Blob" by climatologists after a voracious and incessantly growing alien, the villain of a science fiction film of the same name. It formed south of the Aleutian Islands in the winter of 2013, reached massive proportions in 2014 and continued to grow.

In the summer months of 2014 and 2015, these warm waters, two-thirds the size of the United States, reached the west coast of North America and caused thousands of animal deaths. California beaches were covered with countless orange-red red-tuna crabs, which normally live much farther south, as well as the tropical hammerhead sharks, marlins, and dolphinfish, which reached further north than ever before.

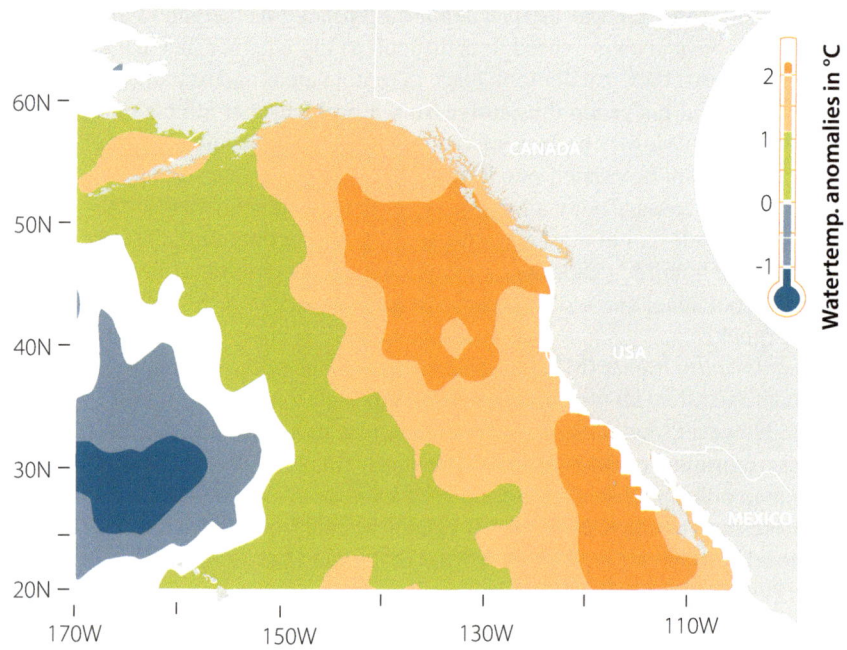

Fig. 5 Temperature distribution during *The Blob* thermal anomaly: clearly two regions with water warmer than 2 degrees. (Data from Bond, 2015; NOAA ESRL, 2015)

As far as is known, the sequence of events followed a classic bottom-up pattern. (Piatt et al., 2020) The trigger was apparently the widespread absence of winter storms, which normally ensure the mixing of warm surface water with cold and nutrient-rich water masses from deeper layers. The resulting nutrient deficiency first became apparent in 2014 in the form of an exceptionally low chlorophyll content, the lowest since measurements began at the end of the twentieth century. The lack of phytoplankton was followed by a collapse in zooplankton, which was dominated by nutrient-poor small crustaceans that had recently migrated from the south. Starvation continued through the prey fish species (sand eels and capelins) to the top predators of the ecosystem. Tens of thousands of dead seabirds washed up along six thousand kilometers of coastline all the way down to California. Experts estimate that up to 1.2 million guillemots died, 10–20% of the total population in the North Pacific, plus other seabirds and thousands of sea lions and fur seals (Fig. 6).

Cod and pollock stocks collapsed by 70%. Off Hawaii, sightings of mother humpback whales with calves dropped by more than 70%, "unexplained" whale

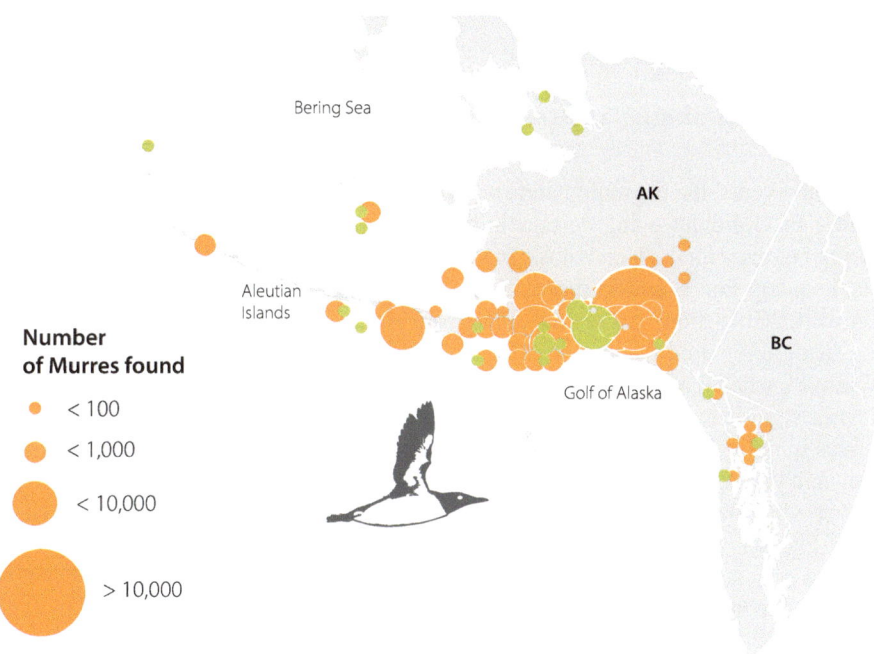

Fig. 6 Victims at the top of the food chain: anomalously warm water has led to the starvation of thousands of guillemots on many coasts around the Gulf of Alaska. (Data from Piatt et al., 2020)

strandings shocked people in Canada and Alaska. The animals all appeared healthy, but like the lost birds and seals, were severely malnourished.

In addition to the *bottom-up process,* there was also a *top-down effect,* because the predatory fish at the top of the food chain develop an enormous appetite when water temperatures rise. Just 2 degrees more and their food requirements increase by 63%. Not much is left for seabirds. The fauna of the Northeast Pacific suffered as a result of the *blob* starvation.

It was not until the winter of 2015/2016 that the heat wave began to dissipate. The water masses had warmed by a maximum of 2.5 degrees above the long-term average during the *blob*, to temperatures that are expected to be normal in the North Pacific for the second half of the twenty-first century in the worst case (ICCP scenario RCP 8.5) according to computer modelling.

The frequency, duration, extent, and intensity of heat waves will continue to increase with climate change. The statements of marine biologists and oceanographers sound anything but confident: "Many ocean areas will reach a state of almost permanent marine heat waves in the late 21st century. […] We have moved away from the conditions under which marine heat waves naturally occurred, from a condition that has shaped the distribution of marine species and the structure and function of ecosystems for millennia" (Oliver et al., 2019).

Shifted Phenologies (*Mismatch*)

For many years, the beginning and end of the phenological seasons have been determined in Germany using certain indicator plants. Thus, the phenological early spring begins with the hazel blossom, and the summer starts with the blossom of the black elder. For Bavaria, these long-term records (Bayerisches Landesamt für Umwelt, 2014) show that the hazel blossom has come 23 days earlier as a result of the rising temperatures between 1961 and 2010, while the start of summer is now 17 days earlier. These shifts, which can be observed all over the world, naturally have a significant impact on the organism world, which depends on certain processes in nature being synchronized in time. This ensures that predators meet their prey and young hatch or are born when they find optimal food in nature. Plants must flower when their pollinators are active, parasites must meet their hosts at the right time (Fig. 7).

However, studies from Switzerland now show that living beings react very differently to the shifts (Vitasse et al., 2021). In the meantime, many cases have been documented worldwide in which temporal synchrony is increasingly lost—*mismatch* occurs. Greenlandic caribou start a long migration, triggered by the length of the day, to arrive in the tundra exactly when the especially nutritious fresh green appears there. However, the plants now sprout much earlier due to the drastic rise in temperatures in the Arctic, the caribou arrive too late, and this impacts the quality and quantity of milk produced by the dams. Since 1993, the number of their calves has dropped to a quarter (Post & Forchhammer, 2008). The deer in the Champagne

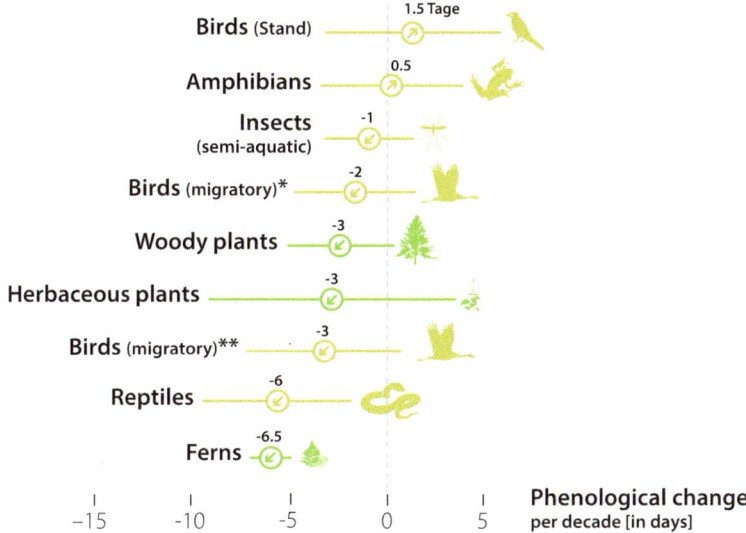

Fig. 7 Things are happening earlier: life cycle changes are occurring across a broad spectrum of species. In the Alps reptiles, for example, became active 6 days earlier on average per decade (*long-distance/**short-distance migrants). (Data from Vitasse et al., 2021)

region of France are suffering a similar fate (Plard et al., 2014). In the Baltic Sea, hatchling young herring increasingly miss the early onset of the algal bloom.

Migratory birds adapt by shortening their migratory routes or becoming sedentary birds in order to arrive at their breeding grounds earlier in spring. This is impossible for long-distance migrants. Their numbers are declining because they find the best breeding sites already occupied by birds that were there earlier, a trend that can only be discerned after 1990. Apparently, the development has accelerated since then (Møller et al., 2008).

In the Netherlands, populations of the pied flycatcher, which winters in tropical Africa, have plummeted by a catastrophic 90% because caterpillars, their preferred prey, now hatch too early for them. Moreover, they are in fatal competition with great tits, which fiercely defend their nesting holes. More and more often, dead pied flycatchers are found in them, mostly inexperienced males that have arrived very late after their long flight (Both et al., 2006; Samplonius & Both, 2019).

Cuckoos are also long-distance migrants and have problems because their offspring have to hatch before the chicks of the host birds. The animals that have specialized in sedentary birds or short-distance migrants are increasingly arriving too late and will therefore probably become extinct. Since the breeding parasites amazingly match the coloration of their eggs to that of their hosts, they cannot simply change bird species. Cuckoos, which lay their eggs in the nests of long-distance migrants, fare better. It can already be observed that early-breeding species such as robins and wagtails are increasingly relieved of the burden of cuckoos, while late-arriving host species are increasingly burdened. The presence of breeding parasites in the nests of reed warblers has more than doubled in the last 30 years (Saino et al., 2009).

Extinction and Defaunation

Climate change and the *range and regime shifts, mismatches*, heat waves and other extreme weather events it triggers will massively increase the pressure on flora and fauna. However, the global crisis of biodiversity that we are currently experiencing is not an effect of climate change but caused by massive habitat destruction due to agriculture and urbanization. In addition, there is the large-scale use of pesticides, the introduction of invasive species, as well as hunting, fishing, and poaching. Currently, between 11,000 and 58,000 species die out each year, a loss that exceeds the natural extinction rate by orders of magnitude (Dirzo et al., 2014).

In recent years, nature conservation has focused primarily on rare and endangered species. However, recent data shows that the dramatic nature of the crisis has been vastly underestimated in the process, as wildlife is dying across the board. A recent study by British and Czech researchers puts the loss of bird individuals in the European Union at 17–19% of the total population since 1980. Yet, there are also winners such as blackcaps with a population increase of 55 million birds, followed by chiffchaffs, blackbirds, and wrens (Fig. 8). Overall, however, the total number of birds has shrunk by 560–620 million, with species of agricultural landscapes being especially affected (Burns et al., 2021). The decline primarily affects common species such as Yellow Wagtail, Starling, Skylark, and House Sparrow, whose populations have halved. The so-called insect mortality, which was discovered by a group of entomologists in Krefeld and which has since been confirmed for other countries, is certainly connected to this (Hallmann et al., 2017, 2019; Powney et al., 2019).

Other studies show a collapse of the global seabird population by almost 70% since 1950. (Paleczny et al., 2015) As a result of hunting and poaching, more than half of the remaining tropical forests are practically without larger mammals. The horror image of the *empty forests* is making the rounds (Benítez-López et al., 2019).

The *Living Planet Report* of the World Wildlife Federation and the Zoological Society of London, based on more than 20,000 monitored populations of vertebrates around the world, also laments a 68% decline since 1970—in individuals, mind you, not in species (WWF, 2020). To describe the full extent of this historically unprecedented faunal destruction, terms such as defaunation or *biological annihilation* have been coined (Dirzo et al., 2014).

So far, climate change has contributed little to this development. In the future, however, it will intensify it and, in doing so, will encounter an animal and plant world that is already under enormous stress and, through the loss of countless individuals, in danger of losing the one thing that could ensure its survival in a rapidly changing environment: its genetic diversity.

Nature conservation has therefore never been as important as it is today. The IPBES (Inter-governmental Science-Policy Platform on Biodiversity and Ecosystem Services) also demands that biodiversity and its threats must play a more important role in the future. Climate protection and nature conservation should go hand in

**Changes in bird
numbers (million)**

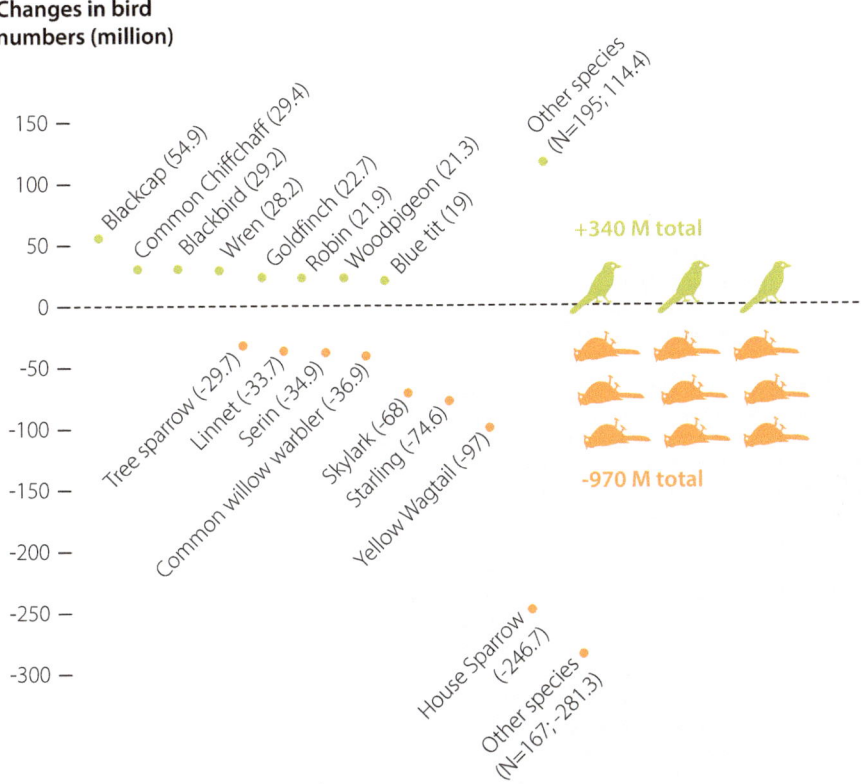

Fig. 8 Loss of bird individuals in the EU: profiteers such as the blackcap stand in contrast to dramatic declines in once "commonplace species" such as the starling and sparrow. (Data from Burns et al., 2021)

hand. More than 100 countries, including Brazil and Russia, have just decided in Glasgow to stop the global destruction of forests by 2030. Colombia, Ecuador, Costa Rica, and Panama are creating a huge protected area off their coasts that also includes the Galapagos Archipelago, a "marine corridor in the eastern tropical Pacific." These are important signals that hopefully will finally be implemented and must be followed by others. One of the most prominent biologists of our time, the recently deceased ant and biodiversity researcher E. O. Wilson, called for half of the earth to be placed under protection; the UN is aiming for 30% by 2030 (Wilson, 2016). The destruction and fragmentation of habitats must be stopped, and the rise in the average global temperature limited, otherwise there is a risk of losing countless animal and plant species and vital ecosystem functions, the end of the biosphere as we know it.

References

All_palaeotemps_G2.svg. upload.wikimedia.org/wikipedia/commons/5/5b/All_palaeotemps_G2.svg

Barnosky, A. D., et al. (2012). Approaching a state shift in Earth's biosphere. *Nature, 486,* 52–58.

Bayerisches Landesamt für Umwelt. (2014). Bayerisches Landesamt für Umwelt (Ed.): *Beeinflusst der Klimawandel die Jahreszeiten in Bayern? Antworten der Phänologie.* LfU Bayern.

Benítez-López, A., et al. (2019). Intact but empty forests? Patterns of hunting-induced mammal defaunation in the tropics. *PLoS Biology, 17,* e3000247.

Bond, J. (2015). *What is the "warm blob" in the Pacific and what can it tell us about our future climate?* https://theconversation.com

Both, C., et al. (2006). Climate change and population declines in a long-distance migratory bird. *Nature, 441,* 81–83.

Burns, F., et al. (2021). Abundance decline in the avifauna of the European Union reveals cross-continental similarities in biodiversity change. *Ecology and Evolution, 2021,* 1–14.

Campbell, D., et al. (2015). Climate-related spatial and temporal variation in bill morphology over the past century in Australian parrots. *Journal of Biogeography, 42,* 1163–1175.

Chen, I. C., et al. (2011). Rapid range shifts of species associated with high levels of climate warming. *Science, 333,* 1024–1026.

Cunze, S., Leiblein, M. C., & Tackenberg, O. (2013). Range expansion of Ambrosia artemisiifolia in Europe is promoted by climate change. *Hindawy, 13,* 610126.

Dirzo, R., et al. (2014). Defaunation in the Anthropocene. *Science, 345,* 401–406.

Dunne, D. (2017, October 9). Hyperthermals: What can they tell us about modern global warming? In *Carbon brief.* www.carbonbrief.org/hyperthermals-what-can-they-tell-us-about-modern-global-warming

Fey, S. B., et al. (2015). Recent shifts in the occurrence, cause, and magnitude of animal mass mortality events. *PNAS, 112,* 1083–1088.

George, J.-P., et al. (2021). Long-term forest monitoring unravels constant mortality rise in European forests. *bioRxiv preprint.* https://doi.org/10.1101/2021.11.01.466723

Hallmann, C. A., et al. (2017). More than 75 percent decline over 27 years in total flying insect biomass in protected areas. *PLoS One, 12,* e0185809.

Hallmann, C. A., et al. (2019). Declining abundance of beetles, moths and caddisflies in the Netherlands. *Insect Conservation and Diversity, 13,* 127–139.

Hannah, L. (2011). *Climate change biology* (2nd ed.). Academic.

Hickling, R., et al. (2006). The distributions of a wide range of taxonomic groups are expanding polewards. *Global Change Biology, 12,* 450–455.

Holbrook, N. J., et al. (2019). A global assessment of marine heatwaves and their drivers. *Nature Communications, 10,* 2624.

Hubau, W., et al. (2020). Asynchronous carbon sink saturation in African and Amazonian tropical forests. *Nature, 579,* 80–87.

Jirinec, V., et al. (2021). Morphological consequences of climate change for resident birds in intact Amazonian rainforest. *Science Advances, 7,* eabk1743.

Kegel, B. (2013). *Die Ameise als Tramp. Von biologischen Invasionen.* DuMont Buchverlag.

Kegel, B. (2021). Die Natur der Zukunft. In *Tier- und Pflanzenwelt in Zeiten des Klimawandels.* DuMont Buchverlag.

Kennett, J. P., & Scott, L. D. (1991). Abrupt deep-sea warming, palaeoceanographic changes and benthic extinctions at the end of the Palaeocene. *Nature, 353,* 225–229.

Kleinbauer, I., et al. (2010). *Ausbreitungspotenzial ausgewählter neophytischer Gefäßpflanzen unter Klimawandel in Deutschland und Österreich.* Bundesamt für Naturschutz.

Martin, J. M., et al. (2018). Bison body size and climate change. *Ecology and Evolution, 8,* 4564–4574.

Massimino, D., et al. (2015). The geographical range of British birds expands during 15 years of warming. *Bird Study, 62,* 523–534.

McInerney, F. A., & Wing, S. L. (2011). The Paleocene-Eocene Thermal Maximum: A perturbation of carbon cycle, climate, and biosphere with implications for the future. *Annual Review of Earth and Planetary Sciences, 39*, 489–516.

Møller, A. P., Rubolini, D., & Lehikoinen, E. (2008). Populations of migratory bird species that did not show a phenological response to climate change are declining. *PNAS, 105*, 16195–16200.

Moret, P., et al. (2019). Humboldt's *Tableau Physique* revisited. *PNAS, 116*, 12889–12894. alamy stock, Science History Images.

Newman, J. A., et al. (2011). *Climate change biology.* CABI.

Nikiforuk, A. (2011). *Empire of the beetle. How human folly and a tiny bug are killing North America's great forests.* Greystone Books.

NOAA/ESRL Physical Sciences Division. (2015). *Surface SST © composite anomaly 1981–2010.*

Oliver, E. C. J., et al. (2018). Longer and more frequent marine heatwaves over the past century. *Nature Communications, 9*, 1324.

Oliver, E. C. J., et al. (2019). Projected marine heatwaves in the 21st century and the potential for ecological impact. *Frontiers in Marine Science, 6*, Article 734.

Paleczny, M., et al. (2015). Population trend of the world's monitored seabirds, 1950–2010. *PLoS One, 10*, e0129342.

Parmesan, C. (2019). Range and abundance changes. In T. E. Lovejoy & L. Hannah (Eds.), *Biodiversity and climate change. Transforming the biosphere* (pp. 25–38). Yale University Press.

Pearce, F. (2015). *The new wild. Why invasive species will be nature's salvation.* Beacon Press.

Pecl, G. T., et al. (2017). Biodiversity redistribution under climate change: Impacts on ecosystems and human well-being. *Science, 355*, eaai9214.

Piatt, J. F., et al. (2020). Extreme mortality and reproductive failure of common murres resulting from the Northeast Pacific marine heatwave of 2014-2016. *PLoS One, 15*, e0226087.

Pinsky, M. L., et al. (2019). Greater vulnerability to warming of marine versus terrestrial ectotherms. *Nature, 569*, 108–111.

Plard, F., et al. (2014). Mismatch between birth date and vegetation phenology slows the demography of roe deer. *PLoS Biology, 12*, e1001828.

Post, E., & Forchhammer, M. C. (2008). Climate change reduces reproductive success of an Arctic herbivore through trophic mismatch. *Philosophical Transactions of the Royal Society B, 363*, 2369–2373.

Powney, G. D., et al. (2019). Widespread losses of pollinating insects in Britain. *Nature Communications, 10*, 1018.

Saino, N., et al. (2009). Climate change effects on migration phenology may mismatch brood parasitic cuckoos and their hosts. *Biology Letters, 5*, 539–541.

Samplonius, J. M., & Both, C. (2019). Climate change may affect fatal competition between two bird species. *Current Biology, 29*, 327–331.e2.

Secord, R., et al. (2012). Evolution of the earliest horses driven by climate change in the Paleocene-Eocene Thermal Maximum. *Science, 335*, 959–962.

Simberloff, D. (2013). *Invasive species. What everyone needs to know.* Oxford University Press.

Sluijs, A., et al. (2007). The Paleocene-Eocene Thermal Maximum super greenhouse: Biotic and geochemical signatures, age models and mechanisms of global change. In M. Williams et al. (Eds.), *Deep-time perspectives on climate change: Marrying the signal from computer models and biological proxies* (pp. 323–349). The Micropalaeontological Society, Special Publications/The Geological Society.

Smith, J. J., et al. (2009). Transient dwarfism of soil fauna during the Paleocene-Eocene Thermal Maximum. *PNAS, 106*, 17655–17660.

Smol, J. P., et al. (2005). Climate-driven regime shifts in the biological communities of arctic lakes. *PNAS, 102*, 4397–4402.

Somero, G. N. (2010). The physiology of climate change: How potentials for acclimatization and genetic adaptation will determine 'winners' and 'losers'. *Journal of Experimental Biology, 213*, 912–920.

Somero, G. N. (2012). The physiology of global change: Linking patterns to mechanisms. *Annual Review of Marine Science, 4*, 39–61.

Tinner, W., & Lotter, A. (2001). Central European vegetation response to abrupt climate change at 8.2 ka. *Geology, 29*, 551–554.

van der Kooij, J., Engelhard, G. H., & Righton, D. A. (2016). Climate change and squid range expansion in the North Sea. *Journal of Biogeography, 43*, 2285–2298.

Vitasse, Y., et al. (2021). Phenological and elevational shifts of plants, animals and fungi under climate change in the European Alps. *Biological Reviews*. https://doi.org/10.1111/brv.12727

von Humboldt, A., & Bonpland, A. (1807). *Essai sur la géographie des plantes, accompagné d'un tableau physique des régions équinoxiales*. Levrault & Schoell.

Welbergen, J. A., et al. (2008). Climate change and the effects of temperature extremes on Australian flying-foxes. *Proceedings of the Biological Sciences, 275*, 419–425.

Williams, J. W., & Burke, K. D. (2019). Past abrupt changes in climate and terrestrial ecosystems. In T. E. Lovejoy & L. Hannah (Eds.), *Biodiversity and climate change. Transforming the biosphere* (pp. 128–141). Yale University Press.

Williams, J. W., & Jackson, S. T. (2007). Novel climates, no-analog communities, and ecological surprises. *Frontiers in Ecology and the Environment, 5*, 475–482.

Wilson, E. O. (2016). *Half-earth: Our planet's fight for life*. Norton & Company.

Wing, S. L., et al. (2005). Transient floral change and rapid global warming at the Paleocene-Eocene boundary. *Science, 310*, 393–396.

WWF. (2020). *Living planet report 2020—Bending the curve of biodiversity loss*. WWF.

Agriculture in a Hot World

Why Efficiency Improvements Are Not Enough to Secure Our Food Supply

Ralf Seppelt, Stefan Klotz, Edgar Peiter, and Martin Volk

As this text takes shape, there are about 8 billion people living on our planet. There is no need to go to bed hungry these days. Healthy, sufficient, and diverse food could in principle be available in sufficient quantities to everybody. In fact, global food production averages an energy equivalent of 5000 kilocalories (kcal) per person per day. Per capita consumption differs by 430% between the richest countries (Australia, Germany, Canada, …) with more than 8000 kcal per person and day and the poorest countries (Chad, Congo, Niger, …) with about 2000 kcal (Tilman et al., 2011).

At the same time, about 800 million people are still undernourished. 250 million children under the age of five are either malnourished, have reduced height growth or are significantly over-nourished (Fig. 1). Children in countries of the global South are especially affected by malnutrition. In these regions, people must get by on less than $8 per day, even though the situation has improved significantly in recent decades: Child mortality halved between 1990 and 2017, and the number of people living in extreme poverty fell to 736 million in 2015.

However, there are regional deviations or setbacks: Between 2013 and 2015, child mortality in sub-Saharan Africa rose slightly again (Rosling et al., 2018).

R. Seppelt (✉) · M. Volk
Department Computational Landscape Ecology, Helmholtz Centre for Environmental Research (UFZ), Leipzig, Germany
e-mail: ralf.seppelt@ufz.de; Martin.volk@ufz.de

S. Klotz
Department Community Ecology, Helmholtz Centre for Environmental Research (UFZ), Halle, Germany
e-mail: tefan.klotz@ufz.de

E. Peiter
Institute of Agricultural and Nutritional Sciences, Martin Luther University Halle-Wittenberg, Halle, Germany
e-mail: plant.nutrition@landw.uni-halle.de

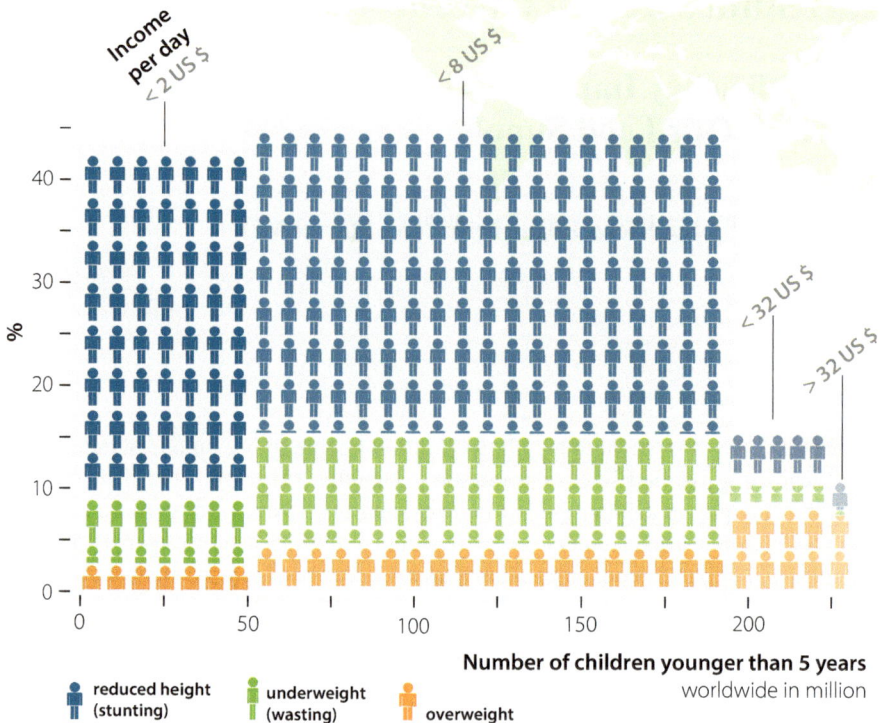

Fig. 1 Malnutrition has three manifestations: reduced height (blue), underweight (green) and overweight (orange). Sorted by income groups—from extreme poverty with less than $2 per person per day to the rich states with more than $32—the absolute numbers of affected children under 5 years of age and their percentage share in this population group are shown here. (Data from UNICEF, WHO)

Africa is repeatedly shaken by individual regional hunger crises, such as in Madagascar in 2020. These are only two examples that show that a development that is quite positive on a global average can turn into the opposite on a regional level. It is to be feared that such crises and setbacks will become more frequent due to climate change.

Food security is not exclusively a question of production quantity. Besides stable production at a sufficiently high level, it is a question of availability, access, distribution, and use of food. This makes it clear that agricultural production and food security are very complex issues, of which only the most important aspects can be highlighted here in relation to the supply of food to the world's population. What our food supply will look like if climate change continues unchecked is therefore not easy to answer, but nonetheless a clear (rather bleak) picture emerges.

According to UN/WHO figures, there will be between 8.8 and 11.6 billion people on the planet by 2060. These people are expected to have a higher income and their eating habits will have shifted towards more energy-rich diets. Accordingly, the demand for food will increase by a further 59–98% by 2050 (Valin et al., 2014).

At the same time, climate change will cause annual average temperatures to rise and extreme events to increase in frequency and intensity (Hansen & Cramer, 2015). Climate change will lead to changes in the distribution of water, earlier flowering dates, longer growing seasons, more frequent crop failures, increased pest infestations, and much more. Yield patterns will change, and it is very likely that agricultural yields will not increase, but rather decrease, as 52% of agricultural land is already considered degraded (WWF, 2020).

To get a sense of how our food supply will evolve in the near future, we will here examine three issues: First, we have a look at the Anthropocene and show how our supply of agricultural products has improved so much that no one should go hungry today, which unfortunately is not the case. We will then, second, also look at where we come from, at what has shaped our basic beliefs about our food system: namely a process of productivity optimization of agricultural crops that has been going on since about 12,000 BC, evolving in the stable climate of the Holocene and optimized through domestication and breeding (Doebley et al., 2006). Finally, we are trying to understand the complex ecological interactions that are crucial for sufficient and stable production of healthy food. Recognizing that, if average temperatures rise by (much) more than 2 degrees, our world will look very different. Business as usual is by no means an option. Investing in innovation and optimization of food production might, on the contrary, be just one part of the story. In the end, we are probably left with the conclusion that a different, healthier way of life has many positive side effects, while at the same time, new technologies and breeding successes must be part of the solution (Löwenstein, 2017; Willett et al., 2019).

Our Recent Past in the Anthropocene

Since 1960, global food production has risen so much that, despite a growing world population, the amount of food produced per capita has increased (Fuglie et al., 2019; Fuglie, 2021). This increase resulted only partly from the expansion of the area under agricultural cultivation. This is not to ignore that increasing deforestation of tropical rainforests is primarily driven by demand for agricultural products. But the expansion of agriculture does not explain a 2.5-fold increase in production. The main reason why agricultural production has increased so much is the higher intensity of land use. The yield potential of our crops has increased greatly in the course of domestication. For example, the Green Revolution in the 1960s let to varieties that, due to their compact growth, could convert higher amounts of nitrogen fertilizer into produce (Bailey-Serres et al., 2019). On agricultural land, more fertilizer, more pesticides, but also more water are used to achieve this increased yield potential. What plants lack in the field, the so-called limiting factors, are supplied by humans.

Mitscherlich and Liebig had already developed a general law in the nineteenth century (Mitscherlich, 1909). Today we remember it with the term "Liebig's Minimum Principle" (Liebscher, 1895). It says that yields can be increased to a maximum by eliminating all limiting factors—but not indefinitely. Mitscherlich also formulates the principle of diminishing returns: the higher the yield, the more likely it is that further increases in yield, for example by adding fertilizers, will have only a minimal positive effect on yield. So-called external effects for the environment (e.g., pollution of groundwater and the species community in the receiving water) are then all the greater. One way of reducing external effects is to increase the nutrient efficiency of crops through breeding, i.e., more efficient uptake and utilization of the given amount of nutrients. Such strategies benefit from an improved understanding of how the crop functions, such as the nitrogen uptake of rice (Hu et al., 2015). Mechanisms of wild plants, which are often highly efficient in taking up nutrients such as phosphate (Lambers et al., 2015), can also be transferred to crops (Fig. 2).

Interim conclusion: the surface area on our planet is obviously limited. Humans have already anthropogenically shaped 70% of the earth's surface, be it through the creation of pastureland, crop farming, or the construction of roads and settlements (Watson et al., 2016; Díaz et al., 2019b). The question of the extent to which yields can be further increased through further technological developments and ultimately intensification is highly topical and open: what is certain is that the answer to this question varies greatly from region to region. In regions with a severe undersupply of nutrients, even small amounts of fertilizer lead to large increases in yields, while other regions suffer from severe oversupply, for example with phosphate (MacDonald et al., 2011). A look at current discussions of agricultural production in Europe or in Germany shows that it is difficult to achieve further increases in yield and thus higher incomes for agriculture with the currently cultivated crops on the given areas,

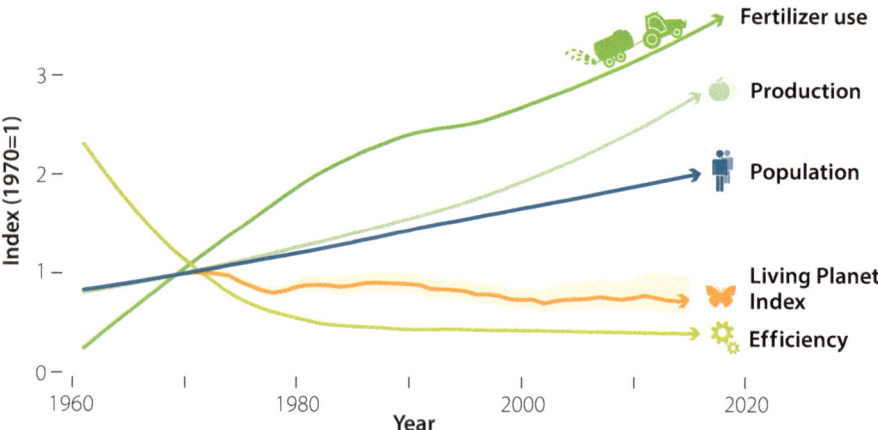

Fig. 2 Global agricultural production since 1960 in index values (all data for 1970 are normalized to 1). While total agricultural production increases by a factor of 2.8, the population grows by a factor of two. As an example of a factor that made this increase in production possible, the amount of fertilizer applied worldwide is shown, which has increased by more than 3.5 times. Over the same period, the ratio of goods produced per amount of fertilizer used ("efficiency") has fallen continuously. The index for the integrity of ecosystems ("Living Planet Index") shows that this is continuously decreasing. (Data from WHO, FAOStat, WWF)

precisely because the areas and the farms are already farmed very intensively. In Europe we are already achieving maximum yields—but is this also true globally?

An analysis of time series of the production of different renewable resources answers the question of whether these time series are subject to continuous growth or whether they tend toward saturation. In the case of fossil resources, we are (still) experiencing a continuous increase in production: mankind's hunger for energy is so great that more and more fossil fuels are being extracted, with the well-known consequence that these continue to pollute the atmosphere and lead to a further increase in CO_2 concentrations worldwide. In contrast, for many renewable resources, it is clear that the point of maximum yield increase has long been exceeded globally (Fig. 3).

However, this analysis of globalized agriculture also shows the process of intensification described above. If we look at the time series of the means of production used ("resources" in Fig. 3), we see that first the rate of increase of agricultural land (the expansion) reached its peak and slows down since 1950. Then the increase in irrigated land reached its maximum in 1978, and finally, the increase in nitrogen fertilizer use peaked in 1983. Liebig's principle can be observed well here. First, the most fertile land was put under the plough; then land use was extended to less fertile

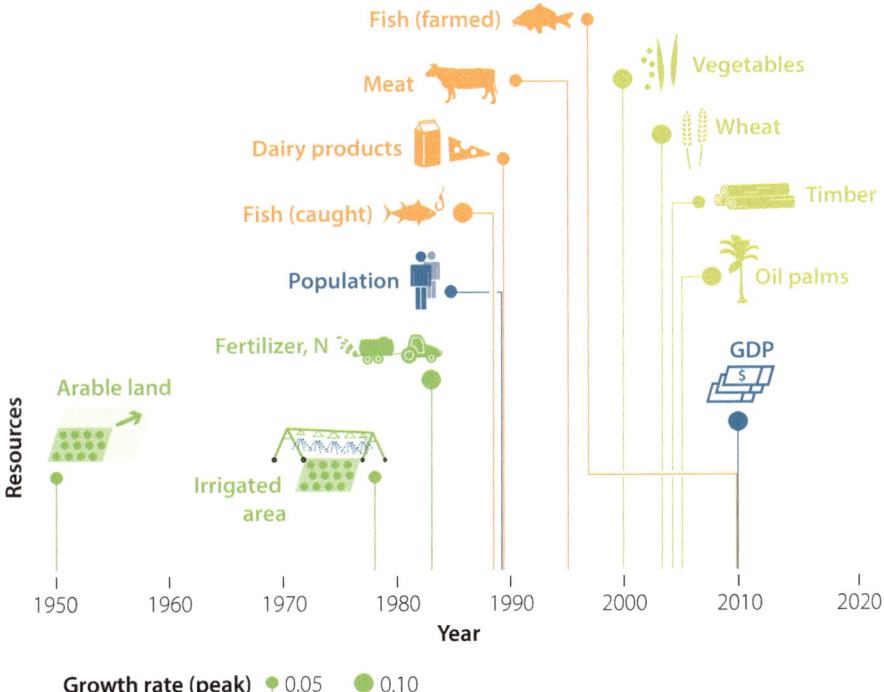

Fig. 3 Results of the determination of years of maximum yield or production growth *(peak year)* of products from animal husbandry (orange), arable farming (green) and the use of means of production and socio-economic variables (blue). The dot shows the respective year of maximum production growth, the size of the dot corresponds to the growth rate in the respective year, see also www.ufz.de/global-agriculture. (Seppelt et al., 2014)

areas; and finally, the limiting factors (first water, then nutrients) were targeted, which made further increases in yield possible. In the end, the message remains: the maximum increase in yields of all renewable resources falls almost synchronously in the period from 1989 to 2008 (median: 2006). We are managing a finite planet.

However, these increases were not linear but, with the emergence of innovations like the Green Revolution, in spurts. The question arises whether there will be innovations in the near future that will bring about a new push in yield gains. Recent breakthroughs in our functional understanding of plants hold out the prospect of such a disruptive development (Bailey-Serres et al., 2019). New possibilities of genome editing will massively accelerate the further development of crops, as precise changes can be made now, for the first time, in the genome of adapted genotypes (Chen et al., 2019). It will even be possible to make wild plants usable within a few generations, which is equivalent to domestication in fast motion (Zsögön et al., 2018).

But what might be causing the slowing of production growth in many renewable resources? It is unlikely that declining consumption and demand are the causes. Although global population growth is slower, this trend is more than offset by increasing consumption due to growing prosperity and by a shift toward more energy-rich (meat-heavy) dietary patterns is taking place.

Before we come back to how sensitive the global system of agriculture is, let us look back at how we have reacted to bottlenecks in the past millennia. This should make it clear that so far, we have always found creative responses to problems: agricultural production in the Holocene actually seems to be a continuous success story. This can be encouraging. But the retrospective also makes clear how drastic the changes coming our way are and how short the time span is in which the right decisions must be made.

Agricultural Activity in the Holocene

With the Neolithic Revolution, about 10,000 years before our era, Homo Sapiens began to leave behind his hunter-gatherer stage becoming sedentary. This has brought many changes to the in these days still very small societies, through which our species has literally grown to become the most successful, or rather most influential, species on the planet. Humans consume 25% of the annually growing net primary production (Krausmann et al., 2013). No other species in Earth's history has ever had such a high global turnover of matter, not even the dinosaurs. We are indeed in the process of (over)shaping the Earth's ecosystem, which is why we now speak of the "Anthropocene".

In the beginning, one of the most important innovations for us was the domestication of the first arable crops, such as emmer, barley, lentils and rice, and of the first animals (Doebley et al., 2006). This happened in China, in New Guinea, in Central America and of course in the Fertile Crescent in what is now the Middle East (see Fig. 4a). About 2600 years before our era, about 50% of all domesticated plants and animals were already known and used (Fig. 4c). After that, it took another 4000 years until this way of cultivation spread to almost the entire surface of the earth and displaced other ways of life of the genus Homo (which were still surviving then and even today). New domesticated plants such as cotton and maize arrived from Central and South America and from regions in Africa. Much of the land used today was settled and tilled by humans at that time. Until the end of the Middle Ages, agriculture continued to expand across Europe, Asia, Africa, Central and South America, new forms of cultivation such as three-field crop rotation were invented, and later four-field rotation was added as a further innovation. Industrialization made the use of machines possible, large-scale synthesis of mineral nitrogen fertilizer made agricultural work vastly easier, less physically demanding, and thereby facilitated the cultivation of ever larger areas. By now, we have modified almost all the fertile land on the planet in some way. Since 1950, productivity has been further optimized through breeding of more efficient varieties to genetically modified crops and the

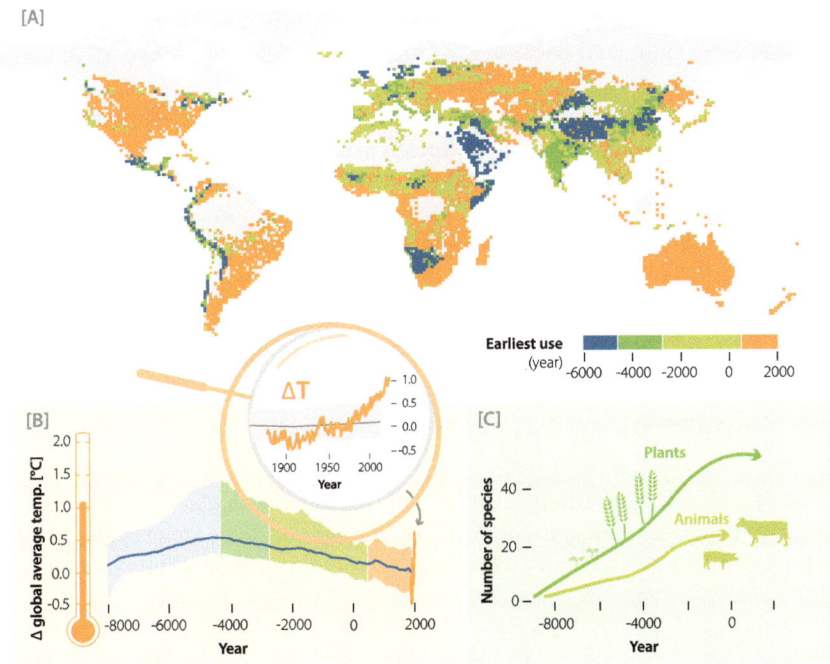

Fig. 4 Agriculture developed (Map **a**) about 10,000 years ago in the fertile areas of the Middle East, China, but also Central America and the mountainous regions of South America (agricultural cultures of the Incas). During this time, the climate was surprisingly stable, and temperatures fluctuated by only a few degrees—a pattern that, in the last 100 years or so, we seem determined to leave behind (**b**). Many animal and plant species were quickly domesticated and cultivated (**c**). The expansion of the genus *Homo* over the subsequent 10,000 years has led to the worldwide spread of agricultural production methods, along with the associated cultivated species, which were (and are) not always adapted to their new locations. (Data from Ellis et al., 2010, 2013; Seppelt et al., 2014; Kaufman et al., 2020)

development and application of sophisticated crop protection products. The application of nitrogen, phosphate, and pesticides multiplied between 1960 and 1990 (Fig. 2) (Foley et al., 2011).

In the relatively stable climate of the Holocene, these innovations spread wherever possible. At the same time, homogenization took place. Since the conquest of the world in the late Middle Ages, an exchange of agricultural cultures took place, with advantages and disadvantages. On the positive side, crops that were easy to cultivate became widespread. Cereal crops from the Middle East conquered the world, as did the "American" potato and maize. On the negative side, introduced alien species have drastically changed entire ecosystems. Many species that had no natural enemies until the appearance of humans were hunted to extinction (Díaz et al., 2019a, b). The main crops cultivated by humans are 1-year grasses: cereals such as wheat, rye, barley, rice, maize, millet, etc. This homogenization has led to a one-sided and uniform production system in parts of the world (Khoury et al., 2014, 2016).

Innovation is thus a guiding principle throughout our entire development. In the course of the Holocene, the model of cattle breeders and arable farmers prevailed—and could be optimized because environmental conditions, above all a stable climate, allowed it. Of course, there were always setbacks. The decline of the Fertile Crescent, for example, was a result of both overuse and climatically induced changes in water availability. Between the fifteenth and seventeenth centuries, the so-called Little Ice Age—although it reduced the average temperature by merely 0.1 °C—brought Europe repeated crop failures due to cooler summers. But the bottom line remains the same: A stable climate and human ingenuity have created an extremely successful system of global resource use, which is now slowly reaching its production limits.

Agriculture—Still a Very Sensitive System

Even though a drastic homogenization of production took place, especially with the onset of the modern era, agricultural methods and practices are exceedingly diverse around the world—especially when one takes into account not only biophysical and climatic aspects, but also socio-economic ones.

Yield Gaps

Identifying yield gaps is one way of finding out to what extent yield increases can be achieved in the future and in which regions this is most likely to be feasible. All locations where agriculture can be practised are compared in terms of their biogeographical and climatic factors and grouped into categories. Within these categories, the locations where yields are highest are identified, and it is then assumed that this yield can also be achieved at all comparable locations, which thus are presumed to have yield gaps. The values given in the literature vary, but it can be assumed that, under current climatic conditions, the closing of all yield gaps would raise aggregate yield by 58%. If the option of growing other, more efficiently producible crops in some locations is included, the potential yield increase rises to 148% (Mauser et al., 2015). But how easy (or difficult?) is it to realize these gains in practice? Do we just have to manage agricultural land more intensively, for example, by applying higher amounts of fertilizer, and everything will be fine?

It is obvious that in some regions the necessary means of production are not available or cannot be financed. This becomes clear as soon as social and economic factors are taken into account, resulting in so-called land use systems (also called "land system archetypes" or LSA) (Fig. 5).

Land Use Systems or Land System Archetypes (LSA)
To better assess the effects of land use and to understand the diversity of land use systems around the world, a world map of land use was created and more than 30 indicators on agriculture, environment, climate, and the socio-economic situation were evaluated for this purpose.

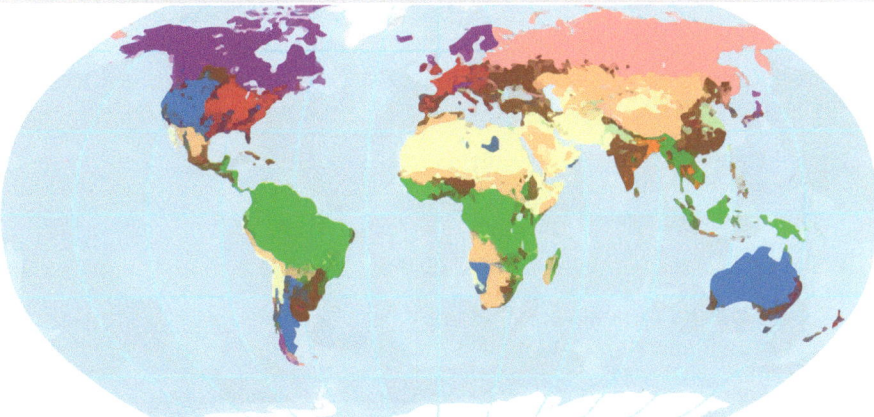

Fig. 5 World map of different land system archetypes (LSA). As a result of an analysis of biophysical, climatic, and socio-economic data, twelve characteristic archetypes can be distinguished, which account for between 0.1% (urban areas) and up to 20% (boreal regions in the east) of the global land area. (Václavík et al., 2013)

Figure 5 (and the corresponding legend) shows for various regions whether maximum intensive cultivation is already practiced (LSA 10) or agriculture can still be intensified (LSA 7)—as well as what climatic conditions are expected and whether more or fewer people will need to be supplied in the future. In large parts of China, India, and Eastern Europe, for example, "traditional agriculture with high labor input" (LSA 7) is practiced. These areas are important because they are classic yield gap regions, meaning areas where farmers could still increase yields because little fertilizer is being applied, the infrastructure is poorly developed, or agriculture still receives little state support.

These findings also support evidence-based conclusions on what specific measures can be taken to counteract the negative effects of land use. In parts of Latin America and Southeast Asia, for example, soil erosion is extremely high (LSA 2). Since agriculture plays an important role in the national economy, the ecological problem becomes a socio-economic one and should be remedied with the highest priority. This would not only remedy environmental harms but also increase yields and returns from agriculture.

1. Forest systems of the tropics (14%)

   high biodiversity,
   increase in arable and pasture land,
   comparatively high climate anomalies

2. Degraded forest and pasture systems of the tropics (0.35%)

   high erosion
   high share of agriculture in GDP
   low political stability

3. Boreal forests of the western world (14%)

   high GDP and political stability
   poor accessibility
   low land use intensity

4. Boreal forests of the eastern world (20%)

   like (3), but low political stability
5. Urban agglomerations (0.1%)

   well above-average population density, rising total population
   a wide range of environmental conditions

6. Rice cropping systems with high yield potential (1%)

   high proportion of arable land and high species richness
   agriculture accounts for a large share of GDP
   high population density

7. Traditional agriculture with high labor input (11%)

   large and increasing share of arable and pasture land
   agriculture accounts for a large share of GDP
   good accessibility
   a wide range of environmental and climatic conditions

8. Grazing systems (13%)

   above-average and increasing share of pasture land
   agriculture has a large share in GDP
   below-average population density, rising population figures

9. Irrigation field cultivation (2%)

   far above-average share of arable land and associated energy use
   high rice yields
   agriculture accounts for a large share of GDP
   well above-average population density, rising population figures

10. Intensive agriculture (5%)

   well above average, but decreasing share of arable land
   far above-average energy use (fertilizer, pesticides)
   agriculture accounts for a rather small share of GDP
   good accessibility and political stability
   temperate climate

11. Marginal land in developed countries (9%)

   High GDP and low population density
   Little to no arable yields, but some pasture land

12. Deserts and wastelands in the countries of the Global South and emerging
    economies (11%)

   high temperatures, little precipitation
   little to no pasture and arable land
   low GDP

Biodiversity—A Threatened Production Factor

The question how we can feed all now and in the future is often answered by saying that a higher intensity of cultivation leads increased production. Irrespective of whether this is feasible everywhere, more intensive land use (and expansion into still semi-natural ecosystems) has negative effects, even though increased application of production inputs such as fertilizers or chemical pesticides did facilitate enormous gains in global agricultural production.

A critical effect of this is the clear downward trend in biodiversity indicators: the so-called Living Planet Index has declined by about 25% since 1970. The Intergovernmental Platform on Biodiversity and Ecosystem Services (IPBES) states in its first global assessment (Díaz et al., 2019a) that our way of land use, from the intensity of resource use to the homogenization of landscapes, is the key driver of biodiversity loss. If we continue on this path, we can expect the extinction of 500,000 to one million species can be expected (Díaz et al., 2019a, b).

We cannot answer the question of possible future scenarios for agriculture under a significantly warmer climate without also considering the aspect of biodiversity. Agricultural production depends on functioning ecosystems, and these are not possible without maintaining biodiversity. Biodiversity is a crucial agricultural production factor: birds and insects fight potential pests (herbivores) and thus help stabilize yields. About 70% of all crops depend on pollination services, which in turn are provided by insects and birds or bats. Almost all agricultural production is based on sufficiently deep and fertile soils, which in turn are due to the interaction of fungi and microorganisms. Currently, soil degradation has reduced the productivity of 23% of the world's land area, and the loss of pollinators threatens $235–$577 billion worth of annual global crop yields (Díaz et al., 2019a).

Intensification is therefore a double-edged sword: it eliminates limiting factors and thereby often leads to a yield increase (Beckmann et al., 2019), but it also decreases biodiversity (Newbold et al., 2015), thereby jeopardizing yields in the long run. This is critical in agricultural systems that have been used rather intensively for a long time (especially in archetypes LSA 6 and 7, Fig. 5).

Here, yield increases of up to 80% are contrasted with a decline in biodiversity of up to 30%, and we recognize that such systems are highly sensitive and do not react linearly (Fig. 6). Therefore, it does not seem wise to pursue agriculture economically optimised at the stress limit, i.e., to clear landscapes to the maximum and to manage them with maximum intensity—especially when one considers that climate change will increase disturbances and weather extremes. We should therefore aim to achieve stable crop production at a sufficient level with a reduced supply of energy and other inputs. Crops that are pathogen-resistant and make better use of nutrients can contribute to this. Through more diverse, small-scale land management and the continuous development of crops adapted to changing conditions, agro-ecosystems and biodiversity can be preserved and yields stabilized (Egli et al., 2020, 2021a, b).

What Is Efficient Agricultural Production?

Conventional intensification is also questionable in other respects. This is due to how "efficiency" is understood. The basic idea of agricultural production is (or rather was) to use the fertility of the soil to convert solar energy into biomass via photosynthesis. While this led to a (net) gain in energy for centuries, conventional agriculture is now a loss-making business in terms of energy. A measure of this relationship is so-called Total Factor Productivity (TFP) which, due to anthropogenic climate change, has deteriorated by 21% since 1961 (Ortiz-Bobea et al., 2021). For countries in Africa and Latin America (incl. The Caribbean), TFP has even declined by 26–34%. Taking increasing yields as a given, this means that more and more energy, labor and effort have had and will have to be expended on further

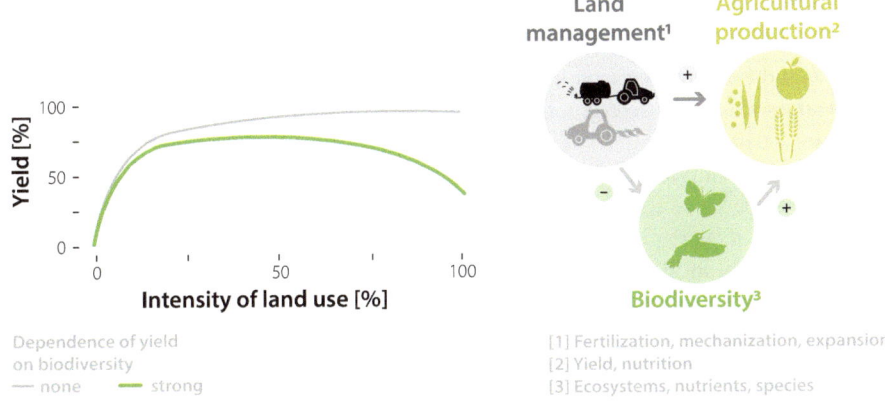

Fig. 6 Taking into account that biodiversity is a production factor, agricultural yields can decline again when a certain intensity of land use is reached (left). Biodiversity positively influences the level and stability of yields but is negatively influenced by the intensity of land use or management (right). This is especially true in cases where crops are sensitively dependent on functioning ecosystems or a stable high level of biodiversity and technical substitutions are difficult to achieve. This *trade-off has* been shown in a meta-analysis for forestry, arable and pasture systems, and various intensification measures. (Graphics adapted from Beckmann et al., 2019; Seppelt et al., 2020)

yield increases. Agricultural production has never been as inefficient as it is today with respect to energy balances (Fig. 2).

In parallel, nutrient withdrawals have never been so high. Never before have we used so many synthetic substances which interact in correspondingly diverse and uncontrollable ways. We have moved far away from the kind of circular economy that was customary in particular through the classical combination of animal and plant production. Large quantities of plant biomass are used as feed for meat production; global meat consumption has risen apace and far exceeds World Health Organization (WHO) recommendations (Willett et al., 2019). A lower-meat diet would not only be healthier, but also more energy-efficient, free up land for food production for humans (without the diversions via animals) and reduce greenhouse gas emissions.

Small-scale agriculture continues to be the main contributor to food security (Tscharntke et al., 2012). Numerous studies show that larger farms are more productive than smaller ones, especially when looking at labor productivity. However, small-scale, sustainable agriculture that relies on polyculture can produce more food per unit of land (Deolalikar, 1981). It is obvious that small-scale farming will prevail wherever there is a lack of energy (as contained in synthetic fertilizers, for example) combined with abundant labor, as in China (Li et al., 2020) or in Andhra Pradesh's zero budget natural farming (Grefe, 2020). In subsidy systems based on the area under cultivation, still prevalent in the European Union (Pe'er et al., 2019), such forms of farming are not even considered.

Agriculture Under a Changed Climate

What would be the effects on agricultural production if the climate protection targets of the Paris Climate Agreement were missed and the temperature rose by, for example, 3 degrees?

Due to the complexity of the issue, we must be pragmatic and simplify things. A possible temperature increase of 3 degrees is the result of different scenarios. The "worst-case" scenario RCP 8.5 of the IPCC presumably reaches this temperature increase already in 2060, whereas it would "only" occur in 2100 in the scenarios RCP 6 and RCP 4.5 (Deutsch et al., 2018). A warming by 3 degrees can therefore become real sooner or later; it will also do so at different times in different places—and there will be very many regions that will heat up by even more than the global average increase.

As we have seen in our considerations of the past, the following aspects of the socio-ecological system "agriculture" are relevant:

- the available area,
- the physiology of arable crops (growth),
- the eco-factors of an agricultural landscape, and
- the people cultivating these areas.

Sounds complicated, and it is, especially if we expand our question to ask whether future food production will suffice to feed everyone. We must then take two additional unknowns into consideration:

- the number of people—the WHO predicts 8.8–11.6 billion people for 2060 and 7.3–15.6 billion for 2100, and
- our future dietary habits.

Available Area and Possible Production Increases

Various studies have looked at potential changes in arable land under rising temperatures. Based on model calculations, it could be shown that for maize in North America and Europe, land gains of 10–20% can be expected under a warmer climate, whereas for Africa, South America and Oceania, losses of up to 40% are to be expected (Ramirez-Cabral et al., 2017). A global analysis of the basic suitability for the cultivation of agricultural crops shows a similarly uneven picture and predicts an average land gain of about 3%, whereby the suitability of newly acquired land is described as rather moderate (Zabel et al., 2014). Expansion agriculture will happen in Northern latitudes, characterized by shorter day length. Some expansion of cropland into previously unusable areas will also be possible through the development of drought-adapted or salt-tolerant crops (Fig. 7a).

The extent to which an increase in the CO_2 concentration has a positive effect on plant growth has been explored at many agricultural test stations (Fig. 7c). Especially

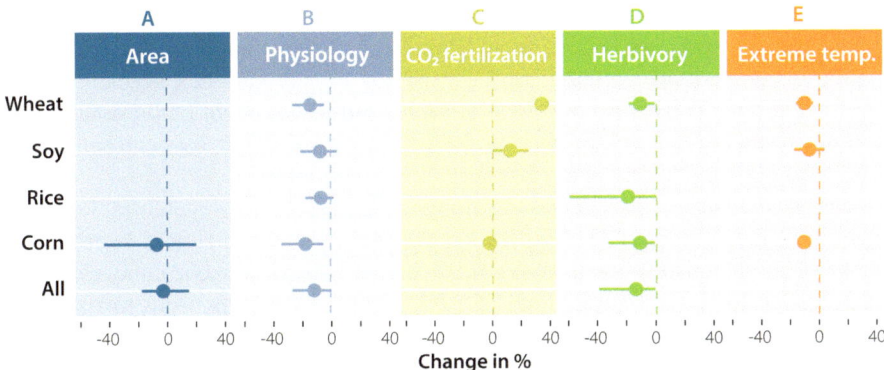

Fig. 7 Overview of possible individual effects of higher global temperatures on available area (**a**) and on yields due to temperature-induced changes in physiology (**b**), CO_2 fertilization effects (**c**), herbivory (**d**) and extreme temperatures (**e**) for different crops. The figures given in the text are in each case mean values of the expected changes under a warmer global climate with a 3 degrees higher mean temperature (the error bars show the existing uncertainties). (Data from Deryng et al., 2011; Zabel et al., 2014; Zhao et al., 2017; Ramirez-Cabral et al., 2017; Deutsch et al., 2018; Jägermeyr et al., 2021)

for plants with less efficient photosynthesis (so-called C3 plants) of the mid-latitudes, such as soy, cassava, rice, potatoes and wheat, a fertilization effect and thus yield increases of 10%, 20%, and 25%, respectively, were observed (Bishop et al., 2015). But these increases can be achieved only if other potentially limiting factors, such as water and nitrogen, are available in sufficient quantities. The atmospheric CO_2 concentrations in these experiments were, however, higher than necessary for a 3 degree warmer world. In more realistic experimental set-ups, there would be significant interactions with parameters such as the availability of nitrogen and water, and clear conclusions would then be more difficult to draw. Within the framework of a model study that assumes a drastic climate scenario with a temperature increase of 6 degrees by 2100 (RCP 8.5), it can be shown for wheat, soy, and rice that a positive CO_2 fertilizer effect is offset by negative effects such as water scarcity, higher ozone levels, or the occurrence of extreme events—with great geographical heterogeneities (Lombardozzi et al., 2018) (Fig. 7e).

Growth of Agricultural Crops in a Changing Climate

Higher temperatures lead to faster growth and earlier maturity of plants. But such a shortened growing season also means that the plant has less time to accumulate biomass, and yields are reduced. The increase in extreme weather events continues to lead to more frequent crop failures. Against this background, it could be shown that the climatic changes already observed between 1980 and 2008 led to a reduction in harvest volumes for maize and wheat by 3.8% and 5.5%, respectively, most strongly in China, Brazil and Russia (Lobell et al., 2011). For wheat, rice, maize and soybean, which provide two-thirds of global calorie demand, losses of 3–7% per degree of temperature increase can be expected for the varieties available today (Zhao et al., 2017). However, 7–18% of such yield losses could be prevented with optimized sowing times alone, as a model simulation for maize, soybean and spring wheat has shown (Deryng et al., 2011).

Less studied are crops that do not belong to the main crops: vegetables, legumes, fruits. For these crops, CO_2 fertilization effects could lead to yield increases of about 22%—but also to yield losses of 9% due to increased ozone concentrations, of 35% due to water scarcity, and of 32% if the temperature were to rise by 4 degrees (Scheelbeek et al., 2020). In addition, these crops depend on intact ecosystem functions even more than annual grasses such as cereals. Pollination, which is so important, could be threatened by temperature-induced shifts in flowering times or changes in insect population dynamics (Fig. 7b).

Breeding new varieties could be another method of choice to achieve higher yields, but no yield increases beyond 1% were observed for wheat, rice, and maize between 1980 and 2008 (Fischer & Edmeades, 2010). If one extrapolating a (quite optimistic) 1% breeding-induced annual yield increase from 2020 to 2060 (at +3 degrees in the RCP 8.5 scenario), this could translate into 48% higher yields. Such a breeding success would only be realistic, however, if it also aimed at increased resistance to pests and robustness against water shortages and higher temperatures (Tester & Langridge, 2010). The fact that every non-extremophilic organism suffers from absolute temperatures of more than 42 °C is a biological constant that cannot be revised even by the most skillful breeding.

Interactions in the Agroecosystem

A warmer climate is in principle good for all species whose activity (metabolism) depends on temperature. Especially ectothermic organisms that cannot regulate their body temperature themselves (e.g., insects) will clearly benefit from a warmer climate. In our context, this is especially relevant for herbivores (plant-eating pests) (Fig. 7d).

In insects, higher temperatures lead to an increased metabolic rate: they simply eat more. Secondly, higher temperatures lead to increased offspring. Based on these basic ecological principles, there is a temperature-related increase in the activity of pests on rice, maize, and wheat, which can lead to crop losses of 10–25% per degree of temperature increase (Deutsch et al., 2018). Even in a 2-degree warmer world, wheat yield losses of about 18% for Europe and North America and of about 17% for East Asia would be predicted. Rice would suffer losses of around 59% (in South and Southeast Asia) and 32% (East Asia). For maize, the losses would be highest in North America, with up to 32%, and in Europe they would be around 23% (Deutsch et al., 2018).

Lastly, plant diseases are expected to increase in a warmer climate, by up to 25% due to weather conditions. Species introduced in the context of crop replacement could be especially affected (plus 56%) (Anderson et al., 2004). Concrete figures on possible yield losses are not yet available (Fig. 7d).

The Socio-Ecological System

Without farmers there is no agriculture. Accordingly, a focus on pure yield effects and ecological aspects must remain incomplete if we want to find answers to the question of the development of the global food system. The effects on people in two of the most endangered regions are largely ignored: Sub-Saharan Africa and Southeast Asia. With a global warming of 3 degrees, agricultural labor capacity could drop by 30–50% due to heat stress. Food prices would rise, many more people would have to work in agriculture to maintain current levels. The global welfare loss at this level of warming could be as much as US$ 136 billion annually (de Lima et al., 2021). If one takes the total economic output into account, there would be a loss in gross world product of 23% (Burke et al., 2015, 2018). As a result of the development of regional heat centers, currently densely populated areas will no longer be suitable for agriculture and perhaps even become uninhabitable. This will lead to unpredictable migration movements (see also chapter "Escape from Heat, Drought, and Extreme Weather" Chazalnoel & Ionesco, (2024)), and even to the collapse of societies—on both sides of the migration movement (Kang & Eltahir, 2018).

Conclusion and Outlook

Under current climatic conditions, yield increases can be achieved by closing yield gaps. Only moderately intensified use allows yield increases of up to 80% on land that has been cultivated moderately intensively up to now (Beckmann et al., 2019). But as we have seen, these potentials are rather poor in a world that is 2–3 degrees warmer. Many theoretically achievable successes will be offset by expected negative side effects. However, it is also possible that the effects will negatively reinforce one another. Finally, different regions will be affected to very different degrees.

Against the backdrop of these uncertainties, simply tabulating future income losses and gains cannot be done in a reasonable way. The general trend, however, is clear and indisputable: a world that is 3 degrees warmer is confronted with the risk of massive yield losses; in extreme cases, large land areas might become desolate or uninhabitable.

Climate impacts affect agriculture all over the world: rich countries like Australia, rising nations like China (Kang & Eltahir, 2018), but also regions where a large part of the population must survive on less than $2 a day. There, the effects of climate-related yield reductions can be devastating, and the expected increase in extreme-temperature days can lead to major famines (Battisti & Naylor, 2009). Conflicts over resources are likely to intensify. Migration to the nearest metropolises or beyond will increase. We cannot rely on yields continuing to increase through inten-sification or expanded cultivation, or on global trade leading to a more equitable distribution of available resources. The gaps in supply which we haven't been able to close in recent decades are likely to widen.

In a situation where we already produce twice as much food as we need, where approx. 40% is lost, most of it as food waste on, only 15% is due to harvest losses, a large part of the world's population eats a diet far too rich in energy, we could today feed even a population of 10–11 billion. Thus, the solution does not call for more production. It is a collective task to mitigate climate change while ensuring agricultural yields under rapidly changing climatic conditions (Löwenstein, 2017). The solution rests on four pillars:

1. Yield security and stabilization with a high level of biodiversity must be ensured by means of agro-ecological principles and diverse crops of useful plants, whereby the latter must be adapted to future climates through innovative breeding;
2. Crop and food losses must be drastically reduced, ideally eliminated completely;
3. Eating habits must change toward a more conscious, lower-energy and healthier lifestyle;
4. Trade must serve to distribute food fairly and to compensate for possible climate-related yield losses, and must not displace locally adapted cultivation methods.

The question of how many people would be able to live and feed themselves healthily and sufficiently under the new climatic conditions of a 3-degree warmer planet cannot be answered with a sufficiently high degree of certainty. Of course, scenarios can be constructed, and well-researched contributions have been pro-duced and is still worthwhile (ABC News, 2009). But it is not the scientific uncer-tainties of the individual effects compiled here in the chapter and the unclear effects of their combination that make an answer to this question seem speculative. More important is the uncertainty concerning our own decisions in the coming years: what resources we will consume in our societies, and in what quantities. It is our decisions that make forecasts uncertain but also provide some grounds for optimism.

References

ABC News. (2009). "Earth 2100": The final century of civilization? In *ABC News*. https://abc-news.go.com/Technology/Earth2100/story?id=7697237&page=1. Accessed 16 Nov 2021.
Anderson, P. K., Cunningham, A. A., Patel, N. G., et al. (2004). Emerging infectious diseases of plants: Pathogen pollution, climate change and agrotechnology drivers. *Trends in Ecology & Evolution, 19*, 535–544. https://doi.org/10.1016/j.tree.2004.07.021

Bailey-Serres, J., Parker, J. E., Ainsworth, E. A., et al. (2019). Genetic strategies for improving crop yields. *Nature, 575*, 109–118. https://doi.org/10.1038/s41586-019-1679-0

Battisti, D. S., & Naylor, R. L. (2009). Historical warnings of future food insecurity with unprecedented seasonal heat. *Science, 323*, 240–244. https://doi.org/10.1126/science.1164363

Beckmann, M., Gerstner, K., Akin, M., et al. (2019). Conventional land-use intensification reduces species richness and increases production: A global meta-analysis. *Global Change Biology, 17*. https://doi.org/10.1111/gcb.14606

Bishop, K. A., Betzelberger, A. M., Long, S. P., & Ainsworth, E. A. (2015). Is there potential to adapt soybean (Glycine max Merr.) to future [CO2]? An analysis of the yield response of 18 genotypes in free-air CO2 enrichment. *Plant, Cell & Environment, 38*, 1765–1774.

Burke, M., Hsiang, S. M., & Miguel, E. (2015). Global non-linear effect of temperature on economic production. *Nature, 527*, 235–239. https://doi.org/10.1038/nature15725

Burke, M., Davis, W. M., & Diffenbaugh, N. S. (2018). Large potential reduction in economic damages under UN mitigation targets. *Nature, 557*, 549–553. https://doi.org/10.1038/s41586-018-0071-9

Chazalnoel, M. T., & Ionesco, D. (2024). Escape from heat, drought, and extreme weather. In *Hot season ahead: What a 3-degree warmer world looks like*. Springer.

Chen, K., Wang, Y., Zhang, R., et al. (2019). CRISPR/Cas genome editing and precision plant breeding in agriculture. *Annual Review of Plant Biology, 70*, 667–697. https://doi.org/10.1146/annurev-arplant-050718-100049

de Lima, C. Z., Buzan, J. R., Moore, F. C., et al. (2021). Heat stress on agricultural workers exacerbates crop impacts of climate change. *Environmental Research Letters, 16*, 044020. https://doi.org/10.1088/1748-9326/abeb9f

Deolalikar, A. B. (1981). The inverse relationship between productivity and farm size: A test using regional data from India. *American Journal of Agricultural Economics, 63*, 275–279. https://doi.org/10.2307/1239565

Deryng, D., Sacks, W. J., Barford, C. C., & Ramankutty, N. (2011). Simulating the effects of climate and agricultural management practices on global crop yield. *Global Biogeochemical Cycles, 25*. https://doi.org/10/c3p72k

Deutsch, C. A., Tewksbury, J. J., Tigchelaar, M., et al. (2018). Increase in crop losses to insect pests in a warming climate. *Science, 361*, 916–919. https://doi.org/10.1126/science.aat3466

Díaz, S., Settele, J., Brondizio, E., et al. (2019a). *Summary for policymakers of the global assessment report on biodiversity and ecosystem services of the intergovernmental science-policy platform on biodiversity and ecosystem services*. Intergovernmental Science-Policy Platform on Biodiversity and Ecosystem Services (IPBES) Secretariat.

Díaz, S., Settele, J., Brondízio, E. S., et al. (2019b). Pervasive human-driven decline of life on Earth points to the need for transformative change. *Science, 366*, eaax3100. https://doi.org/10.1126/science.aax3100

Doebley, J. F., Gaut, B. S., & Smith, B. D. (2006). The molecular genetics of crop domestication. *Cell, 127*, 1309–1321. https://doi.org/10.1016/j.cell.2006.12.006

Egli, L., Schröter, M., Scherber, C., et al. (2020). Crop asynchrony stabilizes food production. *Nature, 588*, E7–E12. https://doi.org/10.1038/s41586-020-2965-6

Egli, L., Mehrabi, Z., & Seppelt, R. (2021a). More farms, less specialized landscapes, and higher crop diversity stabilize food supplies. *Environmental Research Letters, 16*, 055015. https://doi.org/10.1088/1748-9326/abf529

Egli, L., Schröter, M., Scherber, C., et al. (2021b). Crop diversity effects on temporal agricultural production stability across European regions. *Regional Environmental Change, 21*, 96. https://doi.org/10.1007/s10113-021-01832-9

Ellis, E. C., Klein Goldewijk, K., Siebert, S., et al. (2010). Anthropogenic transformation of the biomes, 1700 to 2000. *Global Ecology and Biogeography, 19*. https://doi.org/10.1111/j.1466-8238.2010.00540.x

Ellis, E. C., Kaplan, J. O., Fuller, D. Q., et al. (2013). Used planet: A global history. *Proceedings of the National Academy of Sciences of the United States of America, 110*, 7978–7985. https://doi.org/10.1073/pnas.1217241110

Fischer, R. A. T., & Edmeades, G. O. (2010). Breeding and cereal yield progress. *Crop Science, 50*, S-85–S-98. https://doi.org/10.2135/cropsci2009.10.0564

Foley, J. A., Ramankutty, N., Brauman, K. A., et al. (2011). Solutions for a cultivated planet. *Nature, 478*, 337–342. https://doi.org/10.1038/nature10452

Fuglie, K. (2021). Climate change upsets agriculture. *Nature Climate Change, 11*, 294–295. https://doi.org/10.1038/s41558-021-01017-6

Fuglie, K., Gautam, M., Goyal, A., & Maloney, W. F. (2019). *Harvesting prosperity: Technology and productivity growth in agriculture.* World Bank Publications.

Grefe, C. (2020). *Brahma, Krishna und Öko* (pp. 25–26). Die Zeit.

Hansen, G., & Cramer, W. (2015). Global distribution of observed climate change impacts. *Nature Climate Change, 5*, 182–185. https://doi.org/10.1038/nclimate2529

Hu, B., Wang, W., Ou, S., et al. (2015). Variation in NRT1.1B contributes to nitrate-use divergence between rice subspecies. *Nature Genetics, 47*, 834–838. https://doi.org/10.1038/ng.3337

Jägermeyr, J., Müller, C., Ruane, A. C., et al. (2021). Climate impacts on global agriculture emerge earlier in new generation of climate and crop models. *Nature Food, 2*, 873–885. https://doi.org/10.1038/s43016-021-00400-y

Kang, S., & Eltahir, E. A. B. (2018). North China Plain threatened by deadly heatwaves due to climate change and irrigation. *Nature Communications, 9*, 2894. https://doi.org/10.1038/s41467-018-05252-y

Kaufman, D., McKay, N., Routson, C., et al. (2020). Holocene global mean surface temperature, a multi-method reconstruction approach. *Scientific Data, 7*, 201. https://doi.org/10.1038/s41597-020-0530-7

Khoury, C. K., Bjorkman, A. D., Dempewolf, H., et al. (2014). Increasing homogeneity in global food supplies and the implications for food security. *Proceedings of the National Academy of Sciences, 111*, 4001–4006. https://doi.org/10.1073/pnas.1313490111

Khoury, C. K., Achicanoy, H. A., Bjorkman, A. D., et al. (2016). Origins of food crops connect countries worldwide. *Proceedings of the Royal Society B: Biological Sciences, 283*, 20160792. https://doi.org/10.1098/rspb.2016.0792

Krausmann, F., Erb, K.-H., Gingrich, S., et al. (2013). Global human appropriation of net primary production doubled in the 20th century. *PNAS, 110*, 10324–10329. https://doi.org/10.1073/pnas.1211349110

Lambers, H., Hayes, P. E., Laliberté, E., et al. (2015). Leaf manganese accumulation and phosphorus-acquisition efficiency. *Trends in Plant Science, 20*, 83–90. https://doi.org/10.1016/j.tplants.2014.10.007

Li, C., Hoffland, E., Kuyper, T. W., et al. (2020). Syndromes of production in intercropping impact yield gains. *Nature Plants, 6*, 653–660. https://doi.org/10.1038/s41477-020-0680-9

Liebscher, G. (1895). Untersuchungen über die Bestimmung des Düngerbedürfnisses der Ackerböden und Kul-turpflanzen. In J. von Liebig (Ed.), *Journal für Landwirtschaft* (p. 13). Braunschweig.

Lobell, D. B., Schlenker, W., & Costa-Roberts, J. (2011). Climate trends and global crop production since 1980. *Science, 333*, 616–620. https://doi.org/10.1126/science.1204531

Lombardozzi, D. L., Bonan, G. B., Levis, S., & Lawrence, D. M. (2018). Changes in wood biomass and crop yields in response to projected CO_2, O_3, nitrogen deposition, and climate. *Journal of Geophysical Research, Biogeosciences, 123*, 3262–3282. https://doi.org/10.1029/2018JG004680

Löwenstein, F. (2017). *Food Crash: wir werden uns ökologisch ernähren oder gar nicht mehr.* Erweiterte Taschenbuchausgabe.

MacDonald, G. K., Bennett, E. M., Potter, P. A., & Ramankutty, N. (2011). Agronomic phosphorus imbalances across the world's croplands. *Proceedings of the National Academy of Sciences, 108*, 3086–3091. https://doi.org/10.1073/pnas.1010808108

Mauser, W., Klepper, G., Zabel, F., et al. (2015). Global biomass production potentials exceed expected future demand without the need for cropland expansion. *Nature Communications, 6*, 8946. https://doi.org/10.1038/ncomms9946

Mitscherlich, E. A. (1909). Das Gesetz des Minimums und das Gesetz des abnehmenden Bodenertrags. *Landwirtschaftliche Jahrbücher, 38*, 537–552.

Newbold, T., Hudson, L. N., Hill, S. L. L., et al. (2015). Global effects of land use on local terrestrial biodiversity. *Nature, 520*, 45–50. https://doi.org/10.1038/nature14324

Ortiz-Bobea, A., Ault, T. R., Carrillo, C. M., et al. (2021). Anthropogenic climate change has slowed global agricultural productivity growth. *Nature Climate Change, 11*, 306–312. https://doi.org/10.1038/s41558-021-01000-1

Pe'er, G., Zinngrebe, Y., Moreira, F., et al. (2019). A greener path for the EU common agricultural policy. *Science, 365*, 449–451. https://doi.org/10/gf5qmt

Ramirez-Cabral, N. Y. Z., Kumar, L., & Shabani, F. (2017). Global alterations in areas of suitability for maize production from climate change and using a mechanistic species distribution model (CLIMEX). *Scientific Reports, 7*, 5910. https://doi.org/10.1038/s41598-017-05804-0

Rosling, H., Rosling, O., & Rönnlund, A. R. (2018). *Factfulness: Ten reasons we're wrong about the world—And why things are better than you think.* SCEPTRE.

Scheelbeek, P. F. D., Moss, C., Kastner, T., et al. (2020). United Kingdom's fruit and vegetable supply is increasingly dependent on imports from climate-vulnerable producing countries. *Nature Food, 1*, 705–712. https://doi.org/10.1038/s43016-020-00179-4

Seppelt, R., Manceur, A. M., Liu, J., et al. (2014). Synchronized peak-rate years of global resources use. *Ecology and Society, 19*, art50. https://doi.org/10.5751/ES-07039-190450

Seppelt, R., Arndt, C., Beckmann, M., et al. (2020). Deciphering the biodiversity—Production mutualism in the global food security debate. *Trends in Ecology & Evolution, 35*, 1011–1020. https://doi.org/10.1016/j.tree.2020.06.012

Tester, M., & Langridge, P. (2010). Breeding technologies to increase crop production in a changing world. *Science, 327*, 818–822. https://doi.org/10.1126/science.1183700

Tilman, D., Balzer, C., Hill, J., & Befort, B. L. (2011). Global food demand and the sustainable intensification of agriculture. *Proceedings of the National Academy of Sciences, 108*, 20260–20264. https://doi.org/10.1073/pnas.1116437108

Tscharntke, T., Clough, Y., Wanger, T. C., et al. (2012). Global food security, biodiversity conservation and the future of agricultural intensification. *Biological Conservation, 151*, 53–59. https://doi.org/10.1016/j.biocon.2012.01.068

Václavík, T., Lautenbach, S., Kuemmerle, T., & Seppelt, R. (2013). Mapping global land system archetypes. *Global Environmental Change, 23*, 1637–1647. https://doi.org/10.1016/j.gloenvcha.2013.09.004

Valin, H., Sands, R. D., van der Mensbrugghe, D., et al. (2014). The future of food demand: Understanding differences in global economic models. *Agricultural Economics, 45*, 51–67. https://doi.org/10.1111/agec.12089

Watson, J. E. M., Shanahan, D. F., Di Marco, M., et al. (2016). Catastrophic declines in wilderness areas undermine global environment targets. *Current Biology, 26*, 2929–2934. https://doi.org/10.1016/j.cub.2016.08.049

Willett, W., Rockström, J., Loken, B., et al. (2019). Food in the Anthropocene: The EAT–Lancet Commission on healthy diets from sustainable food systems. *The Lancet, 393*, 447–492. https://doi.org/10/gft25h

WWF. (2020). *Living planet report 2020—Bending the curve of biodiversity loss.* WWF.

Zabel, F., Putzenlechner, B., & Mauser, W. (2014). Global agricultural land resources—A high resolution suitability evaluation and its perspectives until 2100 under climate change conditions. *PLoS One, 9*, e107522. https://doi.org/10.1371/journal.pone.0107522

Zhao, C., Liu, B., Piao, S., et al. (2017). Temperature increase reduces global yields of major crops in four independent estimates. *Proceedings of the National Academy of Sciences of the United States of America, 114*, 9326–9331. https://doi.org/10.1073/pnas.1701762114

Zsögön, A., Čermák, T., Naves, E. R., et al. (2018). De novo domestication of wild tomato using genome editing. *Nature Biotechnology, 36*, 1211–1216. https://doi.org/10.1038/nbt.4272

Escape from Heat, Drought and Extreme Weather

When Local People Have Nothing Left to Survive

Mariam Traore Chazalnoël and Dina Ionesco

Whether in the media or in the negotiating rooms of the United Nations, people compelled to leave their homes due to extreme climate change impacts are more prominent than ever. There is now no doubt that climate change has a direct and indirect effects on when, how and where people migrate. Governments of migrants' countries of origin, of transit and of destination are becoming increasingly aware that policy changes are urgently needed to address current challenges and to deal adequately with future ones. This chapter provides a brief overview of some key issues relevant to climate migration and discusses some of the policy options for shaping future action.

Important Aspects of "Climate Migration"

There is no internationally valid legal definition of people who migrate due to climate and environmental impacts. This has implications for the policies and obligations of states to protect people fleeing extreme climate impacts. For example, climate migrants cannot be classified as "refugees" as their situation is not recognized by the 1951 Refugee Convention. Climate migrants are generally not entitled to specific assistance under international law. Many proposals have been made to close these gaps, including the extension of the Refugee Convention. But few countries currently have an interest in changing the status quo. There are however international legal instruments applicable to climate migration, including human rights law (IOM, 2014a, b). Some states have also developed national instruments to provide protection and assistance to people migrating due to climate impacts and disasters.

M. Traore Chazalnoël (✉) · D. Ionesco
Geneva, Switzerland

© The Author(s) 2024
K. Wiegandt (ed.), *3 Degrees More*,
https://doi.org/10.1007/978-3-031-58144-1_4

63

In this chapter, we will use the definition of climate migrants developed by the International Organization for Migration (IOM) (2019). This definition is broad and includes *people who migrate due to direct or indirect impacts of climate change, both sudden-onset disasters and gradual environmental degradation. Migration can be within or across borders, temporary or permanent, and may take place on a continuum from 'forced' to 'voluntary'.*

The broad scope of this definition makes it possible to capture the diversity of mobility situations in a changing climate. From nomadic populations in the Sahel changing their traditional migration patterns to adapt to the effects of climate change to islanders in the Pacific fleeing the advancing sea, climate migration takes many different forms around the world.

The Effects of Climate Change on Migration Is Manifold

Climate migration is usually multi-causal, as the decision to migrate is often influenced by a combination of different factors. Social, economic or security aspects often overlap and cannot always be sharply separated from purely climate change-related drivers. Various environmental and climate factors can influence the decision or need to migrate, from sudden disasters such as typhoons and floods to slow-onset processes such as sea-level rise or land degradation (Ionesco et al., 2017).

The connection between environmental changes and population movements is relatively easy to recognize, when sudden disasters force large numbers of people to flee immediately. In such situations, people usually migrate to safe(r) nearby locations and return once the emergency is over.

It is more difficult, by contrast, to establish clear connections between slow climate changes and migration movements (Traore Chazalnoël & Randall, 2021; IOM, 2020b). Slow environmental changes usually extend over generations; desertification or ocean acidification directly affect the livelihood of the population as crops can no longer grow or fish can no longer be caught. In such cases, people may consciously decide to move and set out in search of new livelihood opportunities— it is then often no longer possible to make a clear distinction between "economic migration" and "climate migration". Many migrants coming to the United States from Latin America, for example (commonly called "economic migrants") come from areas affected by drought and other climate change impacts (Puscas & Escribano, 2018).

Climate Migration to Europe—Is the Fear of It Justified?
Some studies have found a possible link between global warming and the volume of asylum applications in the European Union (Missirian & Schlenker, 2017). Others have rejected the European narrative attributing international migration primarily to "economic" reasons is inconclusive for failing to consider the importance of climate and environmental factors for migration (Bendandi, 2020). Such narratives need to be challenged because they do not reflect the current factual situation—most people who migrate because of climate change stay in their own countries. Nevertheless, it is important to recognize potential links between climate factors and migration to Europe without classifying it as a mere border security issue. Only with this recognition effective policies can be developed as alternatives to unregulated migration. Such policies include helping countries of origin to adapt to climate change, restoring damaged natural resources and combating poverty.

Understanding possible links between economic migration and the effects of climate change is important for developing policies that mitigate the adverse economic effects of climate change on populations—and for learning how such policies can affect migration. Not yet sufficiently explored, such might become very important in the future. In highly politicized contexts, in which migration is generally seen as negative, such issues would be quite sensitive.

Is Climate Migration a Threat to Security?
Climate migration is often considered under the "security" label and accordingly classified as a threat to political stability. In some contexts, climate change can indeed increase instability and fragility in vulnerable regions, such as in the Lake Chad region, where the combination of political, economic and climatic factors is leading to increased displacement. In contrast, recent studies show that climate migration does not necessarily lead to security problems—even while climate change should definitely be understood as a threat to human security, regardless of whether it triggers migration or not (Baillat & Traore Chazalnoël, 2022).

Numbers and Life Paths of Climate Migrants

Although knowledge on climate migration has grown exponentially in recent years (Université de Neuchatel, 2018), we still lack long-term longitudinal data that capture the extended time periods of slow environmental change, as well as harmonized data sets and disaggregated data. But even if we do not yet have a complete picture to date, it is obvious that action is needed to minimize and manage the effects of climate change on migration patterns.

The exact number of climate migrants is unknown. The lack of a coordinated database and a universal definition makes it difficult to determine the number of people migrating within or across borders due to climate change; different countries and organizations use different criteria and arrive at different figures (Flavell et al., 2020). The annual report of the *Internal Displacement Monitoring Centre* (IDMC), for example, contains information on displacement caused by natural disasters at the country level. According to this report 33.6 million of new displacements due to disasters were registered in 2022 (IDMC, 2023). These figures only refer to people who were displaced within their own country and by a sudden disaster. Less data exists on cross-border migration and migration forced or motivated by slow environmental change.

These gaps are significant, but the people compelled to leave their homes each year due to environmental pressures certainly number to millions. We also know that some regions of the world, such as island states in the Pacific, are disproportionately affected by such migration (IOM & UN-OHRLLS, 2019). The number of migrants from such places may be small in absolute terms, but these migrant flows naturally have a greater impact on small countries than large ones, which are often better placed economically and in terms of infrastructure. So long as people can stay in their own country, they usually choose this option and migrate within their country. Some countries, such as the Kingdom of Morocco, act as a source, transit and destination country for climate migrants at the same time, which is why they need political and legal frameworks to take all these dimensions into account. Finally, it should be mentioned that there are population movements related to climate change in *all* parts of the world, even in wealthy Western countries.

Forecasts on Future Mobility

It is predicted that, absent appropriate policy measures, climate change impacts will force millions of people to flee. The World Bank estimates that by 2050 gradual environmental changes may force 216 million people to migrate within their own country (Clement et al., 2021). Accurate projections require data on how many people live in regions that may suffer severe climate change impacts. A 2017 report estimates that if global temperatures were rise by "only" 1.5 °C, 30–60 million people might live in areas where temperatures in the hottest month will exceed the limits of human endurance (IOM, 2017a). In addition to the island states already mentioned, coastal regions will be especially hard hit because they are heavily populated—and the seas will rise inexorably (McMichael et al., 2020).

These figures paint a picture of a world in which climate-induced migration has the potential to affect not only the daily lives of countless individuals, but also the socio-economic stability of entire states. The portrayal of an apocalyptic future may create a sense of urgency among decision-makers, but it also encourages the creation of narratives that seek to devalue the threat as scaremongering. This makes it even more important to carefully examine what all the numbers really mean. Not all people living in vulnerable areas will be willing and able to migrate in search of a safer home and better living conditions. Projected migration flows will become reality only if no or insufficient policy measures are taken to reduce the climate-related causes of migration.

In the coming decades, governments around the world will need to show that they can look ahead to identify early on where greater migratory pressure might build. Only with such foresight can timely action be taken and the worst outcomes be averted.

Policy Measures for the Future

Information deficits and insufficient legal frameworks can impede the timely development of appropriate measures. But they are no excuse for doing nothing. The data situation is good enough to develop and implement policy responses at global, regional and national levels. The goal of policy action must be to strengthen the resilience of mobile populations and their associated people (Traore Chazalnoël & Ionesco, 2022).

Climate Migration in the Global Political Debate

During the last two decades, the member states of the United Nations have increasingly recognized the problem of climate migration as one of the greatest challenges of our time. Several phases can be distinguished.

The first phase was one of "absence" of the issue on the global stage, with little interest and limited understanding of the causalities. Around 2005, climate migration was recognized as a cross-cutting issue and discussed as part of the agenda on the social impacts of climate change. A second phase of "awareness raising" began thanks to the efforts of countries at risk from climate change and other affected stakeholders. In a third phase ("mainstreaming"), wording on migration was included in texts negotiated at the annual Conferences of the Parties (COP) in the context of the United Nations Framework Convention on Climate Change (UNFCCC). The 2015 Paris Agreement (UNFCCC, 2015) includes passages on the protection of migrants' rights and mandated the establishment of an interdisciplinary task force on displacement. In parallel, the first major projects were launched, such as MECLEP (*Migration, Environment and Climate Change: Evidence for Policy*) (IOM, 2017b) and *Where the Rain Falls* (Warner & Afifi, 2013).

After 2015, a fourth phase of "unification" began, focusing on improving policy coherence and building partnerships between stakeholders, emerging interest groups and donors. During this period, states negotiated and adopted policy statements that recognized the linkages between mobility and environmental issues, such as the 2018 UN *Global Compact for Safe, Orderly and Regular Migration* (GCM) (UNGA, 2018), the recommendations of the UNFCCC Task Force on Displacement (Traore Chazalnoël & Ionesco, 2018a) and migration-related resolutions under the 'Desertification Convention' (the United Nations Convention to Combat Desertification, UNCCD) (IOM & UNCCD, 2019) and the Sendai Agenda for Disaster Risk Reduction (DRR) (IOM, 2018). State-led initiatives such as the *Nansen Agenda for the Protection of Cross-Border Displaced Persons* and the *Climate Vulnerable Forum* also addressed mobility issues in the context of climate and environmental change.

Today we are at a turning point where we must ask ourselves the key question: How can governments translate the progress made in global and regional policy discussions into a new phase of "action" that involves comprehensive policy changes and programmatic action on the ground? This is not an easy question to answer, but states can build on existing instruments that have emerged from strengthened global intergovernmental cooperation.

Strengthening Global Cooperation and Shared Responsibility

Over the past 5 years, countries in the global North and South have demonstrated their political will to discuss climate migration and to develop global guidance for encouraging countries to develop policies that address climate migration at the national level (Traore Chazalnoël & Ionesco, 2018b). This growing political interest has been reflected in the development of global principles adopted by United Nations Member States. The aforementioned 2018 milestones, the 'UN Global Compact for Migration' (GCM) and the recommendations of the *Task Force on Displacement* (UNFCCC, 2019) outline principles and measures that states can develop and implement to counteract the impact of climate change on migration (Traore Chazalnoël & Ionesco, 2018c). These include, in particular, minimizing relevant climate and environmental drivers of migration and facilitating migration, for example through visa options for those affected by climate impacts.

These political advances are important, but they have their limits, because not all "agreements" are binding and their implementation depends on the good will of states. Moreover, the "Global Compact for Migration" only refers to international migration, while most climate migration takes place within countries. Nevertheless, these "declarations" have an important symbolic value and can serve as a starting point for action by governments dealing with the issue. Indeed, we are already seeing some countries (e.g. Peru, 2019a, b; Government of Vanuatu, 2018) making concerted efforts to better align their national policy frameworks with global principles. In parallel to global discussions, regional policy dialogues—on both climate change and migration—are exploring how to develop solutions that will help states manage migration in a changing climate and support the affected population groups (Dal Pra et al., 2021; IGAD, 2020).

Looking ahead to the next decade, we have a better understanding of the challenges and opportunities, as well as a solid global policy foundation and more expertise to take action and achieve tangible results. Global policy discussions can drive engagement at the national level, especially in the countries most affected by climate impacts. Vulnerable states must be supported in developing their own frameworks and programs; at the community level, everything possible must be done to reduce vulnerability and build resilience.

Migration as an Adaptation Strategy to Climate Impacts

In the current debate on climate migration, there is intense discussion about the extent to which migration can be assessed as an adequate adaptation strategy to climate change. In and of itself, migration is not a negative phenomenon—throughout history, people have migrated due to environmental changes and have built prosperous lives and made significant contributions to their new homelands. Yet, when migration occurs out of necessity and/or through irregular channels, serious tragedies can occur, as evidenced by the high numbers of missing persons and deaths (IOM, n.d.). And those who stay behind often suffer, especially children whose parents have emigrated; they are often left to their own devices and have to bear burdens that disrupt or derail their further education (Traore Chazalnoël et al., 2021).

In some contexts, however, migration can also help people, for example through better educational opportunities or more secure and better-paying jobs. Remittances back to the areas of origin increase local resilience to the adverse effects of climate change. It is important that such migrations take place in a safe and orderly manner, as this is the only way that positive effects can outweigh negative ones. Ensuring this raises numerous questions in various policy areas, including above all how to deal with the drivers of migration.

Addressing the Climate-Related Drivers of Migration

Most people would rather stay in their countries of origin to live a decent and good life there. Therefore, supporting climate mitigation and adaptation measures is crucial, especially in countries and regions that have significant outflows. Wealthy countries can play a crucial role by making investments in the most vulnerable countries. Appropriate investments reduce the negative impact of climate change on livelihoods, create alternative, sustainable employment opportunities and reduce potential reasons for leaving. Such investments also contribute to climate justice, as the countries hit hard by climate change are often the least responsible for its causes.

At present, however, migration aspects are still not sufficiently integrated into the instruments for financing climate protection and adaptation measures. Even though some measures explicitly refer to the effects of migration, migration aspects are rarely taken into account when allocating funds, such as those of the *Green Climate Fund* (GCF). Migration-related funds should also increasingly finance programs in areas that are especially vulnerable to climate impacts. Recently, the *Migration Multi-Partner Trust Fund*, established to support the implementation of the *Compact for Safe, Orderly and Regular Migration,* decided to fund joint climate migration projects (UNMN, n.d.). Other migration funds, such as the IOM Development Fund, are financing pilot programs for climate migration in vulnerable areas. It will be interesting to see whether such programs prove successful; at present it is too early to make a final assessment. The political will of donor countries to finance climate migration projects is in any case discernible, so it is all the more important that this remains the case.

Facilitating Migration

There, will be regions that will be so severely affected by climate change that the population there will probably be unable to stay or return. Pacific island states might disappear completely if sea levels continue to rise. In these cases, the only option is to comprehensively facilitate or enable migration from the affected areas.

The "UN Compact for Safe, Orderly and Regular Migration" identifies some ways in which states can facilitate regular international migration in a changing climate, including the development of special visas. There are already countries that grant and have granted visas on humanitarian grounds, such as Brazil, which offered visas to Haitian migrants after the 2010 earthquake (The Nansen Initiative, 2015a, b).

It is still possible for countries to expand the scope of existing migration management measures to help people in vulnerable areas to migrate legally and safely from endangered areas. For example, many countries already have bilateral education, training or labor agreements with partner countries, such as the *Australian Seasonal Workers Program* or the *Pacific Labor Scheme* (Dempster et al., 2021b). These programs might be expanded to offer alternate opportunities for those most impacted by climate change.

Interested countries can also manage climate migration by taking measures to facilitate legal residence for migrants already living in their country. This can be done by issuing residence permits, stopping or reducing deportations and repatriations, and implementing comprehensive regularization programs.

Being Mindful of the Burden of Immobility
Those most impacted by climate change often lack the financial and social means to migrate. These populations are considered "trapped". Policy making should take into account of their inability to migrate.

Another dimension of migration management concerns the return and reintegration of migrants. While return is sometimes possible, it often takes place in areas and countries where climate impacts are more likely to increase. Therefore, voluntary return and reintegration policies and programs should examine the extent to which structures to be established there can ensure that returnees have sustainable employment opportunities and can live there long-term (IOM, 2020a, b, c).

At the regional level, some countries have joined *free movement agreements,* which were examined in a recent study (Dal Pra et al., 2021). To take these policy instruments to the next level, more activity on the political side and increased investment would be needed.

Management of Domestic Migration

In many countries, there are internal migrations related to climate change impacts, migrations from rural regions to large urban centers for example (Baillat et al., 2020). Sometimes migrants can thus improve their living conditions, but it is doubtful whether this will still hold in the future, when climate change will increase migratory pressure. Some countries are therefore trying to incorporate the dimensions of climate migration into urban planning, such as Bangladesh with its initiative to attract internal migrants to secondary cities, which are more climate resilient and migrant-friendly than large cities (Huq et al., 2018).

Another important policy measure is managed resettlement. Such projects are already taking place in over 60 countries worldwide (Bower & Weerasinghe, 2021). Countries such as Fiji are preparing to cope with a growing number of planned resettlements, and other countries may have to follow suit if climate impacts worsen. However, planned resettlement should only be considered as a last resort. The relocation of entire communities is extremely costly, and some, such as Isle of Jean Charles in coastal Louisiana, have been underway for several years.

Strengthening Legal Protections

Even if the status of "climate migrant" or "climate refugee" does not exist, governments could take measures to provide protection and support specifically to migrants affected by climate change. Existing models could inspire the development of targeted protection measures. For example, the United States may grant temporary protected status to nationals of certain countries if conditions in those country temporarily prevent them from returning safely (The Nansen Initiative, 2015a, b). Those eligible for this status may not be deported from the US and are allowed to take up employment. France recently refrained from returning a migrant from Bangladesh to his country of origin because he would have been exposed to extreme pollution there that would have negatively affected his health (Dempster et al., 2021a, b). This type of measure could be extended to people exposed to extreme climatic conditions.

Looking ahead, countries may increasingly need to deal with international legal principles in managing climate migration. In January 2020, the UN Human Rights Committee adopted a decision recognizing that states should refrain from sending people back to areas where the impacts of climate change make it impossible to live in dignity—the principle of non-refoulement (United Nations Human Rights Committee, 2020; IOM, 2020c). This decision might pave the way for national legislative changes in coming decades.

Managing Climate Migration in Wealthy Countries
In recent years, countries such as the United States (The White House, 2021; United States-White House Executive Order, 2021) have commissioned reports examining possible measures to address the negative impacts of climate change on migration. Even though climate-induced migration mostly takes place intranationally, as mentioned several times before, the multicausal nature of climate change means that climatic factors could at least partially influence migration flows to wealthier countries. A recent report makes recommendations on specific measures that high-income countries could consider in this context. These include, for example, extending temporary protection measures to climate migrants, allowing them to work, study and access basic services (e.g. health insurance) (Dempster et al., 2021a); extending labor migration to nationals of climate-vulnerable countries (this would also help reduce skills shortages in receiving countries); and establishing a visa lottery targeting people from climate-vulnerable countries.

Strengthen Monitoring and Evaluation of Existing Policies and Programs

There are some policies and programs that specifically target climate migration, some of which have been analysed (CLIMB, n.d.). However, the status of implementation and long-term impacts of these policies are rarely assessed. Considering that climate migration is a relatively new policy issue, it will be crucial in the next decade to allocate resources to conduct comprehensive evaluations and determine which policy and program options work well. States could use international review processes, such as the quadrennial *International Migration Review Forum*, which reviews the implementation of the GCM, and the annual Climate Conferences of the Parties (COPs) to take stock of progress made and to build political will for future action.

Harnessing the Positive Dimensions of Climate Migration

Climate migration stories are not all negative. It is widely acknowledged that migrants play an important role in destination areas, especially by helping to fill labor gaps or redressing demographic imbalances. Migrants also play an important role in their regions of origin, as they contribute with their remittances to reducing poverty in recipient households and to building resilience.

It would be important to create incentives to make it attractive for migrants to invest in building resilience to climate impacts in their areas of origin. Financial resources can be used to make houses and infrastructure climate-proof; in Senegal, the granting of loans has been made possible (Bendandi & Pauw, 2016) and Mexico also has a program that supports investment by migrants (Villegas Rivera, 2014). A study conducted in Tajikistan, for example, highlights that remittances from migrants are increasingly being used in business start-ups and community-based agriculture, helping to create a more climate-resilient future (Babagaliyeva et al., 2017). A recent report found that aspects of environmental sustainability can provide a common ground for bringing people together across borders and counteracting negative migration narratives (Traore Chazalnoël & Barwise, 2021).

Conclusion: Do People Really Have Nothing Left?

Current knowledge and understanding of the link between climate and migration highlight serious political, social, security and economic challenges. Many people live in areas that are so vulnerable to climate change that they have little hope or prospects. But this can also be seen as a moment of opportunity. Political discussions have shown that states are increasingly interested in addressing the problems. Time is of the essence, and it is crucial that states vigorously develop and implement the policy, legal and programmatic instruments to translate global commitments into concrete action.

There are no tailor-made and one-dimensional solutions that would allow states to quickly "solve" existing and future problems. Policy choices will (have to) vary across different parts of the world. In some cases, facilitating migration can help empower people and reduce their exposure to climate risks while benefiting destination countries. In other cases, migration could become a burdensome experience.

More than ever, states need to consider a two-pronged approach. Countries in the Global North should favor policy options that provide financial, technical and political support for adaptation measures in areas most affected by climate change. However, it is also necessary to discuss the development of regular migration pathways that provide immediate and long-term options for people most affected by climate change. Some countries refuse in principle to discuss how migration policies can help cope with climate change. However, it is important to note that facilitating climate migration does not necessarily lead to more migration to wealthy countries and regions. Rather, the discussion should be about finding ways to better manage current migration flows while implementing measures that support climate adaptation and sustainable development in vulnerable areas. This is the only way of ensuring that people also have the option of not migrating thanks to suitable conditions allowing them to live a dignified life in their homeland.

References

Babagaliyeva, Z., et al. (2017). *Migration, remittances and climate resilience in Tajikistan.* Regional Environmental Centre for Central Asia (CAREC) in cooperation with the Ministry of Labour, Migration and Employment of Population of the Republic of Tajikistan. https://carececo.org/eng_P2%20WP1%20Migration,%20remittances_FIN.pdf

Baillat, A., & Traore Chazalnöel, M. (2022). Anticiper et gérer les migrations dans le contexte des changements climatiques. In N. Regaud, B. Alex, & F. Gemenne (Eds.), *La Guerre Chaude.* Les Presses de Sciences Po.

Baillat, A., Guadagno, L., & Mokhnacheva, D. (2020). L'avenir des villes face aux migrations climatiques. *Revue Urbanisme,* n°417. https://environmentalmigration.iom.int/sites/environmentalmigration/files/URBA%20417_Migrations%20climatiques.pdf

Bendandi, B. (2020). Migration induced by climate change and environmental degradation in the Central Mediterranean Route. In *Migration in West and North Africa and across the Mediterranean.* International Organization for Migration (IOM). https://publications.iom.int/books/migration-west-and-north-africa-and-across-mediterranean-chapter-26

Bendandi, B., & Pauw, P. (2016). *Remittances for adaptation: An 'alternative source' of international climate finance?* https://link.springer.com/chapter/10.1007/978-3-319-42922-9_10

Bower, E., & Weerasinghe, S. (2021). *Leaving place, restoring home: Enhancing the evidence base on planned relocation cases in the context of hazards, disasters and climate change.* https://disasterdisplacement.org/portfolio-item/leaving-place-restoring-home

Clement, V., et al. (2021). *Groundswell part 2: Acting on internal climate migration.* World Bank. https://openknowledge.worldbank.org/handle/10986/36248

CLIMB Database: Human Mobility in the Context of Disasters, Climate Change and Environmental Degradation Database. (n.d.). https://migrationnetwork.un.org/climb

Dal Pra, A., Traore Chazalnoël, M., & Dempster, H. (2021). *Strengthening regional policy frameworks to better respond to environmental migration: Recommendations for the UK Government* (CGD Policy Paper 244). Center for Global Development. https://www.cgdev.org/publication/strengthening-regional-policy-frameworks-better-respond-environmental-migration

Dempster, H., Dal Pra, A., & Traore Chazalnoël, M. (2021a). *How can the UK better facilitate environmental migration?* https://www.cgdev.org/blog/how-can-uk-better-facilitate-environmental-migration

Dempster, H., Dal Pra, A., & Traore Chazalnoël, M. (2021b). *Facilitating environmental migration through humanitarian and labour pathways: Recommendations for the UK Government* (CGD Policy Paper 245). Center for Global Development. https://www.cgdev.org/publication/facilitating-environmental-migration-through-humanitarian-and-labour-pathways

Flavell, A., Milan, A., & Melde, S. (2020). *Migration, environment and climate change: Literature review.* First report in the "Migration, environment and climate" series. On behalf of the German Environment Agency. https://www.umweltbundesamt.de/sites/default/files/medien/1410/publikationen/2020-03-04_texte_42-2020_migration-literature-review_1.pdf

Government of Vanuatu. (2018). *Vanuatu: National policy on climate change and disaster-induced displacement.* https://environmentalmigration.iom.int/vanuatu-national-policy-climate-change-and-disaster-induced-displacement-2018

Huq, S., et al. (2018). *Building climate-resilient, migrant-friendly cities and towns.* International Centre for Climate Change and Development (ICCCAD). http://www.icccad.net/wp-content/uploads/2018/10/Policy-Brief-on-Climate-Migration-and-Cities.pdf

Intergovernmental Authority on Development (IGAD). (2020). *Communique of the sectoral ministerial meeting on the protocol on free movement of persons in the IGAD region.* Khartoum, Republic of Sudan. https://igad.int/attachments/article/2373/Communique%20on%20Endorsement%20of%20the%20Protocol%20of%20Free%20Movement%20of%20Persons.pdf

Internal Displacement Monitoring Centre (IDMC). (2023). *Global Report on Internal Displacement (GRID) Internal Displacement and Foood Insecurity.*

https://api.internal-displacement.org/sites/default/files/publications/documents/ IDMC_GRID_2023_Global_Report_on_Internal_Displacement_LR.pdf

International Organization for Migration (IOM). (2014a). *IOM outlook on migration, environment and climate change.* IOM, Geneva. https://publications.iom.int/system/files/pdf/mecc_outlook.pdf

International Organization for Migration (IOM). (2014b). *Migration, environment and climate change: Evidence for policy.* https://environmentalmigration.iom.int/ migration-environment-and-climate-change-evidence-policy-meclep-2

International Organization for Migration (IOM). (2017a). *IOM MECC info sheet on extreme heat and migration.* https://publications.iom.int/system/files/pdf/mecc_infosheet_heat_and_migration.pdf

International Organization for Migration (IOM). (2017b). *Making mobility work for adaptation to environmental changes: Results from the MECLEP global research.* https://publications.iom.int/system/files/pdf/meclep_comparative_report.pdf

International Organization for Migration (IOM). (2018). *Mapping human mobility (migration, displacement and planned relocation) and climate change in international processes, policies and legal frameworks.* Task Force on Displacement Activity II.2. https://unfccc.int/sites/default/files/resource/WIM%20TFD%20II.2%20Output.pdf

International Organization for Migration (IOM). (2019). *Glossary on migration.* https://publications.iom.int/system/files/pdf/iml_34_glossary.pdf

International Organization for Migration (IOM). (2020a). *Guidance for mainstreaming environmental and climate considerations into reintegration programming.* https://environmentalmigration.iom.int/guidance-mainstreaming-environmental-and-climate-considerations-reintegration-programming

International Organization for Migration (IOM). (2020b). *Internal displacement in the context of the slow-onset adverse effects of climate change.* https://environmentalmigration.iom.int/internal-displacement-context-slow-onset-adverse-effects-climate-change-submission-international

International Organization for Migration (IOM). (2020c). *Position by human rights committee opens possibility for dignified migration in the context of climate change.* https://weblog.iom.int/position-human-rights-committee-opens-possibility-dignified-migration-context-climate-change

International Organization for Migration (IOM). (n.d.). *Missing migrants database.* https://missingmigrants.iom.int/

International Organization for Migration (IOM) & United Nations Convention to Combat Desertification (UNCCD). (2019). *Addressing the land degradation—Migration nexus: The role of the United Nations Convention to Combat Desertification.* IOM, Geneva. https://environmentalmigration.iom.int/sites/environmentalmigration/files/IOM%20UNCCD%20 Desertification%202019%20FINAL.pdf

International Organization for Migration (IOM) & United Nations Office of the High Representative for the Least Developed Countries, Landlocked Developing Countries and Small Island Developing States (UN-OHRLLS). (2019). *Climate change and migration in vulnerable countries, a snapshot of least developed countries, landlocked developing countries and small island developing States.* IOM, Geneva. https://publications.iom.int/system/files/pdf/climate_change_and_migration_in_vulnerable_countries.pdf

Ionesco, D., Mokhnacheva, D., & Gemenne, F. (2017). *The atlas of environmental migration.* Routledge.

McMichael, C., et al. (2020). A review of estimating population exposure to sea-level rise and the relevance for migration. *Environmental Research Letters, 15*(12). https://doi.org/10.1088/1748-9326/abb398

Missirian, A., & Schlenker, W. (2017). Asylum applications respond to temperature fluctuations. *Science, 358*(6370), 1610–1614. https://science.sciencemag.org/content/358/6370/1610

Peru. (2019a). *Ley No. 30754. Ley marco sobre cambio climatico, El Peruano.*

Peru. (2019b). *Decreto Supremo N° 013-2019-MINAM—Decreto Supremo que aprueba el Reglamento de la Ley N° 30754, Ley Marco sobre Cambio Climático, Sistema nacional de Informacion Ambiental.*

Puscas, I., & Escribano, P. (2018). *The environment is changing: Is the migrant caravan a consequence?* https://environmentalmigration.iom.int/fr/node/1460

The Nansen Initiative. (2015a). *Agenda for the protection of cross-border displaced persons in the context of disasters and climate change* (Vol. I). The Nansen Initiative.

The Nansen Initiative. (2015b). *Agenda for the protection of cross-border displaced persons in the context of disasters and climate change* (Vol. II). The Nansen Initiative.

The White House. (2021). *Report on the impact of climate change on migration.* https://www.whitehouse.gov/wp-content/uploads/2021/10/Report-on-the-Impact-of-Climate-Change-on-Migration.pdf

Traore Chazalnoël, M., & Barwise, K. (2021). Impact of environmental issues on the promotion of intercultural dialogue, in the EuroMed region. In *Intercultural trends and social change in the Euro-Mediterranean region* (Anna Lindh report 2021). https://www.annalindhfoundation.org/sites/default/files/documents/page/Anna_Lindh_Report_Eng.pdf

Traore Chazalnoël, M., & Ionesco, D. (2018a). *IOM perspectives on climate change and migration—10 key takeaways from the COP24 recommendations on integrated approaches to address displacement and climate change.* https://environmentalmigration.iom.int/blogs/iom-perspectives-climate-change-and-migration

Traore Chazalnoël, M., & Ionesco, D. (2018b). Chapter 7: Advancing the global governance of climate migration through the United Nations Framework Convention on Climate Change and the Global Compact on Migration: Perspectives from the International Organization for Migration. In S. Behrman & A. Kent (Eds.), *Climate refugees: Beyond the legal impasse?* Routledge.

Traore Chazalnoël, M., & Ionesco, D. (2018c). A moment of opportunity to define the global governance of environmental migration. Perspectives from the International Organization for Migration. In R. McLeman & F. Gemenne (Eds.), *Routledge handbook of environmental displacement and migration.* Routledge International Handbooks.

Traore Chazalnoël, M., & Ionesco, D. (2022). Perspectives from the International Organization for Migration (IOM), breaking new ground on the governance of climate migration. In S. Behrman & A. Kent (Eds.), *Climate refugees, global, local and critical approaches.* Cambridge University Press. https://www.cambridge.org/core/books/abs/climate-refugees/perspectives-from-the-international-organization-for-migration-iom/CB1081ED571ECABB14811A9F7EDA16

Traore Chazalnoël, M., & Randall, A. (2021). Migration and the slow-onset impacts of climate change: Taking stock and taking action. In M. McAuliffe & A. Triandafyllidou (Eds.), *World migration report 2022.* International Organization for Migration (IOM). https://publications.iom.int/books/world-migration-report-2022-chapter-9

Traore Chazalnoël, M., Ionesco, D., & Duca, I. (2021). *Children on the move; Why, where, how?* https://www.unicef.org/globalinsight/media/1821/file/Children%20on%20the%20Move:%20Why,%20Where,%20How?%20.pdf

United Nations Framework Convention on Climate Change (UNFCCC). (2015). *Adoption of the Paris Agreement.* https://unfccc.int/resource/docs/2015/cop21/eng/l09r01.pdf

United Nations Framework Convention on Climate Change (UNFCCC). (2019). Report of the Executive Committee of the Warsaw International Mechanism for Loss and Damage associated with climate change impacts: FCCC/CP/2018/10/Add.1. In *Report of the conference of the parties on its twenty-fourth session, held in Katowice from 2 to 15 December 2018. Part two: Action taken by the conference of the parties at its twenty-fourth session.* FCCC/CP/2018/10/Add.1. https://unfccc.int/sites/default/files/resource/10a1.pdf

United Nations General Assembly. (2018). *Intergovernmental conference to adopt the global compact for safe, orderly and regular migration, outcome of the conference.* A/CONF.231/3

United Nations Human Rights Committee. (2020, January 7). *Ioane Teitiota v. New Zealand (advance unedited version), CCPR/C/127/D/2728/2016.*

United Nations Network on Migration (UNMN). (n.d.). *International Migration Review Forum (IMRF) website.* Available at https://migrationnetwork.un.org/international-migration-review-forum-2022

United States—The White House Executive Order. (2021). *Rebuilding and enhancing programs to resettle refugees and planning for the impact of climate change on migration.* https://www.whitehouse.gov/briefing-room/presidential-actions/2021/02/04/executive-order-on-rebuilding-and-enhancing-programs-to-resettle-refugees-and-planning-for-the-impact-of-climate-change-on-migration/

Université de Neuchâtel. (2018). *CliMig database on migration, climate change and the environment, produced by the Université de Neuchâtel.* https://www.unine.ch/geographie/climig_database

Villegas Rivera, F. (2014). *Impact of Mexico's 3 x 1 program for migrants and collective remittances.* University of Calgary. https://larc.ucalgary.ca/publications/impact-mexicos-3-x-1-program-migrants-and-collective-remittances

Warner, K., & Afifi, T. (2013). Where the rain falls: Evidence from 8 countries on how vulnerable households use migration to manage the risk of rainfall variability and food insecurity. *Climate and Development, 6*(1). https://www.tandfonline.com/doi/full/10.1080/17565529.2013.835707

Economic Impacts

The Economic Consequences of the Climate Crisis

Leonie Wenz and Friderike Kuik

In order to avert catastrophic climate change, the international community committed to the Paris Agreement, with the goal to limit "the increase in the global average temperature to well below 2 °C above pre-industrial levels" and pursue efforts "to limit the temperature increase to 1.5 °C above pre-industrial levels." (United Nations, 2015). In an optimistic case the recent emission reduction commitments of individual countries would be roughly enough to meet the 2 °C-goal of the Paris Agreement. But the emission reduction measures that are currently already implemented still fall short of these voluntary commitments: According to the United Nations Emissions Gap Report 2021, we are currently heading for a warming of about 2.7 degrees (UNEP and INEP DTU Partnership, 2021).

In this chapter, we shed light on the economic damages that might be expected in a world in which temperatures are 3 degrees higher than in pre-industrial times. Based on the current state of science, we present and discuss various transmission channels from climate change to the economy (see also section "Climate Change Affects All Sectors of the Economy") and outline possible consequences for the economy as a whole (see also section "The Costs of Climate Change"). A guiding question for these discussions is how future damages can credibly be estimated. The past years have given a glimpse of the high humanitarian and economic costs that

The views expressed herein are those of the author and do not necessarily reflect those of the European Central Bank.

L. Wenz (✉)
Complexity Science Department, Potsdam Institute for Climate Impact Research, Potsdam, Germany
e-mail: leonie.wenz@pik-potsdam.de

F. Kuik
European Central Bank, Frankfurt am Main, Germany
e-mail: friderike.kuik@ecb.europa.eu

© The Author(s) 2024
K. Wiegandt (ed.), *3 Degrees More*,
https://doi.org/10.1007/978-3-031-58144-1_5

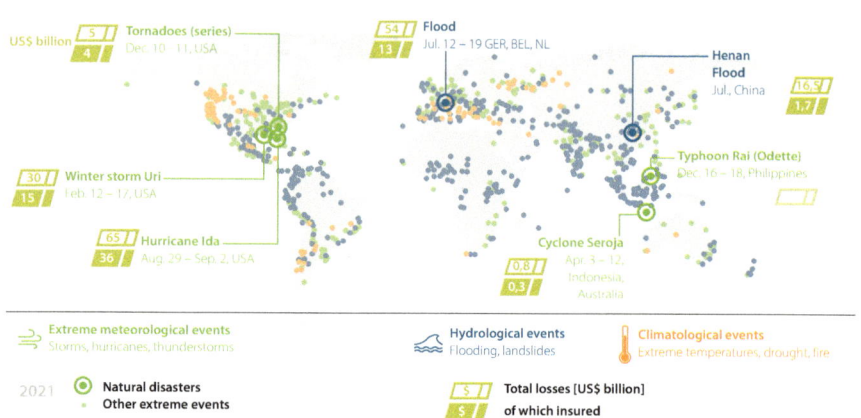

Fig. 1 Overview of some of the most expensive natural catastrophes in 2021. The figure only includes catastrophes related to weather or climate extremes, not earthquakes or volcanic eruptions. (Data from Munich Re, 2023)

climate change can cause. Figure 1 gives an illustration of some of the most severe events in 2021—the series of extreme weather events has continued since then.

From Degrees Dollars: How to Measure the Cost of Climate Change?

Weather extremes—which are becoming more frequent and more intense due to climate change—do not only lead to high human costs but also cost us a lot of money and harm the economy. But what exactly is the cost of climate change— what are plausible economic damages in a world in which warming reaches 3 degrees? Is it even possible to provide a credible estimate, and why do we need such precise cost estimates?

The biophysical effects of climate change, such as rising temperatures, changing precipitation patterns, melting glaciers, rising sea levels, increasing weather extremes, and ocean acidification are well understood for the most part and can, thanks to sophisticated computer models, be estimated with increasing precision.

These climate impacts can affect the economy in many ways, either directly or indirectly through different channels and—in many cases complex—interactions. Some impacts are easily expressed in monetary terms, others, such as losses in biodiversity or human lives, are difficult or impossible to put a price tag on. Yet other effects are difficult to foresee and assess, such as those that can occur when dangerous tipping points are crossed. Ultimately, the Earth's climate system as well as our economy are highly complex: when strongly disturbed—as is the case with a changing climate—we may be facing effects that are so far not anticipated.

Cost-Benefit Arguments Or: What's the Price Tag?

If it is complex to assess the damages from climate change, and fraught with uncertainty, why should we even bother to express them in dollars or euros? Shouldn't it be sufficient to understand the biophysical effects of climate change, to come to the conclusion that the global community of states must urgently act to stop the emission of greenhouse gases?

In fact, cost arguments do play an important role in public and political discourse. However, it is the costs of financing the energy and climate transition that seem rather concrete, whereas perceptions of the costs induced by climate change damages often still remain rather vague.

This prevents a fair and robust comparison of the costs of protecting our climate against the benefits, where the latter consist in averted climate damages and adaptation expenses. Such a comparison can be conducted formally via cost-benefit analyses or rather informally via public perception. A key figure in this context is the so-called *"social cost of carbon"*—a figure that, roughly speaking, expresses the social cost (in US dollars) of emitting one additional metric ton of CO_2. To estimate this metric, we need a good understanding of which climate damages are likely and how much they would cost. This is also important for efficiently planning adaptation measures as well as for climate justice considerations.

Integrated Models

In a cost-benefit analysis, 2018 Nobel laureate William Nordhaus calculated that a temperature target of "+3.5 °C" would be optimal from a purely economic point of view, because it would minimize the sum of the monetary costs of climate protection and climate damages (Nordhaus, 2018; Hänsel et al., 2020). For his calculations, Nordhaus used the DICE model he had been developing since the 1990s, a so-called *Integrated Assessment Model* (IAM) (Nordhaus & Boyer, 2000).

IAMs map the interactions between the economic, energy and climate systems in a simplified way to estimate the costs and benefits of climate policy (Stern, 2007). Other well-known IAMs are, for example, the PAGE model on which the *Stern Report*[1] from 2006 is based on the FUND model. In these models, cost estimates for specific climate impacts are based on one or more damage functions, which

---

[1] The British economics professor and former chief economist of the World Bank Nicholas Stern published a comprehensive report on the economic effects of climate change in 2006, which he had prepared on behalf of the British government.

are informed by empirical estimates. In addition to IAMs, other types of structural or semi-structural models are increasingly being used to estimate the economic impacts of climate change (Gallic & Vermandel, 2020).[2]

Nordhaus' calculations and model have given rise to discussion and criticism since first presented. One aspect criticized is that too little weight was given to climate damages occurring on longer horizons, based on the assumption that future generation would be better offer, which would, for example, facilitate adaptation (discounting) (Azar & Sterner, 1996; Stern, 2007). IAMs have also been criticized for not representing potentially catastrophic climate impacts (Weitzman, 2009; Pindyck, 2013). The most important criticism, however, is related to damage functions, which—for a long time—were only based on a few empirical studies, many of them dating back to the 1990s (Greenstone, 2016; Howard & Sterner, 2017; Auffhammer, 2018).

Since then, our knowledge of socio-economic climate damages has improved significantly(Carleton & Hsiang, 2016). In the last 10 to 15 years, there has been a vast amount of new empirical studies. Various scientific teams have integrated these recent empirical findings into Nordhaus' DICE model or other IAMs—and, based on this new knowledge, now conclude that the Paris Agreement is optimal also from an economic point of view, as the economic costs of additional warming would be much higher than the costs required to meet the goals of the Paris Agreement (Glanemann et al., 2020; Hänsel et al., 2020; Ueckerdt et al., 2019).

Empirical Models: Learning from the Past to Predict the Future

The rapid expansion of empirical literature on the climate-economy relationship that we have seen in the last 10 to 15 years benefitted from several different developments. First, the amount of available data and new data sources continues to grow, such as climate data collected by satellites, but also data on social and economic indicators as obtained from social media, nighttime light measurements, or GPS trackers. Second, increased computing capacities make it possible to process and analyze these data. Finally, methods for deriving robust conclusions from data have also continuously evolved and improved. These methods come primarily from statistics and econometrics and are increasingly being complemented with machine learning algorithms.

The core idea of these empirical approaches is to explore the impact of past climatic conditions and weather extremes on economically relevant factors, as a basis to derive estimates of future damages (Dell et al., 2014; Hsiang, 2016; Kolstad & Moore, 2020). For example, one might look at how extreme temperatures have

---

[2] Here, for example, the development of structural macroeconomic models should be mentioned, such as so-called dynamic stochastic general equilibrium models (DSGE models). Compared to IAMs, these models focus on a more detailed description of the macroeconomic adjustment after the occurrence of climate impacts.

historically affected labor productivity in order to estimate future productivity losses from rising temperatures. This can be done both for individual economically relevant sectors and variables such as labor productivity, agriculture, or electricity demand (bottom-up; see also section "Climate Change Affects All Sectors of the Economy") and directly at the macroeconomic level with respect to impacts on economic output (top-down; see also section "The Costs of Climate Change").

Roughly speaking, two methodological approaches can be distinguished, with more and more hybrid variants emerging. One approach compares countries or regions with different climatic conditions, to explore the influence of the prevalent conditions on economically relevant factors (cross-sectional analysis). An obvious problem with this approach is that there are many other economically relevant differences among countries. Some of these can be controlled for by also measuring them and including them in the statistical model. Others are not directly observable or correlate with both climate and the economic variable under consideration and thus distort the actual effect that climate has on the economy.

Another approach is to compare a country or region with itself at different points in time (time series analysis), i.e., to examine whether economic performance was lower in an especially hot or especially wet year than in a year with average or moderate weather. The advantage of this approach is that all factors that are specific to a country or region and that have not changed over the observation period can be eliminated from the calculation. If the analysis can be carried out for several countries at the same time (panel analysis), all factors that are specific to a particular year and might have influenced economic performance in that year can also be accounted for. These might be global economic shocks, such as financial crises or a pandemic, or climate phenomena such as El Niño. Controlling for these country- and year-specific "fixed effects" then enables very robust conclusions to be drawn about plausibly causal relationships (Kolstad & Moore, 2020; Auffhammer, 2018).

The disadvantage of this approach is that the effects of short-term weather shocks may only be partly informative about damages induced by long-term climatic changes (Kolstad & Moore, 2020); this is especially problematic with regard to extreme weather events of previously unknown strength, frequency and simultaneity as well as the crossing of dangerous tipping points. In a similar vein, adaptation measures that may not yet have been observable in the past, but seem plausible for the future, are not accounted for.

Risk and Adaptation

In addition to climate change damages or the *social cost of carbon*, some studies focus on a risk-based approach, i.e., assessing the risk of a sector, region, or country of being affected by climate change (IPCC, 2022a, b). This risk depends on the biophysical effects of climate change itself *(hazard)*, but also on how much one is *exposed* to it *(exposure)*, and on how vulnerable one is to damage *(vulnerability)*. For example, a region's risk of economic damage from forest fires may be higher if the area at risk of fire is close to settlements, cities, or industrial plants. Another example is flooding: a region is at a higher risk of economic damage from flooding if populated areas, infrastructure, or industrial facilities are located in the flood zone.

The vulnerability and hence the economic damage caused by climate change can theoretically be reduced through adaptation measures such as (as in the last example) flood protection. Other examples of such measures are increased coastal protection, better water management, an expanded extreme weather warning network, or infrastructure adapted to climate change (Feyen et al., 2020).

Adaptation measures, however, entail costs of their own—financial resources that must be mobilized and that, in the absence of climate change, could have been invested in other, more productive ways. Furthermore, there is a limit to the possibility to adapt: with increasingly severe climate change, adaptation will not be sufficient to avert all economic damages. Already now, according to estimates by the Intergovernmental Panel on Climate Change (IPCC), some weather and climate extremes have led to irreversible damage (IPCC, 2022a, b).

If adaptation measures are insufficient or impossible, people will have to use another one of the above-mentioned three levers: their exposure. This can mean, for example, resettlement or migration—with economic effects that are very difficult to assess. Climate and weather extremes are already leading to increased migration (IPCC, 2022a, b). Conversely, an increase in exposure—for example, expansion of settlements or industrial areas in a region threatened by climate impacts—can entail increased risk. This phenomenon has likely contributed to rising costs from extreme events in recent decades.

Climate Change Affects All Sectors of the Economy

Many biophysical climate impacts have a direct and immediate effect on the economy. For example, extreme weather events such as tropical cyclones or floods can destroy houses, factories, or important infrastructure and can disrupt transportation routes. Dying coral reefs affect the tourism and fishing industries, forest fires harm the forestry business, and droughts destroy harvests, thereby raising food prices.

Besides these direct effects, there may be other, less obvious, indirect, or interacting effects. In fact, the ways in which climate change affects the economy are numerous and may be complex and interconnected. For a classification and quantification of climate change damages, an overview of different transmission channels is helpful. For example, the main risks from climate change in Europe, according to the IPCC (Kovats et al., 2014), include:

- Extreme heat, impacting health and well-being as well as ecosystems;
- Extreme heat and drought, impacting agricultural yields;
- Water scarcity, impacting various areas of economic and daily life;
- Flooding near rivers or coasts, impacting people, the economy, and infrastructure.

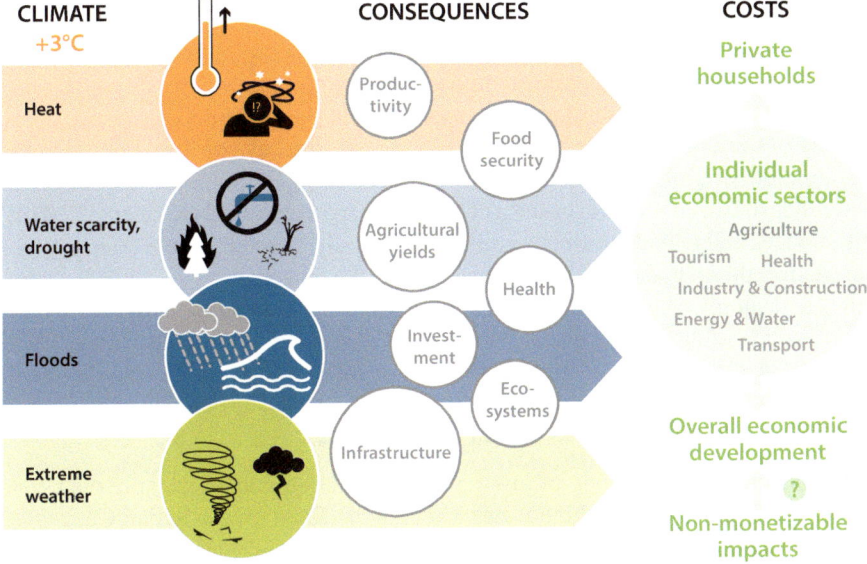

Fig. 2 The diagram shows examples of transmission channels through which biophysical climate impacts can cause economic damages. The categorization of biophysical impacts is illustrative, the individual categories overlap and are not mutually exclusive. The transmission channels are complex and interlinked, and the effects may mutually influence one another (Wenz & Kuik)

In the following we present examples of transmission channels that are related to some of these risks (Fig. 2). It should be noted that most of our examples concern Europe—yet the world's poorer countries are generally likely to be more strongly affected and will have fewer means to adapt, with severe consequences for their food security and human health (Stern, 2007). For example, the IPCC concludes in its Sixth Assessment Report that climate change contributes to humanitarian crises in especially vulnerable areas and that extreme weather events have a greater impact on economic growth in developing than in industrialized countries (IPCC, 2022a, b). The examples we give stem from a variety of studies, some of which are based on different warming scenarios. Therefore, some of the examples do not refer to a global mean temperature increase of 3 degrees, but to an even more pessimistic scenario, leading to warming of more than 4 degrees by the end of the twenty-first century.[3]

---

[3] In order to facilitate the comparability of studies and results, the scientific community has agreed on a common set of scenarios that describe different emission pathways and the respective increases in global mean temperature by the end of the twenty-first century. Since the IPCC's Fifth Assessment Report 2013/14, these have been the so-called *Representative Concentration Pathways* (RCPs) (van Vuuren et al., 2011) which were supplemented by socio-economic narratives (Shared Socioeconomic Pathways; SSPs) with the Sixth Assessment Report 2021/22. Some of the examples mentioned here are based on the RCP 8.5 scenario. This corresponds to an increase in global mean temperature of about 4.4 degrees by the end of the century (2081–2100) compared to pre-

Main Transmission Channels

Too Much Water Many transmission channels are associated with changes in the availability of water—either due to extreme precipitation and flooding, or due to the absence of rain. For example, a higher number of rainy days or days with extreme precipitation within a year were found to reduce economic output—especially in richer industrialized countries such as Germany, Japan or the US (Kotz et al., 2022). The services and manufacturing sectors are especially affected, where effects might materialize through damage to infrastructure, the interruption of transportation routes and supply chains, planning uncertainties or health effects. Extreme precipitation is increasing almost everywhere in the world due to climate change (Min et al., 2011). For example, in a world that is 2 degrees warmer, the probability of an extreme rainfall event, which currently occurs once in 20 years, increases by 45% in northern Europe and by 37% in central Europe (Kharin et al., 2018).

Moreover, with a warming of just 1.5 degrees, there is already a more than 40% higher chance of extremely high water levels in the Rhine or the Indian Ganges, to name some examples (Paltan et al., 2018). In a world that is 3 degrees warmer, river floods in Europe might lead to damages amounting to €40 billion (Feyen et al., 2020)—about as much as the costs caused by the 2021 flood disaster in Western Germany, the Netherlands and Belgium, but recurring every year. The cost of coastal flooding might even reach €238 billion per year in Europe—significantly exceeding, year after year, the cost of Hurricane Katrina, the most expensive natural disaster in US history to date (NOAA, 2021). Other estimates suggest that in a 4 degree warmer world, economic losses from river flooding in Germany would increase more than ten-fold—and by as much as 3214% in Bangladesh (Alfieri et al., 2017). In the examples mentioned, the total costs may be considerably reduced with adequate adaptation measures.

Too Little Water More than two thirds of global freshwater resources are used for irrigation and food production (in some countries of the Global South even up to 95%), one fifth for industry and energy, and only about 12% directly by households (Zhongming et al., 2021). Water scarcity in a world that is 3 degrees warmer thus affects food security in particular, but also industry and the energy sector. In a world that is 3 to 4 degrees warmer, the proportion of the earth affected by extreme drought could increase from 3% at the beginning of the millennium to 30% by the end of the twenty-first century (Burke et al., 2006). In southern Europe, droughts that statistically occurred once every 100 years at the beginning of the millennium might recur roughly every 10 years (Lehner et al., 2001). The European Commission estimates that droughts in a 3 degree warmer world would cost Europe €45 million per year, compared to €9 million per year today (Feyen et al., 2020). The same study estimates

---

industrial times (1850–1900). Other examples mentioned assume less pessimistic scenarios or explicitly estimate the effect at that point in time at which, according to climate models, a warming of 3 degrees is likely to be reached.

that 13 million more people in Europe would live in regions at risk of water scarcity. Changing weather conditions (especially drought) would also lead to a significantly higher risk of forest fires. This risk would still be highest in Southern Europe—but by no means limited to this region: across Europe, a further 15 million people could be exposed to a similarly high risk of forest fires. In a 3 degrees warmer world, the area potentially affected by forest fires might almost double in a normal Mediterranean summer (Turco et al., 2018).

Heat In a world that is 3 degrees warmer, about half the population of the EU and the UK could be exposed to an intense heatwave every year—an event that without climate change statistically occurs only every 50 years (Feyen et al., 2020). This could cause up to 90,000 additional deaths each year. Heat waves are deadly and expensive weather extremes: the damage heatwaves already cause each year is estimated at around $100 billion for the United States alone (Atlantic Council, 2021). High temperatures affect our well-being, social interactions, and productivity. For example, researchers found that higher temperatures increase the risk of mental health problems, suicides, and individual and group conflicts (Hsiang et al., 2013; Obradovich et al., 2018; Helman & Zaitchik, 2020). They also found that the tone in social networks becomes harsher and that schoolchildren perform worse in warm classrooms (Stechemesser et al., 2021, 2022; Graff et al., 2018). Temperatures above about 25 °C reduce the productivity of workers, an effect that is especially relevant for industries, such as construction or agriculture, which require a lot of outdoor work (Ramsey, 1995; Hsiang, 2010; Dunne et al., 2013; Szewczyk et al., 2021). For the European heat waves of 2003, 2010, and 2015, one study puts the losses in the ten most affected countries at $59 to $90 per worker in agriculture and $41 to $72 per worker in the construction sector (Orlov et al., 2019). For China, another study estimates that heat-related productivity losses resulted in costs of $126 billion in 2017 (Wenjia et al., 2021). Heat stress also causes a variety of health complaints, ranging from skin rashes to muscle cramps and insomnia to heat-related strokes. In addition to the suffering of the people affected, it also causes costs for the general public, for example through increased hospitalization and absenteeism from work (Semenza et al., 1999; Gronlund et al., 2014; Phung et al., 2016; Obradovich et al., 2017; Sherbakov et al., 2018).

Storms, Unstable Weather and More The impact channels described above are not exhaustive but provide a first insight into the multitude of economically relevant damages and costs that materialize in a warming world. In addition, there are the costs of tropical cyclones, such as those that hit North and Central America as well as East and Southeast Asia in 2021 and caused immense damage there (Fig. 1). As warming progresses, such storms could also form at higher latitudes and thus affect millions more people (Studeholme, 2021). But even weather that is "just" more unstable can have a negative economic impact—for example through health effects and agricultural losses, or because it means planning uncertainty for decision-makers and thus paralyses investments (Wheeler et al., 2000; Shi et al., 2015).

Complex Interactions

In a closely interconnected economic world, the effects of climate and weather extremes may not remain local but can propagate along global supply and value chains as well as via price signals—even across national borders (MacKenzie et al., 2012; Wenz & Levermann, 2016; Wenz & Willner, 2022). For example, severe flooding in the Thai capital Bangkok in 2011 resulted in a shortage of hard drives in Europe (Haraguchi & Lall, 2014). In 2021, the timber industry in North America was affected by forest fires, a beetle infestation, and sawmill closures due to the pandemic, so that more timber was imported from Germany. Subsequently, timber became expensive and scarce in Germany (Denkler, 2021). As a result of such "cascading effects", the actual damage from weather extremes can be greater than what is observed only locally—especially if several events occur simultaneously in different regions (Kuhla et al., 2021). For example, not only direct damages from river floods are expected to increase (a global increase of 15% to around US $600 billion within the next 20 years), but also indirect effects could arise along the supply chains (leading to damage of another US $200 billion) (Willner et al., 2018). Such effects are especially critical if they lead to supply shortages, for example of medicines or food (Bren d'Amour et al., 2016).

Interaction effects with other crises such as the Covid-19 pandemic are also relevant, for example if combating them ties up important resources: The IPCC emphasized that the interplay of different climatic and non-climatic risks can lead to risk cascades across sectors and regions (IPCC, 2022a, b). At the same time, adaptation to climate change can also have an impact on other sectors relevant to climate change. One example is the installation of power-hungry air-conditioning systems to prevent heat stress. In emerging economies such as India, Indonesia, and Vietnam, a recent study foresees rising electricity demand due to increasing heat (Rode et al., 2021).[4] Another study shows that rising temperatures will also change electricity in Europe: demand will shift from Northern to Southern countries, and the annual peak load will shift from winter to summer—shifts that might pose major challenges to the existing infrastructure (Wenz et al., 2017).

Only a few studies estimate economic damages that might occur if individual tipping points[5] in the climate system are exceeded. What complicates such assessment is that the effects of tipping points are felt on different, sometimes very long time scales. The discounting already mentioned (see also section "Empirical

---

[4] However, many regions of the world could still be too poor at the end of the twenty-first century for their electricity demand to increase drastically due to rising temperatures. Moreover, the additional demand for electricity in emerging economies may be offset globally by reduced heating in countries with colder climates (Rode et al., 2021).

[5] If tipping points are exceeded, large, accelerating and often irreversible changes occur in the climate system which can have serious consequences. It is assumed that some tipping points will already be exceeded with an average warming of 1 to 2 degrees (Lenton et al., 2019). In a world that is three degrees warmer, the risk of additional damage from exceeded tipping points would be substantial.

Models: Learning from the Past to Predict the Future") therefore plays an important role here. A recently published overview study integrates several estimates of economic impacts from the crossing of different tipping points (Dietz et al., 2021), concluding that tipping points increase the Social Cost of Carbon by about 25% (compared to an estimate of economic climate impacts without taking tipping points into account). The study also indicates a probability of about 10% that the integration of tipping points in damage estimates more than doubles the costs.

A Price Tag for Everything?

As already mentioned, some damages can directly be expressed in monetary terms, whereas others are difficult to monetize but still very relevant for the economy and society. This includes, for example, the loss of labor due to migration, illness, or death, or the lost recreational function of burnt forest areas. In addition to the purely economic costs, many climate events cause high humanitarian and social costs, which in turn can have direct and indirect effects on the economy. These include the fact that people may develop mental problems such as depression or anxiety and post-traumatic stress disorders as a result of extreme weather events (Munro et al., 2017; Schwartz et al., 2017), which have to be treated, with corresponding negative effects on their productivity. In addition, things that do not easily carry a price tag may also have value for us, such as the preservation of biodiversity.

An important question is whether and how such damages should be incorporated into estimates of the costs of climate change. Economists have developed various techniques for assigning a monetary value to non-market damages. So-called willingness-to-pay approaches, for example, aim to measure what we are willing to pay to avoid certain damages or to maintain certain features and functions like, for example, ecosystem services. Some studies follow a recommendation of the US Environmental Protection Agency (EPA) and value a statistical human life at $7.4 million (EPA, 2010). This value is controversial: There are voices that argue in favor of greater differentiation, reflecting, for example, that older people contribute less to economic growth (EPA, 2010; Hsiang et al., 2017). Others highlight significant ethical problems—especially if the value is set differently for developed and less developed countries—and recommend that such damages should not be expressed in monetary terms but presented separately (Stern, 2007).

The Cost of Climate Change

As outlined above, climate change will have massive effects on many sectors and areas of life that are economically highly relevant. But what will be the effect on the economy as a whole? A 2021 Reuters survey of climate economists shows a wide range of estimates (Fig. 3). On average, the experts estimated that under a

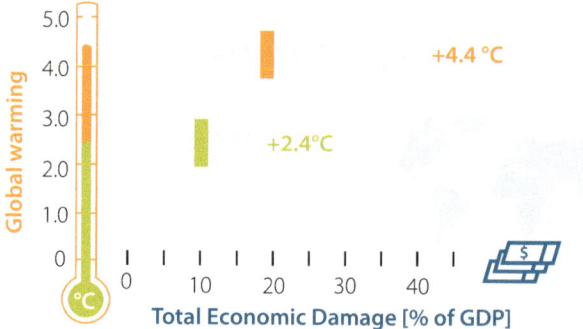

Fig. 3 Total economic damage for different levels of warming. Shown are median (orange), minimum and maximum, based on the assessments of 13 experts. (Data from Reuters, 2021)

pessimistic scenario of unchecked climate change—implying about 2.4 degrees of warming by 2050 and 4.4 degrees by 2100—global economic output (Gross Domestic Product, GDP) would be reduced by about 10% by the middle of the twenty-first century and by about 18% by the end. In Stern's 2006 Review, the damage caused by 3 degrees warming was estimated at 5% to 20% of global GDP, whereby the lower end of the estimate does not take into account damage to health, ecosystems, etc. Due to the wide range of estimates and the difficulty of comparing the underlying methods, the IPCC's Sixth Assessment Report does not include any concrete figures in its summary, but concludes that the damage could be higher than previously assumed (IPCC, 2022a, b).

The cost range can be explained by, amongst others, the diversity of approaches to estimating the macroeconomic damage, which was already discussed in this chapter. Different underlying assumptions also play an important role. In the following three sections, we discuss some of the factors and sources of uncertainty that contribute to the wide range of damage estimates.

From Micro to Macro

One obvious way of estimating the total cost of climate change is to consider the net effect of all individual effects. This is the approach taken by a comprehensive 2017 study for the US. The study focuses on six different sectors and—building on previous research findings– estimates and monetizes the expected damage in each sector and then adds them up (Hsiang et al., 2017). Specifically, the interdisciplinary research team looked at agriculture, crime, storm surges, energy, mortality and labor. They identified the greatest damages due to higher mortality, followed by damages in the agriculture sector, to labor, and in the energy sector. With global warming of 3 degrees, the total direct damage by the end of this century is estimated at about 1.5% to 2% of US gross domestic product, with the costs distributed very unevenly across the US and disproportionately burdening regions in the already poorer South. A report by the Deloitte Economics Institute published in 2021 uses a similar approach to estimate the costs of a 3-degree warmer world for Germany. The damage channels considered there include heat stress, damage to capital stock, loss of agricultural land and agricultural yields, declining tourism revenues, and human health impacts. Based on this, the report concludes that the damage to the German economy could amount to €730 billion over the next 50 years. Such figures would, according to the report, lead to the loss of almost half a million jobs.

Such bottom-up approaches have the advantage that, in addition to estimating the total costs, they also provide a good understanding of processes—for example, how much the individual damages contribute to the total costs. Such insights can play an important role in prioritizing adaptation measures. A shortcoming of bottom-up approaches is that one has to be confident that all mechanisms through which climate change can cause significant economic damage are sufficiently well-known and considered. Furthermore, aggregation may not always be straight-forward due to concerns of double-counting and possible interaction effects between different sectors (Dell et al., 2012).

Climate Change Impacts on Economic Growth

Another possibility is to directly assess the effects of climate change on macroeconomic growth. In this case, changes in economic output observed in the past years are statistically compared against changes in local weather, while accounting for confounding effects. The thus identified effect of changes in temperature and precipitation on the economy is then used to derive possible economic losses under future warming.

Various studies have found a clear non-linear relationship between average annual temperature and productivity (Dell et al., 2012; Burke et al., 2015; Pretis et al., 2018; Kalkuhl & Wenz, 2020; Kahn et al., 2021). If the temperature in a country or region rises from one year to the next, this usually harms the local economy.

An evaluation of climate and economic data of the last 40 years from more than 1500 regions worldwide has shown that an increase in the annual mean temperature of about 1 degree leads to economic losses of 1 to 2% (Kalkuhl & Wenz, 2020). The warmer the region, the greater the losses—though in some regions that were previously very cold, an increase in the mean annual temperature can even be beneficial from an economic point of view. With this approach it also becomes clear that climate change will affect different regions differently. The Earth is warming at different rates regionally, and individual regions' vulnerability to damages also varies.

If we take this observed relationship between temperature and economic output as a starting point for future damage, the following picture emerges: If the Earth were to warm by about 4.4 degrees by the end of the century compared to pre-industrial times (corresponding to the RCP 8.5 scenario described in section "Climate Change Affects All Sectors of the Economy"), this would reduce global GDP by about 14%. In tropical, poorer regions, losses would be even higher,

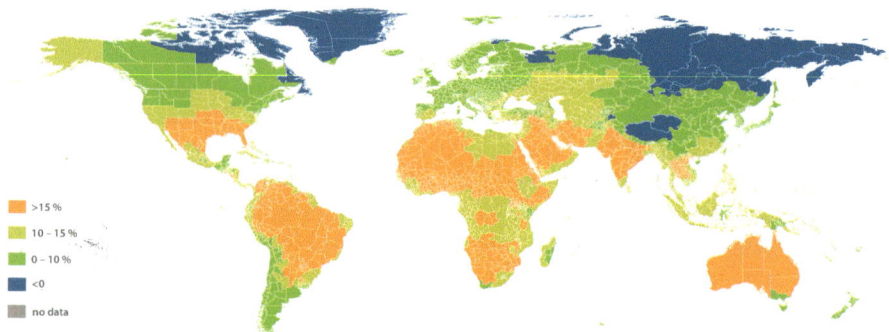

Fig. 4 Regional income losses in the year 2100. Estimates assume a pessimistic scenario (about 4.4 degrees warming compared to pre-industrial times) (Kalkuhl & Wenz, 2020)

possibly over 20%. In the comparatively cooler German regions, they would be about 5% (see Fig. 4). This is comparable to the 5% decline in German output in 2020 in the wake of the Covid-19 pandemic or the impact of the financial crisis of 2008/09 (5.7%) (Tagesschau, 2021).

There is, however, one very important difference to previous economic crises such as the financial crisis or the Covid-19 pandemic: these previous crises were of limited duration. This enables governments to mitigate negative economic and social impacts, for example with government aid and economic stimulus programs. Climate change, however, will not simply recede again, but will be permanent in the best case, or become ever more severe as long as emissions are not reduced—the resulting decline in economic productivity may therefore also be permanent.

Even though the warming scenario used here is more pessimistic than the 3-degree scenario, it does provide a good indication of the massive challenges that await us in a 3 degrees warmer world. And yet, this estimate is a conservative one for various reasons. The 14% reduction in global GDP should be understood as a lower limit for the actual economic losses, because climate change is more than just a gradual increase in the annual mean temperature. The effects of extreme events and sea-level rise, as discussed earlier, are not included in such analyses. The same applies to non-monetary damages such as the loss of biodiversity or health impacts. Current studies also show that looking at annual averages falls short of gauging the actual damages. If, for example, temperatures fluctuate strongly around the monthly average or if there are more rainy days or days with extreme precipitation within a year, this causes additional harm to the economy (Kotz et al., 2021, 2022), resulting in higher economic losses (Waidelich et al., 2024). A recent study that also takes the effects from changes in rainfall and temperature variability into account, as well as the persistence of damages, projects a 19% income reduction on global average in 2050 compared to a world without climate change, irrespective of the emission scenario (Kotz et al., 2024).

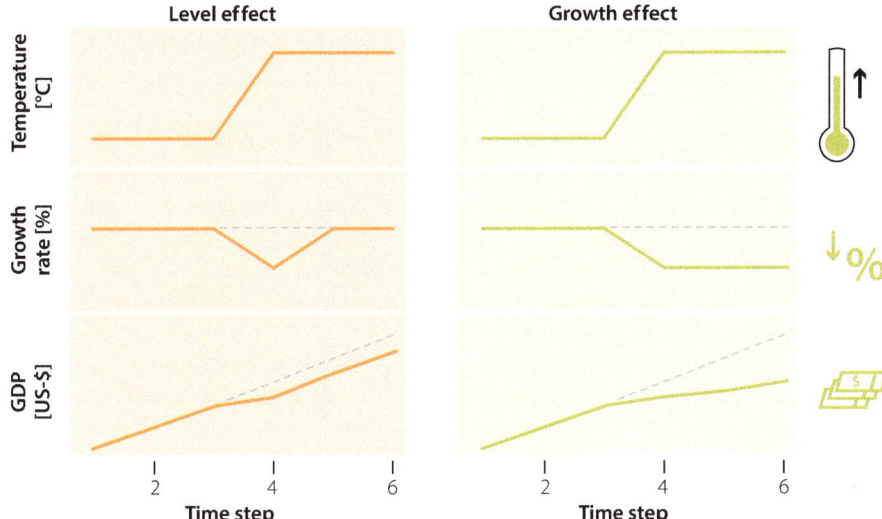

Fig. 5 Level and growth effects. The two columns illustrate how a permanent temperature increase would affect the growth rate of the economy and the gross domestic product—assuming level effects (left column) and growth effects (right column), respectively. In the case of growth effects, economic losses accumulate over time (Wenz & Kuik)

Level or Growth Effects—How Persistent Are the Economic Impacts of Climate Change?

A central question when estimating economic damages is whether the economy is only slowed down in the short term, or whether economic growth is permanently lowered by temperature changes and weather extremes (Fig. 5). Many damage estimates assume so-called level effects (Kalkuhl & Wenz, 2020). The assumption is that a permanent rise in temperature leads only initially to a reduction in economic growth which then returns to its original path. Consequently, economic output is permanently lowered by the same factor.

But there are also reasons to believe that weather extremes can more permanently reduce economic growth. Such long-term growth effects can occur when destroyed capital assets cannot be repaired or replaced for years, when people must give up their education as a result of weather extremes, or when investments cannot be made (Fankhauser et al., 2005; Moore & Diaz, 2015). A recent study shows that tropical cyclones and river flooding can reduce economic growth in affected countries for more than a decade (Krichene et al., 2021).

The right column in Fig. 5 illustrates growth effects for the temperature example. In this case, too, economic output decreases when temperature rises permanently. However, the losses increase with each year as they accumulate. A study from 2015 estimates that unmitigated climate change (scenario RCP 8.5) would reduce global

GDP in 2100 by about 23%, assuming such growth effects (Burke et al., 2015). In many countries of the global South, the losses would be as high as 100%.

A 2021 study demonstrates that the question of damage persistence is a key source of uncertainty in assessing the economic impacts of climate change (Kikstra et al., 2021). Based on an updated version of the PAGE model used in the Stern Review they find that economic output would be reduced by about 50% at the end of the century for a medium warming scenario, if assuming growth effects.[6] With level effects, the loss would "merely" be 6%. The authors consider an intermediate case likely, in which economic growth is slowed down for quite a while, i.e., the harm is partially, but not fully, persistent. In this scenario, the global GDP would be 37% lower, according to this analysis.

The question of level versus growth effects is thus by no means a purely technical one, but has major implications for the magnitude of damages and therefore for estimates of the costs of climate change and optimal climate policy. Accordingly, it is heavily debated in the scientific community and is the subject of active research, as it is statistically challenging to cleanly distinguish between the two effects (Bastien-Olvera et al., 2022; Kikstra et al., 2021; Newell et al., 2021; Kotz et al., 2024).

Climate Damages More Costly Than Climate Protection

The early 2020s have impressively illuminated several possible economic effects of extreme weather—a preview of a world in which global mean temperature could be 3 degrees higher than in pre-industrial times and in which weather extremes would be even more frequent and intense. In such a world, no region or economic sector would be spared from the effects of climate change.

Some costs arise through rather direct impact channels which can be estimated relatively well with conventional methods: the influence of rising temperatures and extreme heat on productivity and health, the effects of droughts and water shortages on agriculture and industry, the effects of heavy rainfall and flooding on buildings and infrastructure. By aggregating damages from these individual sectors or by assessing macroeconomic losses directly using data-intensive empirical approaches we can infer that the costs of a 3-degree warmer world could easily exceed 10% of global GDP. These costs are, moreover, very unevenly distributed globally with regions least responsible for historical climate change and with fewest means to adapt generally hit hardest.

Approaches to estimate these costs are associated with uncertainties: not all transmission channels and interactions can be captured, it is not certain how cost-effectively the world can adapt to a warmer climate, it is unclear how persistent the

---

[6] The RCP 4.5 scenario, which corresponds to a warming of about 2.7 degrees at the end of the twenty-first century compared to pre-industrial times.

damages will be, and not all climate impacts can be expressed in monetary terms. Nonetheless, these approaches give a fairly reliable picture: the economic costs of climate change that is not ambitiously mitigated will be significant.

Of particular concern are complex interactions within and between the climate and economic systems. On the socio-economic side, these include crises, conflicts and migration as well as effects that trickle down complex supply chains. Uncertainties are further aggravated by the possibility of crossing tipping points in the climate system, by potential interaction effects among different climate impacts, or interacting climate-related and non-climate-driven risk factors.

As has become clear in this chapter, many aspects, assumptions, and uncertainties affect estimates of the costs of climate change. These factors also explain why there is a wide range of cost estimates and why it is likely that there is no upper bound to estimates of the economic costs from future climate change. However, one common message arises despite all uncertainties and different methodologies: it is much cheaper to protect our climate than to live with the economic consequences of a 3 degrees warmer world.

References

Alfieri, L., et al. (2017). Global projections of river flood risk in a warmer world. *Earth's Future, 5*(2), 171–182.

Atlantic Council. (2021). *Extreme heat: The economic and social consequences for the United States.*

Auffhammer, M. (2018). Quantifying economic damages from climate change. *Journal of Economic Perspectives, 32*(4), 33–52.

Azar, C., & Sterner, T. (1996). Discounting and distributional considerations in the context of global warming. *Ecological Economics, 19*(2), 169–184.

Bastien-Olvera, B. A., Granella, F., & Moore, F. C. (2022). Persistent effect of temperature on GDP identified from lower frequency temperature variability. *Environmental Research Letters, 17*(8), 084038.

Bren d'Amour, C., et al. (2016). Teleconnected food supply shocks. *Environmental Research Letters, 11*(3).

Burke, E. J., Brown, S. J., & Christidis, N. (2006). Modeling the recent evolution of global drought and projections for the twenty-first century with the Hadley Centre climate model. *Journal of Hydrometeorology, 7*(5), 1113–1125.

Burke, M., et al. (2015). Global non-linear effect of temperature oneconomic production. *Nature, 527*(7577), 235–239.

Carleton, T. A., & Hsiang, S. M. (2016). Social and economic impacts of climate. *Science, 353*(6304).

Dell, M., Jones, B. F., & Olken, B. A. (2012). Temperature shocks and economic growth: Evidence from the last half century. *American Economic Journal: Macroeconomics, 4*(3), 66–95.

Dell, M., Jones, B. F., & Olken, B. A. (2014). What do we learn from the weather? The new climate-economy literature. *Journal of Economic Literature, 52*(3), 740–798.

Denkler, T. (2021). *Kleines Tier, großer Effekt: Ein kanadischer Käfer lässt Holz in Deutschland knapp werden.* Sueddeutsche Zeitung.

Dietz, S., et al. (2021). Economic impacts of tipping points in the climate system. *Proceedings of the National Academy of Sciences, 118*(34).

Dunne, J. P., Stouffer, R. J., & John, J. G. (2013). Reductions in labour capacity from heat stress under climate warming. *Nature Climate Change, 3*(6), 563–566.

EPA, US Environmental Protection Agency. (2010). *Valuing mortality risk reductions for environmental policy: A white paper.*

Fankhauser, S., et al. (2005). On climate change and economic growth. *Resource and Energy Economics, 27*(1), 1–17.

Feyen, L., et al. (2020). *Climate change impacts and adaptation in Europe* (JRC PESETA IV Final Report).

Gallic, E., & Vermandel, G. (2020). Weather shocks. *European Economic Review, 124.*

Glanemann, N., Willner, S. N., & Levermann, A. (2020). Paris Climate Agreement passes the cost-benefit test. *Nature Communications, 11*(1), 110.

Graff, Z. J., Hsiang, S. M., & Neidel, M. (2018). Temperature and human capital in the short and long run. *Journal of the Association of Environmental and Resource Economists, 5*(1), 77–105.

Greenstone, M. (2016). *A new path forward for an empirical social cost of carbon.* Presentation to the National Academies of Sciences.

Gronlund, C. J., et al. (2014). Heat, heat waves, and hospital admissions among the elderly in the United States, 1992–2006. *Environmental Health Perspectives, 122*(11), 1187–1192.

Hänsel, M. C., et al. (2020). Climate economics support for the UN climate targets. *Nature Climate Change, 10*(8), 781–789.

Haraguchi, M., & Lall, U. (2014). Flood risks and impacts: A case study of Thailand's floods in 2011 and research questions for supply chain decision making. *International Journal of Disaster Risk Reduction, 14*(3), 256–272.

Helman, D., & Zaitchik, B. F. (2020). Temperature anomalies affect violent conflicts in African and Middle Eastern warm regions. *Global Environmental Change, 63.*

Howard, P. H., & Sterner, T. (2017). Few and not so far between: A meta-analysis of climate damage estimates. *Environmental and Resource Economics, 68*(1), 197–225.

Hsiang, S. M. (2010). Temperatures and cyclones strongly associated with economic production in the Caribbean and Central America. *Proceedings of the National Academy of Sciences, 107*(35), 15367–15372.

Hsiang, S. M. (2016). Climate econometrics. *Annual Review of Resource Economics, 8*, 43–75.

Hsiang, S. M., Burke, M., & Miguel, E. (2013). Quantifying the influence of climate on human conflict. *Science, 341*(6151).

Hsiang, S. M., et al. (2017). Estimating economic damage from climate change in the United States. *Science, 356*(6345), 1362–1369.

IPCC. (2022a). *Climate change 2022: Impacts, adaptation and vulnerability.* Working Group II Contribution to the IPCC Sixth Assessment Report, Cambridge.

IPCC. (2022b). *Summary for policy makers.*

Kahn, M. E., et al. (2021). Long-term macroeconomic effects of climate change: A cross-country analysis. *Energy Economics, 104*, 105624.

Kalkuhl, M., & Wenz, L. (2020). The impact of climate conditions on economic production. Evidence from a global panel of regions, *103*, 102360.

Kharin, V. V., et al. (2018). Risks from climate extremes change differently from 1.5 C to 2.0 C depending on rarity. *Earth's Futures, 6*(5), 704–715.

Kikstra, J. S., et al. (2021). The social cost of carbon dioxide under climate-economy feedbacks and temperature variability. *Environmental Research Letters, 16*(9).

Kolstad, C. D., & Moore, F. C. (2020). Estimating the economic impacts of climate change using weather observations. *Review of Environmental Economics and Policy, 14*(1), 1–24.

Kotz, M., et al. (2021). Day-to-day temperature variability reduces economic growth. *Nature Climate Change*, 1–7.

Kotz, M., Levermann, A., & Wenz, L. (2022). The effect of rainfall changes on economic production. *Nature, 601*(7892), 223–227.

Kotz, M., Levermann, A., & Wenz, L. (2024). The economic commitment of climate change. *Nature, 628*(8008), 551–557.

Kovats, R., et al. (2014). Europe. In *Climate change 2014: Impacts, adaptation, and vulnerability. Part B: Regional aspects.* Contribution of Working Group II to the Fifth Assessment Report of the Intergovernmental Panel on Climate Change. Bednar-Friedl, B., et al. (2022). Europe. In *Climate change 2022: Impacts, adaptation, and vulnerability.* Contribution of Working Group II to the Sixth Assessment Report of the Intergovernmental Panel on Climate Change.

Krichene, H., et al. (2021). Long-term impacts of tropical cyclones and fluvial floods on economic growth – Empirical evidence on transmission channels at different levels of development. *World Development, 144*, 105475.

Kuhla, K., et al. (2021). Ripple resonance amplifies economic welfare loss from weather extremes. *Environmental Research Letters, 16*(11).

Lehner, B., et al. (2001). *Model-based assessment of European water resources and hydrology in the face of global change.* Centre for Environmental Systems Resources. Universität Kassel.

Lenton, T. M., et al. (2019). *Climate tipping points – Too risky to bet against.* Nature Publishing Group.

MacKenzie, C. A., Santos, J. R., & Barker, K. (2012). Measuring changes in international production from a disruption: Case study of the Japanese earthquake and tsunami. *International Journal of Production Economics, 138*(2), 293–302.

Min, S.-K., et al. (2011). Human contribution to more-intense precipitation extremes. *Nature, 470*(7334), 378–381.

Moore, F. C., & Diaz, D. B. (2015). Temperature impacts on economic growth warrant stringent mitigation policy. *Nature Climate Change, 5*(2), 127–131.

Munich Re. (2023). NatCatSERVICE.

Munro, A., et al. (2017). Effect of evacuation and displacement on the association between flooding and mental health outcomes: A cross-sectional analysis of UK survey data. *The Lancet Planetary Health, 1*(4), e134–e141.

Newell, R. G., Prest, B. C., & Sexton, S. E. (2021). The GDP-temperature relationship: Implications for climate change damages. *Journal of Environmental Economics and Management, 108*.

NOAA Office for Coastal Management. (2021). *Hurricane costs.* https://coast.noaa.gov/states/fast-facts/hurricane-costs.html

Nordhaus, W. (2018). Projections and uncertainties about climate change in an era of minimal climate policies. *American Economic Journal: Economic Policy, 10*(3), 333–360.

Nordhaus, W. D., & Boyer, J. (2000). *Warming the world: Economic models of global warming.*

Obradovich, N., et al. (2017). Nighttime temperature and human sleep loss in a changing climate. *Science Advances, 3*(5), e1601555.

Obradovich, N., et al. (2018). Empirical evidence of mental health risks posed by climate change. *Proceedings of the National Academy of Sciences, 115*(43), 10953–10958.

Orlov, A., et al. (2019). Economic losses of heat-induced reductions in outdoor worker productivity: A case study of Europe. *Economics of Disasters and Climate Change, 3*(3), 191–211.

Paltan, H., et al. (2018). Global implications of 1.5 C and 2 C warmer worlds on extreme river flows. *Environmental Research Letters, 13*(9).

Phung, D., et al. (2016). Ambient temperature and risk of cardiovascular hospitalization: An updated systematic review and metaanalysis. *Science of the Total Environment, 550*, 1084–1102.

Pindyck, R. S. (2013). Climate change policy: What do the models tell us? *Journal of Economic Literature, 21*(3), 860–872.

Pretis, F., et al. (2018). Uncertain impacts on economic growth when stabilizing global temperatures at 1.5 C or 2 C warming. *Philosophical Transactions Royal Society A: Mathematical, Physical and Engineering Sciences, 376*(2119).

Ramsey, J. D. (1995). Task performance in heat: A review. *Ergonomics, 38*(1), 154–165.

Reuters. (2021). Climate inaction costlier than net zero transition: Reuters poll | Reuters.

Rode, A., et al. (2021). Estimating a social cost of carbon for global energy consumption. *Nature, 598*(7880), 308–314.

Schwartz, R. M., et al. (2017). Displacement and mental health after natural disasters. *The Lancet Planetary Health, 1*(8).

Semenza, J. C., et al. (1999). Excess hospital admissions during the July 1995 heat wave in Chicago. *American Journal of Preventive Medicine, 16*(4), 269–277.

Sherbakov, T., et al. (2018). Ambient temperature and added heat wave effects on hospitalizations in California from 1999 to 2009. *Environmental Research, 160*, 83–90.

Shi, L., et al. (2015). Impacts of temperature and its variability on mortality in New England. *Nature Climate Change, 5*(11), 988–991.

Stechemesser, A. H., et al. (2021). Strong increase of racist tweets outside of climate comfort zone in Europe. *Environmental Research Letters, 16*.

Stechemesser, A., Levermann, A., & Wenz, L. (2022). Temperature impacts on hate speech online: Evidence from 4 billion geolocated tweets from the USA. *The Lancet Planetary Health, 6*(9), e714–e725.

Stern, N. (2007). *The economics of climate change: The Stern review.*

Studeholme. (2021). Poleward expansion of tropical cyclone latitudes in warming climates. *Nature Geoscience, 15*(1).

Szewczyk, W., Mongelli, I., & Ciscar, J.-C. (2021). Heat stress, labour productivity and adaptation in Europe – A regional and occupational analysis. *Environmental Research Letters, 16*(10).

Tagesschau. (2021). *Wirtschaft bricht um fünf Prozent ein.*

Turco, M., et al. (2018). Exacerbated fires in Mediterranean Europe due to anthropogenic warming projected with non-stationary climate-fire models. *Nature Communications, 9*(1), 1–9.

Ueckerdt, F. et al. (2019). The economically optimal warming limit of the planet. *Earth System Dynamics 10*(4), 741–763.

UNEP and INEP DTU Partnership. (2021). *Emissions Gap Report 2021: The heat is on – A World of climate promises not yet delivered.*

United Nations. (2015). *Paris Agreement.* https://treaties.un.org/Pages/ViewDetails. aspx?src=TREATY&mtdsg_no=XXVII-7-d&chapter=27&clang=_en

van Vuuren, D., Edmonds, J., & Kainuma, M. (2011). The representative concentration pathways: An overview. *Climatic Change, 109*, 5–13.

Waidelich, P., et al. (2024). Climate damage projections beyond annual temperature. *Nature Climate Change*, 1–8.

Weitzman, M. L. (2009). On modeling and interpreting the economics of catastrophic climate change. *Review of Economics and Statistics, 91*(1), 1–19.

Wenjia, C., et al. (2021). The 2020 China report of the Lancet Countdown on health and climate change. *The Lancet, Public Health, 6*(1), e64–e81.

Wenz, L., & Levermann, A. (2016). Enhanced economic connectivity to foster heat stress–related losses. *Science Advances, 2*(6), e1501026.

Wenz, L., & Willner, S. N. (2022). 18. Climate impacts and global supply chains: An overview. *Handbook on Trade Policy and Climate Change, 290.*

Wenz, L., Levermann, A., & Auffhammer, M. (2017). North-South polarization of European electricity consumption under future warming. *Proceedings of the National Academy of Sciences.*

Wheeler, T. R., et al. (2000). Temperature variability and the yield of annual crops. *Agriculture, Ecosystems & Environment, 82*(1–3), 159–167.

Willner, S. N., Otto, C., & Levermann, A. (2018). Global economic response to river floods. *Nature Climate Change, 8*, 594–598.

Zhongming, Z., et al. (2021). *UN world water development report 2021* (Valuing Water).

Part II
Nature-Based Solutions: How Can We Still Prevent a 3-Degree Warmer World

Stop Rainforest Deforestation

The Most Urgent Way of Combining Climate and Species Protection

Susanne Winter

Clear and undoubted, as the sixth IPCC report of August 2021 confirms, it has to be noted that we humans are causing rapid climate change. Our emissions and destruction of our environment are leading us into a drastic change of life on a global scale. Our treatment of forests transforms them, too, into major sources of emissions. Man-made emissions from the economy, including agriculture and forestry, must be reduced immediately!

Reliable goals and ways of achieving climate neutrality as quickly as possible are the political order of the day. No more delaying, no more wait-and see, no more blame-shifting, no more beating around the bush—we need ambitious action in the right direction. Compliance with the Paris Climate Agreement is the narrow pathway into a future that might resemble the present. We need much stronger commitments and efforts toward keeping the climate roughly as we know it.

Forests at the Center of Multiple Crises

To protect the climate, as well as our health, much greater efforts are needed to preserve the biodiversity and biosphere integrity of our planet. Biodiversity is the backbone of our life (Díaz et al., 2006; Cardinale et al., 2012; Kadykalo et al., 2019). Its loss, which is dangerous for us, has progressed much farther than the climate crisis (Fig. 1), which is a burden on us and is now finally being perceived politically (Rockström et al., 2009; Steffen et al., 2015; Richardson et al., 2023). The biodiversity and biosphere integrity crisis is caused by the unchecked heavy encroachment on, and elimination of habitats that are permanently necessary for the diversity of species. Already 75% of the world's land surface is severely degraded

S. Winter (✉)
WWF Germany, Department Policy and Biodiversity, Berlin, Germany
e-mail: susanne.winter@wwf.de

© The Author(s) 2024
K. Wiegandt (ed.), *3 Degrees More*,
https://doi.org/10.1007/978-3-031-58144-1_6

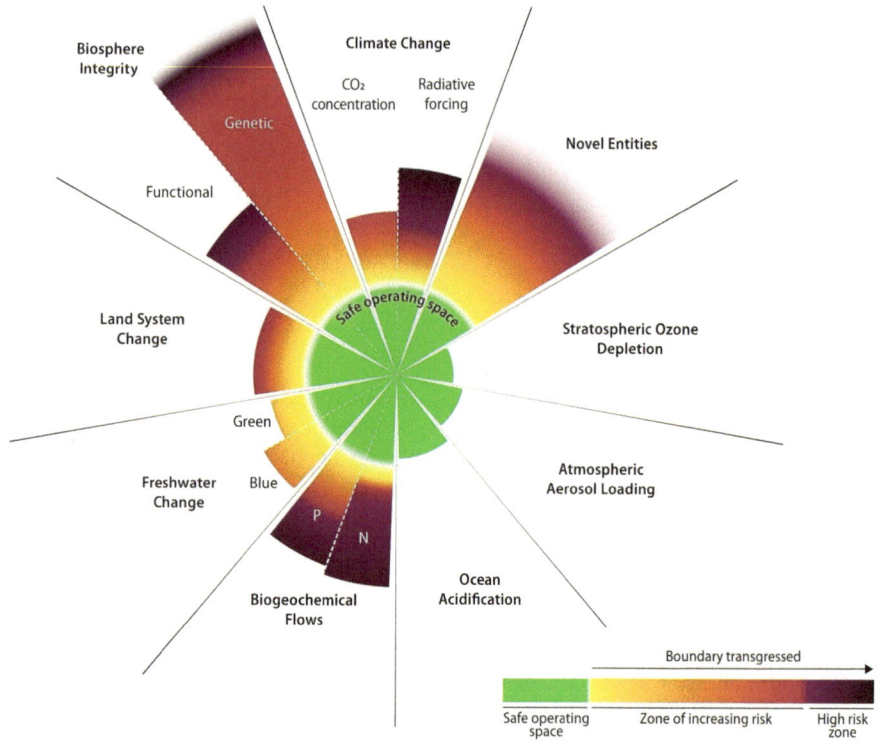

Fig. 1 Triple crisis: biosphere integrity, nitrogen, and climate. Crises that reinforce each other in both the short and long term (Richardson et al., 2023)

(IPBES, 2019). The damage amounts to an estimated $6.3 trillion per year (that's 6.3 times a thousand billion!) and the livelihoods of about half a billion people are increasingly at high risk (Ding et al., 2017). Faced with the threat posed by the damage to nature, with the need to safeguard the Earth's habitats, politics is not sufficiently attentive nor doing enough (Mazor et al., 2018). Political measures are inadequate. Moreover, most people today are completely unfamiliar with the things that must be protected, the still nearly untouched primary forests for instance. Most people live in an environment that has been almost completely degraded and excessively transformed. The loss of biodiversity, manifested in the decline of species and populations, continues unchecked around the world and especially in the tropics (e.g. Barley et al., 2016; Giam, 2017).

We Europeans have reached a level of consumption that entails not merely a very large ecological footprint in our continent, but also burdens for the whole world. In regard to forest destruction caused by international trade, the European Union (EU) plays a leading role, second only to the People's Republic of China (WWF, 2021a). Yet the losses and the comprehensive need for action are still not sufficiently perceived as politically urgent. The opportunity for the EU to become a role model is still there—and hopefully will be seized as soon as possible.

Accounting for over 90% of forest loss, agriculture and forestry are its primary drivers, especially in tropical regions. Political regulations are needed that constrain responsible producers, companies, and traders through rules and obligations that prevent deforestation in their supply chains. In this way, climate change and the downward trend in vertebrate species (as a proxy for all biodiversity) could be slowed and vital biodiversity be rebuilt.

Forest as a Health Precaution
On average, a first transmission of pathogens from animals to humans occurs every four months. This transition is described for Ebola, malaria, avian influenza, Lassa fever and the Niphavirus, among others (Ellwanger et al., 2020; Wolfe et al., 2005; for Ebola: Olivero et al., 2017; for malaria: Brock et al., 2019; for avian influenza: Sehgal, 2010; for Lassa fever: Adetola & Adebisi, 2019; for Niphavirus: Shanko et al., 2015), and is an essential source for developing pandemics. With ongoing decimation, fragmentation, and overexploitation of forests, such spillovers occur ever more frequently as people are getting more often too close to forest animals, so that infections can take place more easily (WWF, 2020b). But it is not only our physical health that is protected by resilient biodiversity and threatened by habitat change (WWF, 2020b; Kilpatrick & Randolph, 2012). Our psyche, our mental balance, is also better nourished when we humans live in a diverse environment and can be in contact with biological diversity (Lovell et al., 2014; Harvey et al., 2020).

The One Health approach (Rabinowitz et al., 2013; Jorwal et al., 2020) shows our dependence on natural ecosystems. Only the preservation and restoration of natural and near-natural ecosystems will maintain our health in the long term. Conservation of—especially natural—forests is a form of protective health care in a context of continued population growth.

Nature's intrinsic value should be an imperative for its preservation in our social and political conduct. But we civilized humans seem to have evolved with a blind spot, a missing development. So long as population density was low, nature seemed an inexhaustible resource for humans. But this is no longer true. We have already trimmed nature to such an extent that its necessary contributions to our lives (from soil fertility to water supply to pastoral care for stressed people) are ever more insecure. Moreover, increasing destruction of nature makes transmission of infectious diseases to humans more likely, thus magnifying the danger of pandemics. We cannot afford to expose what nature remains to further human-technical penetration. However, this is not fully felt nor accepted by the humans.

To avoid inflicting further large-scale damage on ourselves and the planet, we must avert the greatest current crises (Fig. 1), the biodiversity, biochemical,[1] land system change and climate crisis. Forests play a role in these crises in four ways, namely as

- shrinking and degraded ecosystems with immense biodiversity and functional losses;
- endangered utilization systems that, among other things, provide groundwater and drinking water, supply raw materials, and keep us healthy;
- large-scale carbon sources as a result of degradation and loss;
- objects of climate policy insofar as forests, storing carbon and regulating the climate, offer important nature-based solutions to the climate crisis.

While the transformative power of climate policy (away from fossil fuels) and the growing world population are steadily raising the demands on forests, the decline of forests and their quality is increasingly reducing their sustainable usability and thus the security of their services. The passage into the future has become very narrow and will become even narrower with every tenth of a degree of additional global warming due to human-induced emissions, with further deforestation, and with intensified land use. Animal, plant, and fungal species of the forest no longer have a future, and people thereby lose their livelihoods.

Main Message *With the loss of primary forests, global biodiversity, our health, and climate protection are in free fall as well. Without preserving the remaining natural forests, a climate that protects our (human) lives is not possible.*

Forest Loss and Degradation in the Tropics

Some 30% of Germany's land area is covered by forest. During the last decade, we have globally lost natural forest areas of this size every year (FAO, 2020)! More than 80% of this deforestation is due to the spread of agriculture (Kissinger et al., 2012; WWF, 2018a). The three main destroyers are soy cultivation for animal feed, the creation of palm oil plantations and cattle pastures (Fig. 2). Regionally, however, other factors are also strongly involved in deforestation, such as mining and infrastructure projects (dams and roads) in South America (WWF, 2015). Furthermore, the conversion of natural forests into timber plantations is another important factor: forests rich in biodiversity and carbon thus become degraded timber plantations that resemble agricultural land more than real forests. Such plantations are the fourth largest global driver of the destruction of valuable natural forests (WWF, 2021a).

The effects are not adequately captured by adding up lost areas. Thus, the shrinking of the Amazon Forest, for example, leads to drying out with supra-regional

---

[1] Biochemistry cannot here be dealt with in depth. Pesticides and nitrogen are especially damaging to forest biodiversity and resilience (Hofmann et al., 2019; de Vries et al., 2011).

Fig. 2 Between 2005 and 2017, more than 80% of tropical forest destruction was caused by just six commodities: soy, palm oil, beef, wood products, cocoa, and coffee (WWF, 2022)

effects (Leite-Filho et al., 2021; Zemp et al., 2017). Without forests, precipitation may cease in a worst-case scenario, leaving millions of people without sufficient drinking water for extended periods. In São Paulo, the supply of fresh water had to be cut off for months (Le Monde diplomatique, 2015) because fewer clouds formed over the Amazon rainforest. It is expected that such water shortages due to deforestation will become more frequent, especially at the end of the already noticeably lengthened dry seasons (Lima et al. 2014; WWF, 2020c). In Southeast Asia, palm oil production and industry are causing the loss of extremely climate-relevant peat swamp forests (WWF, 2018a). In Africa, where palm oil plantations are also expanding, population pressures on forests are already especially intense, as some 90% of wood production there is used for firewood and charcoal. The overexploited forests can no longer recover and are disintegrating on a large scale (Allen & Barnes, 1985).

Forests are lost through practices that are legal or illegal in the countries in which they occur. Political actors are only rarely committed to the conservation of natural forests. Illegal forest use and destruction are often even legitimized retroactively through the granting of property titles—amounting to governmental promotion of forest loss. Examples are known not only from Brazil, but from many other countries, including the European region—and such action is typically linked to corruption. Illegal logging, the largest sector of global environmental crime, ranks third in organized crime worldwide after drug trafficking and product counterfeiting (Nellemann et al., 2018). Its rise is only possible through "looking the other way" or "tolerating" at government level, German ministries included (WWF, 2021b). Illegally felled timber and timber products from overexploitation reach us on a large scale via China, for example; even barbecue charcoal, which seems harmless, represents the destruction of nature and misdeclaration (Haag et al., 2020; WWF, 2020d). Analyses by the WWF have uncovered serious consumer deception. Barbecue charcoal suppliers have advertised with the claim "no tropical wood," but in fact exclusively used charcoal that could be traced back to tropical wood (WWF, 2018b).

Much criminal energy is devoted to extracting short-term benefits from forests. Some 15% to 30% of timber is harvested illegally (Nellemann & Nellemann, 2012). This can be curbed only through decisive action by state and society. Long-term protection of forests will be achieved only when we no longer treat environmental crime as a trivial offence and structure global supply chains in a thoroughly sustainable way. Appropriate action requires that we recognize such blind spots and take responsibility for forest loss and degradation. For this we need politicians, companies, and consumers who tackle forest protection and prioritize it on their long-term agendas.

Main Message *We know about forest loss and degradation in the tropics and the vital importance of forests—so why don't we stop the loss of forests? Our civilization's evolutionary blind spot, which makes us regard nature as inexhaustible, consistently leads to large and small myopic economic and utilization decisions that downgrade forests to mere resources. There is a danger that the importance of forest protection as a necessary basis of the economy will be recognized only once decisive tipping points will have been passed.*

Forests as Centers of Life

Forests are the most important creators and preservers of biological wealth. Around 80% of all species living on land depend on forests as a form of vegetation, even though many of these species do not spend their entire lives there. As natural forests shrink, so does global diversity. The WWF report "Below-the-canopy" (WWF, 2019) uses data from 1970 to 2010 to show that the distribution of wild vertebrates has shrunk by more than half (53%). On average, there are only 47 individuals left for every 100. Imagine a village whose population has shrunk from 100 to 47 in just 40 years. The monitoring of 455 populations of 268 forest vertebrate species carried out since 1970 shows that the loss has so far continued unabated. The decline of forest species is disproportionately steep compared to the overall dataset (including non-forest species) (WWF, 2016a). Data updated to 2016 already show losses of 68% for all vertebrate species (WWF, 2020a). At this rate, we must fear that the forest-dependent vertebrate population today is only 20–30% of what it was in 1970.

We should also recognize that this negative trend goes back to well before 1970. Whereas humans and their farm animals accounted for only a few percent of the biomass of land-based vertebrates a few centuries ago, this share has now grown to around 95% (Bar-On et al., 2018; Harari, 2015). This increase is as good an illustration as any of the transformation process: away from nature, into the new age of the Anthropocene. Less than 5% of the biomass of land-based vertebrates (including here the excessive wild stocks in our forests!) are now "living free" (Fig. 3). Our

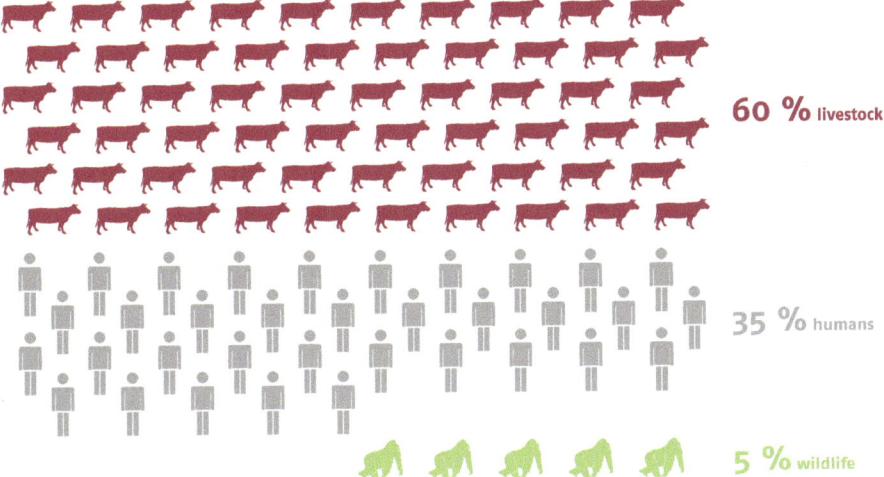

Fig. 3 Humans and their livestock now make up over 95% of the terrestrial vertebrate biomass. (Data from Bar-On et al., 2018)

entire biodiversity of eagles, antelopes, macaws, capercaillies, and monkeys to cheetahs, wildebeests, and giraffes to lions, orangutans, tigers, bison, and zebras hardly registers today in terms of biomass. Of all land-based vertebrates, 90% are born only to provide food for us humans.

The international environmental agreements adopted so far are far from sufficient to turn this around. What we need now is a transformation of our thinking, decision-making, conduct, and way of life. The UN Decade from 2021 to 2030 has therefore been proclaimed as the time of the necessary restoration of natural habitats. Only if we loosen the massive land-use pressure and give wild animals more space will they have a chance to survive the climate change that is already underway.

For "civilized" societies, forests are primarily an economic sector. Timber extraction is the dominant use, which in no way does justice to the importance of the forest for human life. Because the two great human tasks of protecting the climate and preserving biodiversity have so far barely been considered in macro- and micro-economic decisions concerning forests, current economic and forest-policy goals are exacerbating the "twin crises."

Only an understanding of forests and nature as a value system for health, nutrient circulation, climate, and biodiversity (Kadykalo et al., 2019) can lead to the urgently needed sound forest-policy and nature-conservation decisions and to measures that protect and restore natural forests.

Main Message *Is it really so bad that forests lose their biological wealth? Yes, because 80% of biodiversity depends on protecting the forests. Some two-thirds of wildlife populations have already disappeared.*

Biodiversity Protection: Not a Conduct-Relevant Concern

Countries with stagnating populations and without hunger theoretically offer the best prerequisites for enabling the preservation of biodiversity in an exemplary manner, especially on publicly owned land. Given this fact, how can it be that in the state of Brandenburg (Germany) today 581 animal species are on the brink of extinction and around 6,000 of 15,000 species are endangered (MOZ, 2021)? The main reasons cited for the threat to biodiversity are (1) intensive agriculture, (2) drying up of streams, lakes, and wetlands, and (3) urban sprawl and fragmentation of the countryside.

This shows that German politics (Brandenburg can serve as an example because the situation is similar in Germany's other extended states) is not achieving its tasks. The needed continuous safeguarding of necessary habitats for animal, plant, and fungal species is evidently not a sufficient concern for politicians (and hence also for us citizens).

Germany is representative for the other 195 countries of this world. Societies, though they know better, show little willingness seriously to pursue the preservation of biological diversity (Ding et al., 2017). Companies as well as business and political lobbyists succeed again and again at preventing such pursuit. Moreover, the laboriously negotiated agreements, directives, conventions, laws, and regulations for the protection of nature are not sufficiently implemented. Even a country like Germany, with wealth, a (social) market economy, and democracy, fails to protect biodiversity. Instead, we leave an additional immense footprint outside our national borders.

Large forest landscapes are transformed more (e.g., Russia and Canada) or less (many countries in the tropics) systematically. Important narrow corridors of life, such as those between the peat swamp forests of Sebangau and the "Heart of Borneo," the central forests of the Indonesian island, are not being deliberately preserved. However, continuous habitats are needed by the over 2 million known species as well as by the species not concretely known today, which may number over 10 or 20 million.

Continuity in nature is not a rigid concept. Even a dynamic river landscape, with its floods, deposits, accumulations, and overlays of sediment, continuously provides a great variety of habitats that emerge anew again and again. The decomposition of dead trees and the regrowth of young ones is part of one continuity, so long as both are connected on a small scale within the forest's life cycle and are not separated from each other on a large scale to the detriment of biodiversity. Even natural fires are part of this in boreal forests, whereas today's frequent and very extensive fires, well over 90% of which are caused by humans, prevent continuity.

The preservation of biodiversity absolutely requires spatial and temporal continuity and can be facilitated today only through reliable political action. Only effectively managed protected areas and forest management that is close to nature and worthy of the name (DNR-Forest management guideline, 2021) can reliably achieve biodiversity preservation. Yet the biodiversity-and-climate crisis is fueled anew each day by unsustainable decisions and financial flows. Protestors of all kinds are insulted, threatened, persecuted, or even murdered. With political reason and a well-functioning executive, this could be changed immediately and in an exemplary manner.

Main Message *We are setting the wrong priorities and are thus recklessly putting biodiversity at risk at all levels—as if we did not need it! Diversity requires space and restraint in our politics and daily living.*

Timber Use and Forest Loss—A Climate Problem

If the forests were a state, they would currently be the third largest carbon emitter after China and the United States of America. This is due to its man-made destruction, as forests would naturally be a carbon sink: around 50% of the carbon bound to land is bound in forests (WWF, 2020c). Overexploitation of forests, forestry that pays no attention to the carbon storage in forest landscapes, water drainage, slash-and-burn agriculture, out-of-control forest fires, conversion of forests to other uses (especially agricultural land), biologically degraded timber plantations, and the expansion of infrastructure and mining are continuously releasing sequestered carbon (WWF, 2015). A combination of local human activities and local to global political economic decisions are putting pressure on the forest system. This puts a massive strain on our climate. On average, 15% of annual global CO_2 emissions are caused by deforestation and forest degradation (WWF, 2016b; Smith et al., 2014). At 8%, tropical forest degradation accounts for more than half of these losses.

Not only forest loss through slash-and-burn, but also the use of wood itself poses a problem. Any use of wood biomass for energy and heat or for short-lived wood products leads to a further burden on the climate through carbon release. Swedish forestry and wood use has been calculated to release more CO_2 than all other sectors of Swedish society (Protect the Forest & Greenpeace Nordic, 2021).

The Forest Declaration signed at the Climate Change Conference in Glasgow (COP 26) will hardly stop the loss of forests by 2030. Instead, it may even fuel the conversion of carbon- and biodiversity-rich forests and their soils. German and EU policy must see the forest for what it is: one of the largest terrestrial CO_2 reservoirs and guarantor of diverse life on earth—and also, for some time already, a main source of emissions and therefore cause of immense climate harms. This must be considered for the immediate initiation of appropriate preservation measures.

Main Message *Considered as a state, forests are the third-largest emitter of carbon after China and the United States. Wood use and forest loss constitute immense climate problems!*

Forests as Carbon Reservoir—Urgently Needed

Forests store about 638 gigatonnes (Gt) of carbon, 283 Gt in living biomass above ground, 38 Gt in deadwood, and 317 Gt in the soil (top 30 cm; FAO, 2011), but with a decreasing trend due to global forest losses. Due to the growth of vegetation, the primary forests of the boreal and temperate zones alone can store around one Gt of carbon per year on average (1.3 ± 0.5 Gt per year). Primary forests that are over 200 years old store an additional 2.4 metric tons of carbon per hectare per year on average (Luyssaert et al., 2008), reason enough to preserve them at all costs. Tropical forests store around 200 to 300 Gt of carbon. About 60% of photosynthesis world-wide takes place in tropical forests—they absorb about 72 Gt of carbon each year and release normally a little less again through respiration (Pan et al., 2011). The world's tropical forests have already been weak net emitters of carbon between 2000 and 2014. In the dry El Niño year of 2015, they were calculated to have released up to 2.7 Gt (Mitchard, 2018).

Natural forest loss, degradation, and fires result in forest and savannah land-scapes being net carbon emitters, despite their just outlined sequestration capacities. Savannah and forest fires annually release 1.7 to 4.1 Gt of carbon into the atmosphere worldwide—plus an estimated 39 million metric tons of methane (with a much higher global warming potential than CO_2), 20.7 million metric tons of nitrous oxides (NO_x), and 3.5 million metric tons of sulphur dioxide (SO_2). The difference is over 3 billion metric tons of carbon equivalents, to the detriment of the—ever decreasing!—storage capacity (WWF, 2016b).

Ideally, this flow should be reversed. Since many commercial forests are now very young, their stored carbon stocks can be increased very quickly by a longer rotation period, that is, a higher tree age at felling. Carbon sequestered in the forest does not harm the climate. This insight could drive action, if it were not for the forestry and timber industry's pronounced will to exploit the forest: "Only people outside the forestry world believe that the forest is about nature. It is about growth, felling and stock" (Fokken, 2021). Forestry worldwide primarily aims at felling and then using wood to benefit financially from forests and wood plantations. The narrative of carbon storage is summarily rewritten: forest storage becomes wood storage. It is claimed that what matters is not the carbon stored in the forest, but primarily the carbon stored in the harvested and processed (wood) products. Thinking in this way, one can harvest at will, because intensive use produces a large stock of wood and thus constitutes practical climate protection—but without attending equally to the degradation of carbon storage in live forests.

However, such carbon storage in wood works only if this wood is preserved. Yet, presently less than 20% of the wood processed in Germany ends up in long-lasting wood products such as furniture or parts of houses (based on Mantau, 2012). And even this figure is only of limited significance, because many processes for manu-facturing wood products (pressboard, MDH panels, etc.) require much energy, whose generation is associated with carbon dioxide emissions. More than three quarters of the 135 million cubic meters of raw wood used thus lead to the release of over 10 million metric tons of carbon dioxide per year. In China, the wood prod-ucts industry has already become a net contributor of carbon (Ji et al., 2013).

Main Message *Forests are urgently needed as a carbon sink. The one-sided narra-tive of carbon storage in wood products and of the uses of wood as a substitute material thus detracts from sound and necessary action for climate and biodiversity protection.*

Using Wood for Energy—Globally Not Climate Neutral

The strategists of the forestry and timber industry have meanwhile recognized the problem and developed another narrative. According to them, short-term wood products (such as paper towels and cardboard boxes) and the energetic use of wood (burning for energy and heat) are also a gain for the climate: burning wood is climate-neutral and moreover can replace more energy-intensive fossil fuels. But both claims are wrong. The use of wood for energy is not climate-neutral just because new trees can grow in the forest. The forest's carbon cycle has been mas-sively altered and disturbed by humans; due to loss and degradation, the global forest area is in decline in extent as well as in ecological quality. So far, 2 billion hectares (about one-third) of the global forest area and most of its carbon storage capacity have been lost. Any additional energetic use of wood thus places an addi-tional burden on the climate through the release of carbon. Forests cannot compen-sate for their own emissions because they are losing ground around the world and are becoming patchier and younger. Just between 2005 and 2017, the EU's interna-tional trade was responsible for 3.5 million hectares of forest destruction in the tropics (WWF, 2021a).

The substitution effect of wood used for energy is not present, as the efficiency of using wood for energy is low. Compared to the combustion of coal or gas, wood combustion even releases more carbon (Fig. 4). Moreover, in view of the agreed

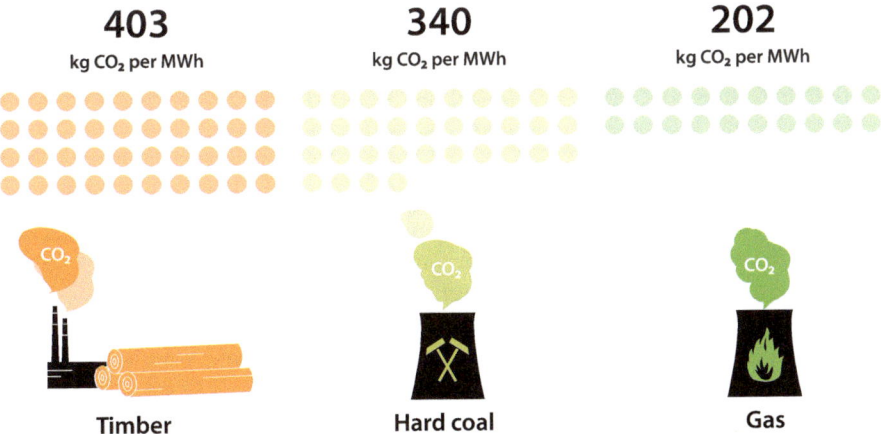

Fig. 4 Carbon release during the combustion of wood, hard coal, and gas (Mwh: megawatt hour; based on IPCC, 2006 in Kuhlmann & Gerhardt, 2017)

decarbonization of the economy by 2050 at the latest, any substitution effects existing today (dpa, 2021) are rapidly declining; with the transformation of the economy and the achievement of climate neutrality in the land use sector, they may disappear completely.

Main Message *The "regenerative" sustainable energy source forest is too small. The current use intensity is aggravating our acute climate crisis. The use of wood for energy is no more climate neutral than a forest fire.*

Forestry—Not Climate Neutral

The most commonly used harvesting and silviculture methods worldwide are clear-cutting and clear-cut-like logging. Clear-cutting is economically most lucrative in the short term, as the wood biomass can be used in only one operation. This creates an area of open space. Clear-cut areas become significantly warmer than the original forest area due to unhindered solar radiation and have no forest interior climate. The higher temperatures cause soil carbon to be released through increased activity of soil organisms and accelerated chemical processes (Wieting & Leversee, 2019). The faster turnover of substances reduces the humus content and soil fertility. Clear-cut areas are destroyed forest ecosystems whose material cycle can only be closed again very slowly. Increased carbon release can often be detected in such areas even more than 10 years after clear-cutting (Wieting & Leversee, 2019). Clearcutting is not similar to a natural disturbance, neither windthrow nor fire, as biomass is removed to a much higher degree. The frequency, spatial structure and extent of clear-cutting are also not remotely comparable to natural disturbance dynamics.

To make forestry more climate-friendly, management must be adapted accordingly. Soil-conserving management methods that preserve the forest climate are just as imperative as adapted game densities. A move away from clear-cutting towards forest use that maintains the cooling effect of forests, the improved water balance (groundwater supply; Flade & Winter, 2021), and the carbon storage of vegetation and soil is most likely to compensate for the biomass losses caused by wood harvesting and can be described as climate- and biodiversity-friendly. The realization that—due to the necessary forest paths and aisles, the removal of wood, and the reduction of thermoregulation—wood harvesting always leads to an impairment of forest ecosystems should motivate politics to minimize these impairments. Plans—concerning the Central European beech forests for example—exist (Winter et al., 2015) and should serve as blueprints for forest management here.

Forest growing stock is the key to carbon storage, and this can sustainably take place only in a near-natural forest ecosystem. The circulation of carbon within the forest, that is, the decomposition of dead trees and the renewed carbon fixation in young trees, can thus represent an immanent balanced part of the storage medium and the growing storage capital. In a beech forest—the natural vegetation in over two thirds of Germany's forest area—dead older trees need up to 30 years to become humus. Even dead trees thus bind carbon within the forest ecosystem over a longer term than the period straightway necessary to implement climate protection measures. Even in spruce forests that have died due to drought and bark beetle infestation, the carbon remains sequestered on the surface longer than when the damaged wood is used for energy. The clear-cutting-like clearing and soil cultivation of planted areas again releases (soil) carbon for years and in large quantities (for example Nõgu, 2014; Wieting & Leversee, 2019). Young forests, including young commercial forests, have an immense potential within the next decades to absorb carbon into the living tree population, which could remain stored for many decades—so long as people do not overuse, clear, burn, or degrade the forest.

Main Message *While local perspectives can be misleading, from a global perspective, clear-cutting and clear-cutting-like interventions are endangering the climate and biodiversity. Allowing trees to age, preserving deadwood, and allowing natural regeneration with adapted game populations would have a positive effect on the climate and help mitigate forest damage. This would be active climate and biodiversity protection and should be promoted politically.*

What If We Reached a Global Increase by 3 Degrees?

For about two decades, the temperature increase in the Northern hemisphere has been exceeding that in the Southern. With an average global temperature increase of three degrees, the rise in the Northern latitudes would most likely be significantly larger (Feulner et al., 2013). The forest ecosystems of Europe's boreal, cool temperate, and Mediterranean climate zones can neither adapt to this development nor migrate with it (Garamvölgyi & Hufnagel, 2013; Hufnagel & Garamvölgyi, 2014; on forest fragmentation: Bacles & Jump, 2011). Water shortages and droughts are already putting pressure on forests, and it can be assumed that this will intensify as temperatures rise. Extreme weather, such as storms, heavy rainfall and extreme temperatures, will increasingly destabilize forests. The boreal forest belt will shift Northwards, especially in Russia.

Tropical forests will suffer greatly due to increasing use intensity and higher temperatures. Their self-sustaining nutrient and water cycles will no longer be maintainable if further damage occurs. In addition to fires, the tropical forests are increasingly becoming a source of carbon due to desiccation processes. This has been proven for the first time in 2020 for parts of the Brazilian Amazon region (Houghton et al., 2000). These emissions further increase the approximately 20% share of anthropogenic carbon emissions from land use in the tropics (Le Quéré et al., 2016).

Dry years are already leading to increased carbon releases from the tropics (Liu et al., 2017). Further loss of forest area will accelerate the drying process (Baccini et al., 2017). A so far only slightly higher global mean temperature (Fig. 5) has already led to the dry season there lasting up to 6 weeks longer in some years and to

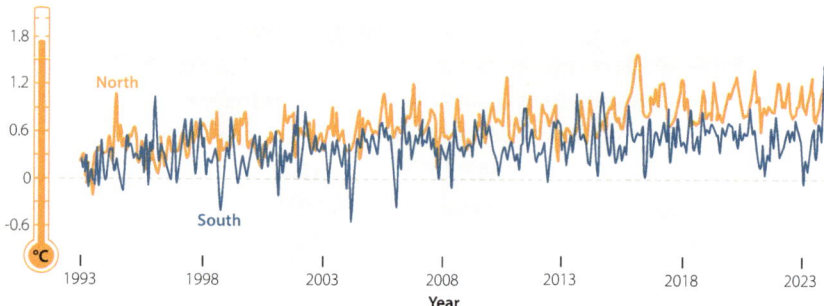

Fig. 5 Monthly values of the global temperature deviation on the land surface over the last 30 years (NCDC, 2022). Temperatures in the Northern hemisphere (orange) are rising more strongly than those in the Southern hemisphere (blue)

a further significant increase in the risk of forest fires, especially at the end of the dry season (WWF, 2020c). It can be assumed that, with a 3-degree higher annual mean temperature, the tropical forests with their extensive services described above will become largely unavailable to us humans. Fifteen different climate models based on the two emission scenarios A2 and B1 (with, respectively, 550 and 860 ppm CO_2 in the atmosphere in 2100) show that the tropical forests of South America could very quickly turn into savannas by the end of the century. The 20% loss of forest area in the Amazon basin by 2020 could increase by at least another 30% due to climate change. Here, we are not looking at the continued destruction by humans, but at the disintegration of tropical forests due to periods of drought, which cause the conversion of tropical forests into savannah (Salazar et al., 2007). The biological diversity of the South American tropics will accordingly also be lost (Senior et al., 2019). A study for Central America calculates in a scenario for the tropical forests there with up to 3.5 degrees more (RCP4.5 scenario) an average decrease in precipitation of up to 4 millimeters per day and a dry season that is around 40 days longer. As for South America, the scenario for the end of the twenty-first century shows a large-scale desertification of the tropical forests there (Lyra et al., 2017).

A scientifically valid prediction or detailed description of a future climate, however, is hardly possible. Since the term climate always refers to a time span of 30 years, this term can in any case not be reliably used in the context of rapidly changing weather patterns. Forests have never been static and will continue to change in the future. However, since the climate and the associated weather patterns change faster than forests can react, they will be able to fulfill their functions only partially or not at all. Single short-term extreme weather events may be decisive for whether a forest is preserved or lost.

It is important to know that it is not possible to adapt a forest through silviculture to a predicted, later climate because (1) we do not know this climate, because (2) this climate will probably not achieve comparatively stable growth conditions for centuries due to the rapidly changing weather patterns triggered by humans, and because (3) active adaptation by business-oriented humans always takes account of only very small parts of ecosystem functionality. In view of the increasing change-ability of weather patterns and the complex interactions in forest ecosystems, the probability of drawing erroneous conclusions is high. We can only try to strengthen the commercial forest for as long as possible by means of forest ecosystems that are as natural as possible with tree species native to the site and forest management that is close to nature.

By 2023, Germany has suffered extensive damage to (about 600,000 hectares of) its forest area for the first time—and this after merely 3 years of drought with a mean temperature increase of merely 1 °C. So far, this damage has mainly affected spruce (*Picea abies*), which is widely planted but not suited to locations below 600 meters above sea level. The profit-oriented spruce industry has deliberately accepted the associated greater economic risk under "normal" weather conditions. Since the loss incurred was and is largely compensated politically through subsidies, the risk is not borne by the forest owners, but by the taxpayer (and of course by the climate). This large-scale silvicultural mistake has been pointed out by nature conservation associations for decades but was not taken seriously by politicians and forest owners. The damage to native tree species is still minor, especially in near-natural forest management, but it is already noticeable there as well.

It is foreseeable that the climate factor, coming on top of the already existing stress factors (excessive nitrogen and pesticide inputs, excessive exploitation and thinning, excessive game or livestock densities, damage to soils, and impaired water management) will greatly harm forests. The heat threshold for plant cells lies between 35 and 46 °C. And every 10 degrees, a tree's respiration and water loss doubles (Coder, 2011). In our temperate latitudes, conifers already reach their heat limit between 35 and 42 degrees, while for deciduous trees it only occurs between 40 and 45 °C (Profft, 2005). It should be borne in mind that the air's increasing carbon dioxide content can be more problematic for some tree species than for others; for example, it has been shown that birch trees cope much worse with high CO_2 concentrations than poplars (Darbah et al., 2010).

Various adaptation strategies can help during prolonged heat, but only temporarily. When heat is coupled with drought, the cooling possibilities through transpiration are lost (Coder, 2011). More water evaporates through the leaves than can be replaced—the transpiration flow breaks off and the cooling stops. The tree dies a quick death from combined heat and drought, something that has been observed in tropical forests as well (Allen et al., 2010). Storms can also harm forests, especially when preceded by heavy rainfall. Both combinations of factors would become more frequent in a 3-degree warmer world, so that forests will continually struggle for survival in many locations. Since we need forests for climate protection, we must not allow this and must act.

Main Message *With a warming by 3 degrees, many forests are ecologically finished. Forests will increasingly disintegrate. Even where they survive, they will no longer be able to provide their previous ecosystem services.*

The Immediate Program for Forests

Climate and biodiversity protection must become central political goals and instruments—cross-sectoral and authoritative at all political levels. There must be no climate protection at the expense of biodiversity protection.

An immediate program to protect our forests should include or consider the following:

- Deforestation and forest degradation has been stopped by law at the EU level since 2023 (EUDR, 2023). Raw materials and downstream products that are associated with forest destruction are no longer allowed to be imported into the EU, exported from the EU, or traded within the EU. This legislation needs strong implementation with high numbers of controls and deterrent penalties. To prevent leakage effects and the destruction of other nature worth protecting, such as grasslands, savannahs and other wooded lands, to complete the raw material list and to stop deforestation resulting from the finance sector all stakeholders, especially politicians, should support inclusion of other wooded land, maize and the finance sector into the EUDR by 2025 that the regulation can become effective without loopholes and trade-offs.
- To enable deforestation-free sustainable supply chains for agricultural and wood raw materials and their products long-term, support programs are needed for the development of a sustainable agriculture that, in particular, would improve soil fertility and prevent further deforestation.
- Reductions are required in climate-damaging emissions from agriculture and in the excessive use of pesticides and nitrogen fertilizers.
- There must be peatland restoration and protection of peatland soils to achieve climate neutrality at the landscape level and to protect forest landscapes from drying out (see also chapter "Peatland Must Be Wet").
- The burning of wood and the production of short-lived wood products such as paper towels, cardboard boxes and disposable pallets must be drastically reduced in favor of reusable items and long-lasting wood products.
- Protection of forest areas must be increased worldwide, and the management of protected areas (e.g., Natura2000 in the EU) must be significantly improved.
- Rapid implementation of the EU Biodiversity Strategy will create large areas within the EU where biodiversity can be preserved through wilderness development or extensive use. The rapid and undiluted full implementation of the strategy should serve as a positive example that the needed ecological stabilization is finally given its due importance.
- The global forest area should be expanded by up to one billion hectares, among other things because of its importance for thermoregulation and water storage as well as for soil fertility and averting soil erosion. In this effort, land rights for indigenous and local communities and benefit-sharing should play central roles.
- We need extensive forest management that favors native tree species and eschews clear-cutting and pesticides.

- Financial resources for the conservation of primary forests and the creation of tree-rich forest landscapes must be made available in large and sufficient amounts.
- Estimates suggest that more than US$300 billion per year are needed for several years to restore forest landscapes (Ding et al., 2017). With somewhat lower investments of around US$200 billion per year, an additional annual storage of 9 Gt of CO_2 could be facilitated (Felbermeier et al., 2016). This requires about US$150 billion per year for tree plantations outside the forest, $22 billion for the conservation of forest land, and another $20 billion for improved forest management (forest restoration). So, the cost of conserving still existing tropical forests sums up to around $40 to $50 billion per year.
- In line with its responsibility for emissions and deforestation, Germany would have to contribute at least US$8 billion annually. The governments of countries with tropical forests could thus use large parts of the already deforested areas sustainably without converting further natural forests (including indigenous forests) into agricultural land by slash-and-burn. The money would have to compensate for the loss of income, since, on the one hand, fewer wood and agricultural raw materials can be sold and, on the other hand, agricultural land that has already been deforested is after a few years no longer as productive as newly created land that has been well supplied with nutrients through slash-and-burn. What is needed, therefore, is a gentle humus-building method of cultivation that produces somewhat higher costs but does not jeopardize the world's food supply in the coming decades (Erb et al., 2016).
- All legislation and binding guidelines, including those for the financial markets, must respect the purpose of forest conservation, that is, exclude further forest conversion. Exceptions must be reduced to a minimum or should be considered only if the overall quality of the forest improves and the forest carbon stock increases significantly to around 700 Gt.
- Experiences from the efforts to implement REDD+, the concept for reducing emissions from forest loss and degradation (WWF, 2011), and the approach "Restoring Forest Landscapes" (NYDF, 2014) show that the reduction of the deforestation rate quickly ceases when the required funds are not made available. The United Nations are trying to provide the pathways and the necessary funding but needs significantly higher contributions from member states to achieve this goal.

Beyond direct forest protection, the following should be noted:

- Politicians have a duty to achieve the climate targets and to impose clear obligations on enterprises.
- The preservation of an approximately stable climate is non-negotiable. The scientific findings are available, which means that politicians, governments, and entrepreneurs have a duty to act.
- Businesses, including the forestry and timber industry, are well-informed about the climate and biodiversity crises and must be held accountable for faulty conduct.

- The containment of environmental crime immediately needs clear laws with effective enforcement. Short-sighted economic activities at the expense of future generations and beyond planetary capacities should be made illegal and punished.
- The Paris Climate Agreement must be fulfilled in full. The nationally determined contributions (NDCs) must be effectively strengthened, as their formulation and implementation have thus far failed to achieve the limitation of warming to a maximum of 1.5 degrees.
- Rich countries like Germany must become role models in climate and biodiversity protection lest we irresponsibly allow the damage we have already done to continue growing.

To close our evolutionary blind spot, we need irrevocable agreements that preserve the Earth with its habitats and diversity. The Glasgow Declaration on Forests, signed by leaders from 140 countries, should be followed up with further agreements. It should be considered whether the preservation of ecosystems requires that some natural areas be placed outside nation states and their decision-making powers to safeguard their continued flourishing and intergenerational justice against the vagaries of national election cycles. The common forest and climate policies of the community of states should make it possible for forests to be no longer the third largest carbon emitter, but part of the solution that would reduce our current emissions by around 15%.

Main Message *No more delaying, no more wait-and see, no more blame-shifting, no more beating around the bush—we need ambitious action in the right direction. Compliance with the Paris Climate Agreement is the narrow pathway into a future that might resemble the present. Will and efforts to preserve the climate and biodiversity roughly as we know them must be significantly strengthened. Reliable goals and ways of achieving climate neutrality as quickly as possible are the political order of the day. Forests should be no longer the third largest carbon emitter, but part of the solution that would reduce our current emission by around 15%.*

References

Adetola, O. O., & Adebisi, M. A. (2019). Impacts of deforestation on the spread of Mastomys natalensis in Nigeria. *World Scientific News, 130*, 286–296.

Allen, J. C., & Barnes, D. F. (1985). The causes of deforestation in developing countries. *Annals of the Association of American Geographers, 75*(2), 163–184.

Allen, C. D., Macalady, A. K., Chenchouni, H., Bachelet, D., Mcdowell, N., Cobb, N., et al. (2010). A global overview of drought and heat-induced tree mortality reveals emerging climate change risks for forests. *Forest Ecology and Management, 259*(4), 660–684. https://doi.org/10.1016/j.foreco.2009.09.001

Baccini, A., Walker, W., Carvalho, L., Farina, M., Sulla-Menashe, D., & Houghton, R. A. (2017). Tropical forests are a net carbon source based on aboveground measurements of gain and loss. *Science, 358*(6360), 230–234.

Bacles, C. F., & Jump, A. S. (2011). Taking a tree's perspective on forest fragmentation genetics. *Trends in Plant Science, 16*(1), 13–18.

Barley, J., Lennox, G. D., Ferreira, J., Berenguer, E., Lees, A. C., Mac Nally, R., Thomson, J. R., de Barros, F., Ferraz, S., Louzada, J., Fonseca Oliveira, V. H., Parry, L., de Castro, R., Solar, R., Vieira, I. C. G., Aragão, L. E. O. C., Begotti, R. A., Braga, R. F., Cardoso, T. M., Cosme de Oliveira, R., Jr., Souza, C. M., Jr., Moura, N. G., Nunes, S. S., Siqueira, J. V., Pardini, R., Silveira, J. M., Vaz-de-Mello, F. Z., Stulpen Veiga, R. C., Venturieri, A., & Gardner, T. A. (2016). Anthropogenic disturbance in tropical forests can double biodiversity loss from deforestation. *Nature, 535,* 144–147.

Bar-On, Y. M., Phillips, R., & Milo, R. (2018). The biomass distribution on Earth. *PNAS, 115*(25), 6506–6511. https://doi.org/10.1073/pnas.1711842115

Brock, P. M., Fornace, K. M., Grigg, M. J., Anstey, N. M., William, T., Cox, J., Drakeley, C. J., Ferguson, H. M., & Kao, R. R. (2019). Predictive analysis across spatial scales links zoonotic malaria to deforestation. *Proceedings of the Royal Society B, 286,* 20182351. https://doi.org/10.1098/rspb.2018.2351

Cardinale, B. J., Duffy, E., Gonzalez, A., Hooper, D. U., Perrings, C., Venail, P., Narwani, A., Mace, G. M., Tilman, D., Wardle, D. A., Kinzig, A. P., Daily, G. C., Loreau, M., Grace, J. B., Larigauderie, A., Srivastava, D., & Naeem, S. (2012). Biodiversity loss and its impact on humanity. *Nature, 486*(7401), 59–67. https://doi.org/10.1038/nature11148

Coder, K. D. (2011). *Drought, heat stress & trees.* Warnell School of Forestry & Natural Resources. University of Georgia. https://bugwoodcloud.org/resource/files/15109.pdf

Darbah, J. N., Sharkey, T. D., Calfapietra, C., & Karnosky, D. F. (2010). Differential response of aspen and birch trees to heat stress under elevated carbon dioxide. *Environmental Pollution, 158*(4), 1008–1014.

de Vries, W., Erisman, J. W., Spranger, T., Stevens, C. J., & van den Berg, L. (2011). Nitrogen as a threat to European terrestrial biodiversity. *The European Nitrogen Assessment: Sources, Effects and Policy Perspectives*, 436–494.

Díaz, S., Fargione, J., Chapin, F. S., III, & Tilman, D. (2006). Biodiversity loss threatens human well-being. *PLoS Biology, 4*(8), e277. https://doi.org/10.1371/journal.pbio.0040277

Ding, H., Altamirano, J. C., Anchondo, A., Faruqi, S., Verdone, M., Wu, A., Ortega, A. A., Zamora Cristales, R., Chazdon, R., & Vergara, W. (2017). *Roots of prosperity: The economics and finance of restoring land.* World Resources Institute. Summary: https://files.wri.org/d8/s3fs-public/roots-of-prosperity-exec-summary-english.pdf

DNR-Forest management guideline. (2021). Von der Waldkrise zur nachhaltig ökologischen und generationengerechten Waldwende. Forderungen von Natur- und Umweltschutzorganisationen im DNR zur Waldpolitik. Deutscher Naturschutzring. https://backend.dnr.de/sites/default/files/2021-12/20211213_DNR-Waldposition.pdf

dpa. (2021) 21.12.2021: dpa-infocom, dpa:211221-99-460587/2 https://www.zeit.de/news/2021-12/21/klimaneutralitaet-stahlbranche-pocht-auf-paradigmenwechsel

Ellwanger, J. H., Kulmann-Leal, B., Kaminski, V. L., Valverde-Villegas, J., Veiga, A. B. G., Spilki, F. R., et al. (2020). Beyond diversity loss and climate change: Impacts of Amazon deforestation on infectious diseases and public health. *Anais da Academia Brasileira de Ciências, 92.* https://doi.org/10.1590/0001-3765202020191375

Erb, K.-H., Lauk, C., Kastner, T., Mayer, A., Theuri, M. C., & Haberl, H. (2016). Exploring the biophysical option space for feeding the world without deforestation. *Nature Communications.* https://doi.org/10.1038/ncomms11382. https://www.nature.com/articles/ncomms11382

EUDR. (2023). Regulation (EU) 2023/1115 of the European Parliament and of the Council of 31 May 2023 on the making available on the Union market and the export from the Union of certain commodities and products associated with deforestation and forest degradation and repealing Regulation (EU) No 995/2010. https://eur-lex.europa.eu/legal-content/EN/TXT/PDF/?uri=CELEX:32023R1115

FAO. (2011). *State of the world'' forests.* http://www.fao.org/3/i2000e/i2000e.pdf

FAO. (2020). *Global Forest Resources Assessment 2020* (Main Report). Food and Agriculture Organization in the United Nations, 164 pp.

Felbermeier, B., Weber, M., & Mosandl, R. (2016). *Zur Machbarkeit eines weltweiten Aufforstungsprogramms—eine Kurzstudie.* Technische Universität München. 25 p. https://www.forum-fuer-verantwortung.de/wp-content/uploads/2016/06/akt_mzn_waldoptionen-kurzstudie.pdf

Feulner, G., Rahmstorf, S., Levermann, A., & Volkwardt, S. (2013). On the origin of the surface air temperature difference between the hemispheres in earth's present-day climate. *Journal of Climate*, 7136–7150. https://doi.org/10.1175/JCLI-D-12-00636.1

Flade, M., & Winter, S. (2021). Wirkungen von Baumartenwahl und Bestockungstyp auf den Landschaftswasserhaushalt. In H. D. Knapp, S. Klaus, & L. Fähser (Eds.), *Der Holzweg – Wald im Widerstreit der Interessen* (pp. 235–242). Oekom.

Fokken, U. (2021). Tote Fichten—Der Harz bildet die Avantgarde des ökologischen Zusammenbruchs in Zeiten der Klimawandels. *Die Tageszeitung, 4.*

Garamvölgyi, Á., & Hufnagel, L. (2013). Impacts of climate change on vegetation distribution no. 1 climate change induced vegetation shifts in the palearctic region. *Applied Ecology and Environmental Research, 11*(1), 79–122.

Giam, X. (2017). Global biodiversity loss from tropical deforestation. *Proceedings of the National Academy of Sciences.* https://doi.org/10.1073/pnas.1706264114

Haag, V., Zemk, V. Z., Lewandrowski, T., Zahnen, J., Hirschberger, P., Bick, U., & Koch, G. (2020). The European charcoal trade. *IAWA International Association of Wood Anatonists.* https://doi.org/10.1163/22941932-bja10017

Harari, Y. N. (2015). *Eine kurze Geschichte der Menschheit.*

Harvey, D. J., Montgomery, L. N., Harvey, H., Hall, F., Gange, A. C., & Watling, D. (2020). Psychological benefits of a biodiversity-focused outdoor learning program for primary school children. *Journal of Environmental Psychology, 67,* 101381.

Hofmann, F., Schlechtriemen, U., Kruse-Plaß, M., & Wosniok, W. (2019). *Biomonitoring der Pestizid-Belastung der Luft mittels Luftgüte-Rindenmonitoring und Multi-Analytik auf >500 Wirkstoffe inklusive Glyphosat 2014–2018.* https://www.enkeltauglich.bio/wp-content/uploads/2019/02/Bericht-H18-Rinde-20190210-1518-1.pdf

Houghton, R. A., Skole, D. L., Nobre, C. A., Hackler, J. L., Lawrence, K. T., & Chomentowski, W. H. (2000). Annual fluxes of carbon from deforestation and regrowth in the Brazilian Amazon. *Nature, 403,* 301–304.

Hufnagel, L., & Garamvölgyi, Á. (2014). Impacts of climate change on vegetation distribution No. 2-climate change induced vegetation shifts in the new world. *Applied Ecology and Environmental Research, 12*(2), 355–422.

IPBES Intergovernmental Platform on Biodiversity and Ecosystem Services. (2019). *Summary for policymakers of the global assessment report on biodiversity and ecosystem services of the Intergovernmental Science-Policy Platform on Biodiversity and Ecosystem Services.* IPBES Secretariat.

IPCC – Intergovernmental Panel on Climate Change. (2006). IPCC Guidelines for National Greenhouse Gas Inventories. Volume 2 Energy. https://www.ipcc-nggip.iges.or.jp/public/2006gl/vol2.html

IPCC – Intergovernmental Panel on Climate Change. (2021). Sixth assessment report. *The Physical Science Basis.* https://www.ipcc.ch/report/ar6/wg1/#outreach

Ji, C., Yang, H., Nie, Y., & Hong, Y. (2013). Carbon sequestration and carbon flow in harvested wood products for China. *International Forestry Review, 15*(2), 160–168. https://doi.org/10.1505/146554813806948530

Jorwal, P., Bharadwaj, S., & Jorwal, P. (2020). One health approach and COVID-19: A perspective. *Journal of Family Medicine and Primary Care, 9*(12), 5888–5891. https://doi.org/10.4103/jfmpc.jfmpc_1058_20

Kadykalo, A. N., López-Rodriguez, M. D., Ainscough, J., Droste, N., Ryu, H., Ávila-Flores, G., Le Clech, S., Munoz, M. C., Nilsson, L., Rana, S., Sarkar, P., Sevenecke, K. J., & Harmáčková,

Z. V. (2019). Disentangling 'ecosystem services' and 'nature's contributions to people. *Ecosystems and People, 15*(1), 269–287. https://doi.org/10.1080/26395916.2019.1669713

Kilpatrick, A. M., & Randolph, S. E. (2012). Drivers, dynamics, and control of emerging vector-borne zoonotic diseases. *The Lancet, 380*, 1946–1955.

Kissinger, G., Herold, M., & De Sy, V. (2012). *Drivers of deforestation and forest degradation: A synthesis report for REDD+ policymakers.* https://www.forestcarbonpartnership.org/sites/fcp/files/DriversOfDeforestation.pdf_N_S.pdf

Kuhlmann, W., & Gerhardt, P. (2017). Kahlschlag für das Klima? Warum das Verbrennen von Holz in Kraftwerken kein Beitrag zum Klimaschutz ist. https://plattform-wald-klima.de/wp-content/uploads/2019/08/Misguided_strategy_burning_wood.pdf

Le Monde diplomatique. (2015). 09.04.2015. https://monde-diplomatique.de/artikel/!200147

Le Quéré, C., Andrew, R. M., Canadell, J. G., Zaehle, S., et al. (2016). Global carbon budget 2016. *Earth System Science Data, 8*(2), 605–649. https://doi.org/10.5194/essd-8-605-2016

Leite-Filho, A. T., Soares-Filho, B. S., Davis, J. L., Abrahão, G. M., & Börner, J. (2021). Deforestation reduces rainfall and agricultural revenues in the Brazilian Amazo. *Nature Communications, 12*, 2591.

Lima, L. S., Coe, M. T., Soares-Filho, B. S., Cuadra, S. V., Dias, L. C. P., Costa, M. H., Lima, L. S., & Rodrigues, H. O. (2014). Feedbacks between deforestation, climate, and hydrology in the Southwestern Amazon: Implications for the provision of ecosystem services. *Landscape Ecology, 29*, 261–274.

Liu, J., Bowman, K. W., Schimel, D. S., Parazoo, N. C., Jiang, Z., Lee, M., Bloom, A. A., Wunch, D., Frankenberg, C., Sun, Y., O'Dell, C. W., Gurney, K. R., Menemenlis, D., Gierach, M., Crisp, D., & Eldering, A. (2017). Contrasting carbon cycle responses of the tropical continents to the 2015–2016 El Niño. *Science, 358*(6360), eaam5690. https://doi.org/10.1126/science.aam5690

Lovell, R., Wheeler, B. W., Higgins, S. L., Irvine, K. N., & Depledge, M. H. (2014). A systematic review of the health and well-being benefits of biodiverse environments. *Journal of Toxicology and Environmental Health – Part B: Critical Reviews, 17*, 1–20. https://doi.org/10.1080/10937404.2013.856361

Luyssaert, S., Schulze, E. D., Börner, A., Knohl, A., Hessenmöller, D., Law, B. E., Ciais, P., & Grace, J. (2008). Old-growth forests as global carbon sinks. *Nature, 455*(7210), 213–215. https://doi.org/10.1038/nature07276

Lyra, A., Imbach, P., Rodriguez, D., Chou, S. C., Georgiou, S., & Garofolo, L. (2017). Projections of climate change impacts on Central America tropical rainforest. *Climatic Change, 141*, 93–105. https://doi.org/10.1007/s10584-016-1790-2

Mantau, U. (2012). *Holzrohstoffbilanz Deutschland, Entwicklungen und Szenarien des Holzaufkommens und der Holzverwendung 1987 bis 2015.* Thünen-Institut.

Mazor, T., Doropoulos, C., Schwarzmueller, F., Gladish, D. W., Kumaran, N., Merkel, K., Di Marco, M., & Gagic, V. (2018). Global mismatch of policy and research on drivers of biodiversity loss. *Nature Ecology & Evolution, 2*, 1071–1074.

Mitchard, E. T. A. (2018). The tropical forest carbon cycle and climate change. *Nature, 559*, 527–534.

MOZ. (2021). 581 Tierarten in Brandenburg vom Aussterben bedroht. Märkische Oderzeitung, 31. Mai 2021.

NCDC. (2022). *National Climatic Data Center der NOAA.* https://meteo.plus/klima-global.php. Accessed 11/02/2023.

Nellemann, C., & Nellemann, C. I. (2012). *Green carbon, black trade: Illegal logging, tax fraud and laundering in the world's tropical forests.*

Nellemann, C., Henriksen, R., Pravettoni, R., Stewart, D., Kotsovou, M., Schlingemann, M. A. J., Shaw, M., & Reitano, T. (Eds.). (2018). *World atlas of illicit flows. A RHIPTO-INTERPOL-GI Assessment.* RHIPTO -Norwegian Center for Global Analyses, INTERPOL and the Global Initiative Against Transnational Organized Crime. www.rhipto.or. www.interpol.int ISBN:978-82-690434-2-6

Nõgu, L. (2014). *The effects of site preparation on carbon fluxes at two clear-cuts in southern Sweden*. Master degree thesis in Physical Geography and Ecosystem Analysis. Department of Physical Geography and Ecosystem Science Lund University. https://lup.lub.lu.se/luur/downlo ad?func=downloadFile&recordOId=4467390&fileOId=4467400

NYDF. (2014). *New York Declaration on Forests*, 2014. https://forestdeclaration.org/images/ uploads/resource/20210628_NYDF_2.0_slide_deck_v0_.7_-_public_website_version_ final_.pdf

Olivero, J., Fa, J. E., Real, R., Márquez, A. L., Farfán, M. A., Vargas, J. M., Gaveau, D., Salim, M. A., Park, D., Suter, J., King, S., Leendertz, S. A., Sheil, D., & Nasi, R. (2017). Recent loss of closed forests is associated with Ebola virus disease outbreaks. *Scientific Reports, 7*, 14291.

Pan, Y., Birdsey, R. A., Fang, J., Houghton, R., Kauppi, P. E., Kurz, W. A., Phillips, O. L., Shvidenko, A., Lewis, S. L., et al. (2011). A large and persistent carbon sink in the world's forests. *Science, 333*, 988–993.

Profft, I. (2005). *Klimawandel und dessen Folgen für den Wald—eine aktuelle Literaturstudie*. http://www.waldundklima.de/klima/klima_docs/wuk_klima_wald_iprofft_01.pdf

Protect the Forest & Greenpeace Nordic. (2021). *More of everything – The Swedish forestry model*. https://www.moreofeverything-film.com/#home

Rabinowitz, P. M., Kock, R., Kachani, M., Kunkel, R., Thomas, J., Gilbert, J., Wallace, R., Blackmore, C., Wong, D., Karesh, W., Natterson, B., Dugas, R., & Rubin, C. (2013). Stone Mountain One Health Proof of Concept Working Group. Toward proof of concept of a one health approach to disease prediction and control. *Emerging Infectious Diseases, 19*(12), e130265. https://doi.org/10.3201/eid1912.130265

Richardson, K., Will, S., Lucht, W., Bendtsen, J., Cornell, S. E., Donges, J. F., Drüke, M., Fetzer, I., Rockström, J., et al. (2023). Earth beyond six of nine planetary boundaries. *Science Advances, 9*(37), eadh2458. https://doi.org/10.1126/sciadv.adh2458

Rockström, J., Steffen, W., Noone, K., Persson, A., Chapin, F. S., Lambin, E. F., Lenton, T. M., Scheffer, M., Folke, C., Schellnhuber, H. J., Nykvist, B., de Wit, C. A., Hughes, T., van der Leeuw, S., Rodhe, H., Sörlin, S., Snyder, P. K., Costanza, R., Svedin, U., Falkenmark, M., Karlberg, L., Corell, R. W., Fabry, V. J., Hansen, J., Walker, B., Liverman, D., Richardson, K., Crutzen, P., & Foley, J. A. (2009). Planetary boundaries: Exploring the safe operating space for humanity. *Ecology and Society, 14*(2), 472–475.

Salazar, L. F., Nobre, C. A., & Oyama, M. D. (2007). Climate change consequences on the biome distribution in tropical South America. *Geophysical Research Letters, 34*, L09708. https://doi. org/10.1029/2007GL029695

Sehgal, R. N. M. (2010). Deforestation and avian infectious diseases. *Journal of Experimental Biology, 213*(6), 955–960. https://doi.org/10.1242/jeb.037663

Senior, R. A., Hill, J. K., & Edwards, D. P. (2019). Global loss of climate connectivity in tropical forests. *Nature Climate Change, 9*, 623–626. https://doi.org/10.1038/s41558-019-0529-2

Shanko, K., Kemal, J., & Kenea, D. (2015). A review on confronting zoonoses: The role of veterinarian and physician. *Veterinary Science & Technology, 6*(2), 1000221. https://doi. org/10.4172/2157-7579.1000221

Smith, P., Bustamante, M., Ahammad, H., Clark, H., Dong, H., Elsiddig, E. A., Haberl, H., Harper, R., House, J., Jafari, M., Masera, O., Mbow, C., Ravindranath, N. H., Rice, C. W., Robledo Abad, C., Romanovskaya, A., Sperling, F., & Tubiello, F. (2014). Agriculture, Forestry and Other Land Use (AFOLU). In O. Edenhofer, R. Pichs-Madruga, Y. Sokona, E. Farahani, S. Kadner, K. Seyboth, A. Adler, I. Baum, S. Brunner, P. Eickemeier, B. Kriemann, J. Savolainen, S. Schlömer, C. von Stechow, T. Zwickel, & J. C. Minx (Eds.), *Mitigation of climate change* (Contribution of Working Group III to the Fifth Assessment Report of the Intergovernmental Panel on Climate Change). Cambridge University Press.

Steffen, W., Richardson, K., Rockström, J., Cornell, S. E., Fetzer, I., Bennett, E. M., Biggs, R., Carpenter, S. R., de Vries, W., de Wit, C. A., Folke, C., Gerten, D., Heinke, J., Mace, G. M., Persson, L. M., Ramanathan, V., Reyers, B., & Sörlin, S. (2015). Planetary boundaries: Guiding human development on a changing planet. *Science, 347*, 6223. https://doi.org/10.1126/ science.1259855

Wieting, J., & Leversee, D. (2019). Clearcut Carbon. Sierra Club BC. https://sierraclub.bc.ca/clearcutcarbon/. "For thirteen years after clearcutting, the carbon released into the atmosphere from decomposing organic matter and exposed soils is more than the carbon captured by the growth of young trees. In other words, it takes thirteen years for young trees to have a net effect of capturing carbon. In the meantime, clearcut areas remain sequestration dead zones".

Winter, S., Begehold, H., Herrmann, M., Lüderitz, M., Möller, G., Rzanny, M., & Flade, M. (2015). *Best practice handbook – Nature conservation in beech forests used for timber.* ISBN:978-3-00-067813-4, 186 pp.

Wolfe, N. D., Daszak, P., Kilpatrick, A. M., & Burke, D. S. (2005). Bushmeat hunting, deforestation, and prediction of zoonotic disease. *Emerging Infectious Diseases, 11*(12), 1822–1827.

WWF. (2011). WWF living forests report: Chapter 3: Forests and climate: REDD+ at a crossroads. https://wwf.panda.org/wwf_news/?202569/Living-Forests-Report

WWF. (2015). WWF living forests report: Chapter 5: Saving forests at risk, 51 pp. https://files.worldwildlife.org/wwfcmsprod/files/Publication/file/5k667rhjnw_Report.pdf?_ga=2.81742739.58627757.1627822934-746113179.1627822934

WWF. (2016a). Living planet report 2016. https://www.wwf.de/fileadmin/fm-wwf/Publikationen-PDF/WWF-LivingPlanetReport-2016-Kurzfassung.pdf

WWF. (2016b). Wälder in Flammen – Ursachen und Folgen der weltweiten Waldbrände. https://www.wwf.de/fileadmin/fm-wwf/Publikationen-PDF/161117_Waldbrandstudie_2016.pdf

WWF. (2018a). Die schwindenden Wälder der Welt. Zustand, Trends und Lösungswege. https://www.wwf.de/fileadmin/user_upload/WWF-Waldbericht-2018.pdf

WWF. (2018b). Marktanalyse Grillkohle 2017 – Das schmutzige Geschäft mit der Grillkohle. https://www.wwf.de/fileadmin/fm-wwf/Publikationen-PDF/WWF_Marktanalyse-Holzkohle_2018.pdf

WWF. (2019). Below the canopy. Plotting global trends in forest wildlife populations. 23 pp. https://www.wwf.org.uk/sites/default/files/2019-08/BelowTheCanopyReport.pdf

WWF. (2020a). Living planet report 2020. Bending the curve of biodiversity loss. https://www.wwf.de/living-planet-report/

WWF. (2020b). The loss of nature and rise of pandemics. https://www.wwf.de/fileadmin/fm-wwf/Publikationen-PDF/WWF-Report_Biodiversity_and_Pandemics.pdf

WWF. (2020c). Fires, forests and the future: A crisis raging out of control? https://wwf.panda.org/wwf_news/?661151/fires2020report

WWF. (2020d). Grillkohle 2020 – Eine EU-Marktanalyse. https://www.wwf.de/fileadmin/fm-wwf/Publikationen-PDF/WWF-EU-Marktanalyse-Grillkohle-2020.pdf

WWF. (2021a). Stepping up? The continuing impact of EU consumption on nature worldwide. Summary report, 7 pp. https://www.wwf.de/fileadmin/fm-wwf/Publikationen-PDF/WWF-Report-Stepping-up-The-continuing-impact-of-EU-consumption-on-nature-worldwide-ExecSummary.pdfundFullreporthttps://www.wwf.de/fileadmin/fm-wwf/Publikationen-PDF/WWF-Report-Stepping-up-The-continuing-impact-of-EU-consumption-on-nature-worldwide-FullReport.pdf

WWF. (2021b). Edles Holz, die Bundeswehr und die Mafia. https://www.wwf.de/wald/gorch-fock-edles-holz-die-bundeswehr-und-die-mafia. Last access 04.10.2021.

WWF. (2022). Cargills böse Welt. https://www.wwf.de/2022/maerz/cargills-boese-welt

Zemp, D. C., Schleussner, C. F., Barbosa, H. M. J., & Rammig, A. (2017). Deforestation effects on Amazon forest resilience. *Geophysical Research Letters, 44*, 6182–6190. https://doi.org/10.1002/2017GL072955

Reforestation in the Tropics and Subtropics

Exploiting the Growth Potential at the Lower Latitudes

Reinhard Mosandl

About 4 billion years ago, the primordial atmosphere was still free of oxygen, but had a very high carbon dioxide (CO_2) content. Over millions of years, large amounts of CO_2 were dissolved from the atmosphere into the oceans and stored in sediments (Schönwiese, 2020). Only through photosynthesis—initially in cyanobacteria, later also by land plants (Hanson & Rice, 2014)—was oxygen added to the atmosphere in large quantities and CO_2 removed to the same extent for the build-up of biomass (Schönwiese, 2020). For photosynthesis, organisms need light. The higher land plants grow, the more sunlight they can use compared to other, nearby plants. Trees gain this competitive advantage by building long, woody stems. Forests therefore dominate the earth's vegetation insofar as the climate allows trees to grow.

Forest and Atmosphere

The existence of trees can be traced back 300 million years (Boenigk & Wodniok, 2014). During this period, there have been repeated major climate fluctuations and corresponding changes in forest area and distribution. Forests and atmosphere are in dynamic equilibrium. If forests expand, large quantities of CO_2 are removed from the atmosphere and stored in the wood of trees. Conversely, when wood rots or burns, CO_2 is released into the atmosphere.

Forests also affect the energy exchange at the Earth's surface and thereby influence the atmosphere's energy balance. Forests take energy from the atmosphere and use it mainly to evaporate water; the air above the forest cools. Conversely, the loss

R. Mosandl (✉)
TUM School of Life Sciences, Technical University of Munich,
Freising-Weihenstephan, Germany
e-mail: mosandl@tum.de

© The Author(s) 2024
K. Wiegandt (ed.), *3 Degrees More*,
https://doi.org/10.1007/978-3-031-58144-1_7

of forests usually leads to a warming of the air layer above the affected areas (Duveiller et al., 2018) (cf. article by Schwarzer starting on p. 255).

After the last ice age, forests had spread over half the global land area. Caused by climate fluctuations, they changed their shape several times, but retained their areal extent for a long time. It was only with the increase in human population and the expansion of agriculture that their decline began. The loss of forests affected the various parts of the world in waves: Europe in the millennium before last, North America in the century before last, and the tropics since the middle of the twentieth century. Today, forests cover 4 billion hectares, or merely one-third of the world's land area (FAO, 2020a). In addition to virgin (primary) forests and commercial forests, degraded forests exist over a large area—in the humid and semi-humid tropics alone they occupy an area nearly the size of Australia (ITTO, 2020). These forests, which have been exploited in the past, are severely altered in structure and reduced in function. They can no longer adequately fulfill their role in the natural balance. In addition, there are roughly 450 million hectares of abandoned agricultural land worldwide (Campbell et al., 2008), most of which was previously forest. All these areas must be taken into consideration for purposes of influencing the CO_2 content of the atmosphere to counteract climate change.

Currently, forests, together with the other land-based ecosystems, bind 31% of anthropogenic CO_2 emissions; the oceans absorb an additional 23%. As in the past millions of years, vegetation and the oceans act as a "giant pump" (Schönwiese, 2020) that removes CO_2 from the atmosphere. In recent decades, therefore, the atmosphere has been polluted with only a portion (46%) of global man-made CO_2 emissions (IPCC, 2021) Even from this simple overview of the Earth's carbon budget (Fig. 1), it is clear that there are essentially only two levers for reducing the anthropogenic CO_2 load on the atmosphere: the burning of fossil fuels can be reduced or carbon reservoirs can be activated, with vegetation (minimizing forest losses and expanding forest area) playing an important role.

Forest Management

For managing existing forests there are the following options:

- Primary forests usually contain very large carbon reservoirs and a high level of biodiversity. These must be safeguarded. Silvicultural interventions in these forests can only be justified if clearing can be prevented through sustainable forestry use or if the adaptation of forests to climate change is actively supported.
- On a regional scale, commercial forests can achieve wood stocks comparable to those of primary forests. They should therefore be managed so that carbon stocks remain stable and good quality wood is sustainably produced. This enables long-term removal of CO_2 from the atmosphere, storing it in wood products. The impact improves with less energy used in further processing, with better raw wood quality, and with longer life and better recyclability of the wood products.

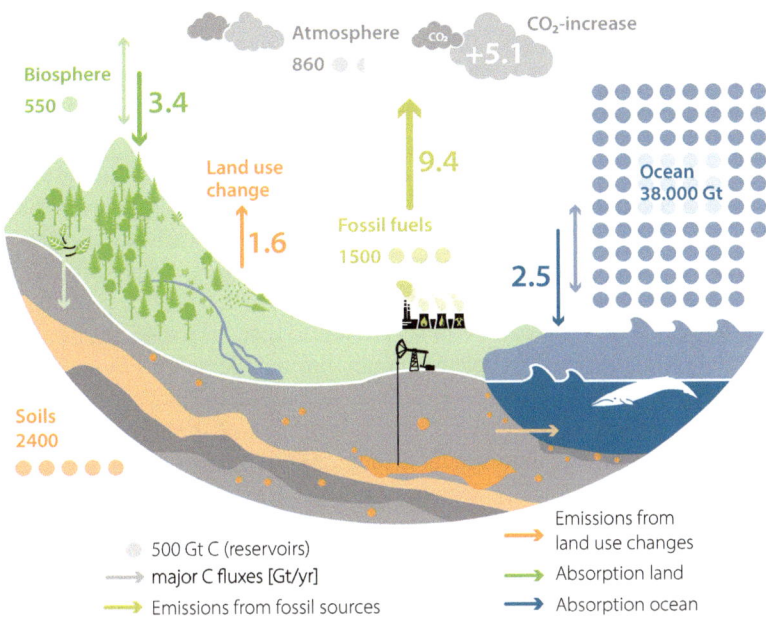

Fig. 1 The global carbon cycle. (Data from Friedlingstein et al., 2020). The largest carbon stocks on earth are found in the sediments, in the ocean, and in fossil deposits. In land-based ecosystems, most carbon is stored in the soil—especially in the permafrost soils of the Northern Hemisphere. More than half of the above-ground carbon is in the biomass of forests in the tropics and subtropics. Mankind pollutes the atmosphere primarily through the combustion of fossil fuels, the production of cement, and through the clearing of forests with carbon emissions, mainly CO_2. Since currently only about half of all these emissions can be stored by vegetation and the oceans, the CO_2 content of the atmosphere is continuously increasing

At the same time, wood can replace other energy-intensive raw materials such as reinforced concrete (see also chapter "Bauhaus Earth").

- In degraded forests, the cause of the impairment (e.g., grazing) must first be eliminated. If these forests are placed under full protection, they will, over an extended period, revert to structured, fully functional forests. Through silvicultural measures, restoration programs can accelerate the restoration of forests and their CO_2 sequestration. Forests can be converted into commercial forests so that the CO_2-ecologically beneficial effects of wood use can be realized. An important management objective for these forests would be to build up the wood and carbon stocks of the stands.

Regarding currently forest-free areas that are to bear forest again, the following possibilities arise:

- On cleared areas, forests will often re-establish themselves sooner or later without human intervention. So-called secondary forests are created through natural seeding and the sprouting of roots and tree trunks. Site conditions and existing landscape structures determine the speed at which natural reforestation and thus CO_2 sequestration take place (Poorter et al., 2016). Silvicultural measures can accelerate natural reforestation, by reducing game feeding on young plants, for example, or by promoting the wider spreading of roots.
- The artificial establishment of forests, when carried out in a professional silvicultural manner, is the fastest form of reforestation. New forest areas are created by planting and choosing the composition of tree species. Most afforestation restores forests in naturally suitable areas. But artificial irrigation makes it possible to create forests also beyond these growth limits.
- Either way, if the newly created forests are left to their own devices, parts of the carbon stored in them will be released again through decomposition and fire. In the case of forestry use, the proportion of wood put to long-term use can be expanded by silvicultural measures, and a large part of the sequestered carbon can be removed from the atmosphere through use of the wood as a raw material. This avoids CO_2 emissions by substituting fossil energies and helps to save climate friendly energies (solar, wind) for reasonable purposes.

Forest in the Tropics and Subtropics

The tropics extend between the latitudes of 23.5 degrees North and 23.5 degrees South. Frost occurs there only in the mountains. At the equator, there is an ever-humid diurnal climate, which means that the temperature differences between day and night exceed those among months of the year. Seasons as we know them do not exist in the ever-humid inner tropics. The natural forests are evergreen deciduous forests with very high productivity and diversity. The further one moves from the equator to the tropics, the longer the dry seasons become. Many trees react to the dry season with leaf fall and an adaptation to the increasing risk of forest fires—through thicker bark, for example. With greater length of the dry season, forests become increasingly unproductive and sparse.

Bordering the tropics to the North and South are the subtropics. There frost may occasionally occur in the lowlands. Large amounts of rain fall on the eastern sides of the continents. This area of the subtropics is humid and well suited for tree growth. In the so-called Mediterranean subtropics, which are located on the Western sides of the continents, precipitation falls mainly in winter. As a result, the summers are dry. Trees must adapt to this dryness in summer and are thereby limited in their productivity. In the center of the continents, the subtropics are dry almost year-round. There forests can often barely survive without artificial irrigation.

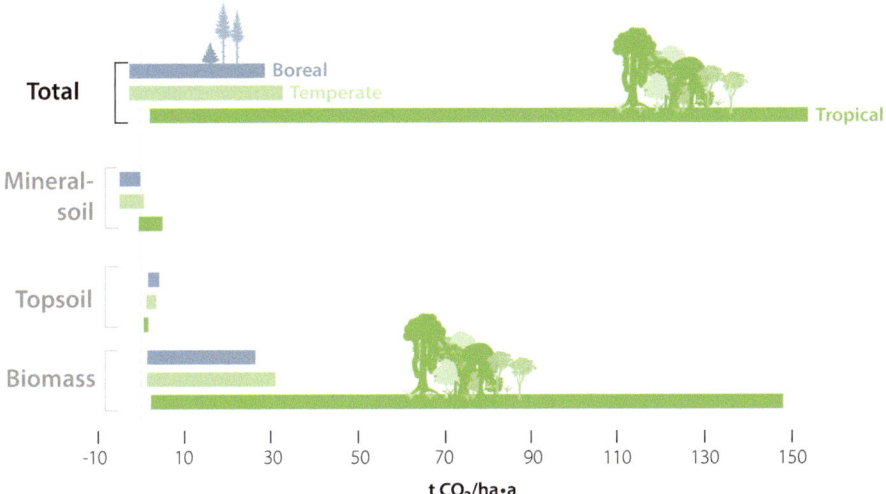

Fig. 2 Range of values for the annual CO_2 reduction (positive values) or release (negative values) in the first 20 years after afforestation. The studies show that a large absorption effect can be expected from the build-up of biomass, especially in the tropics. (Data from Paul et al., 2009)

The greatest destruction of forests is currently taking place in the tropics and subtropics (see also chapter "Stop Rainforest Deforestation"). The release of large amounts of carbon could be prevented if it were possible to stop this forest destruction. Enormous carbon sequestration potential exists where forest areas were cleared but not converted into permanently managed agricultural land. The ever-humid tropics and subtropics support very high productivity, so that reforestation would enable sequestration of large quantities of CO_2 within a few years (Fig. 2). Furthermore, measurements show that reforestation in these regions has cooling effects (Alkama & Cescatti, 2016). Reforestation would therefore make positive biochemical and biophysical contributions to limiting global warming.

With large-scale afforestation, large quantities of valuable timber could be produced in an economically profitable way. There would thus be an economic incentive to promote afforestation even without financial subsidies. Trees could be integrated into both industrial and smallholder agriculture. Due to the large share of agricultural land, the resulting agroforestry systems not only have a large CO_2 sequestration potential, but also offer many other social, ecological, and economic advantages.

For a 3-degree warmer world, climate models predict for the tropics and subtropics a moderate warming and a redistribution of precipitation (Gutiérrez et al., 2021). For the Sahara, higher precipitation is expected, so that parts of this desert may possibly green up (IPCC, 2021). Tropical and subtropical vegetation would spread into the current desert areas. Afforestation could support this process, and unexpected dry periods could be buffered by artificial irrigation. Conversely, precipitation is expected to decrease in the Amazon region. Its current vegetation is not adapted to this, leading to an increase in forest fires with release of large quantities of CO_2. Comprehensive forest management could remedy this situation. Existing forests would have to be protected from man-made fires. Forest development on sensitive sites could be specifically adapted to the expected climate conditions. Climate-adapted afforestation and rehabilitation measures on deforested or degraded forest areas could achieve quite fast reforestation. This would also relieve the pressure of timber use on the still intact tropical forests. Insofar as these activities yield qualitatively valuable wood, the cost of those efforts can be offset, and CO_2 can be sequestered in wood products.

Practical Examples

The foregoing reflections provide the global framework. Planning and implementation of concrete measures require, however, a regional approach due to the highly variable and complex local ecological, technical, and socio-economic conditions. In what follows, reforestation in different regions of the tropics and subtropics is therefore presented in five case studies that were prepared with the Institute of Silviculture at the Technical University of Munich during the time when the author was director of this institute.

Reforestation of Subtropical Forests in Central China

China is home to about one-tenth of the world's tree species, and half of it would by nature be densely forested (Ahrends et al., 2017; Wenhua, 2004). Agriculture developed here more than 10,000 years ago, and forest land gradually became agricultural land. By the twentieth century, the forest area had decreased to about one tenth of the national territory (Ahrends et al., 2017; Miao et al. 2016). In this century, the remaining forests, mostly in mountain areas, were systematically cleared to use the wood for economic development and to create agricultural land for the population (Summa, 2013). On the now predominantly arable land, precipitation increasingly ran off the surface, causing soil erosion. This resulted in greater sediment loads and higher flood peaks in water bodies, with corresponding adverse effects on people, infrastructure, and drinking water supply. Frequent catastrophic floods finally led to a change in China's land use policy in the 1990s, with the aim of compensating for the enormous loss of forest land (Ahrends et al., 2017). Agricultural use of slopes steeper than 25 degrees was outlawed nationwide in 1991 (Liu et al., 2011). The 1998 Forest Conservation Program aimed to rehabilitate degraded forests and stop their exploitation. With the help of afforestation, soil and water conservation were supposed to be promoted and timber production to be increased again (Wenhua, 2004). As a result, China's forest area temporarily expanded by 4 million hectares annually. Today, once again, almost a quarter of the country's area is forested. With 80 million hectares (roughly the area of Turkey), China currently has the largest area of forest plantations in the world (Ahrends et al., 2017). As a result, the carbon stock in the biomass of Chinese forests has since 1990 increased by 4.2 billion metric tons or nearly doubled (FAO, 2020a, b). However, due to the still limited use of wood and the still young forest plantations, there is still a nationwide wood shortage, which the Chinese are trying to mitigate through extensive wood imports. In many cases, other materials are substituted for wood in construction, so that the centuries-old Chinese tradition of timber construction is at present barely practiced.

Located in the center of China, the Qin Ling Mountains have a rich flora and fauna. Among other things, they are home to the famous giant panda (Ailuropoda melanoleuca). The mountain range forms the climatic divide between the dry, temperate Northern parts of China and the humid (sub-)tropical Southern China, which is characterized by the summer monsoon. The mountains are an important water source for Northern China, which tends to be short of water. National resource planning therefore provides that this region supply water to China's economic centers in the dry North.

The abrupt abandonment of agricultural land in steep terrain created in the Qin Ling Mountains large areas that are now mainly covered by grass and shrub vegetation. From older protected areas, where agricultural use had been prohibited for a long time, it is known that rich, stable mixed forests can in due course develop on former agricultural lands. This process of reforestation can be accelerated by afforestation. There is great interest therefore in using afforestation to restore forest to formerly cleared Qin Ling Mountain areas as quickly as possible. This should lead to a more consistent water supply and reduce the high sediment loads in the rivers. One focus is on the creation of stable, near-natural mixed forests of native tree species.

In the course of a German-Chinese cooperation, reforestation areas with native tree species were established in the Qin Ling Mountains on agricultural land that had been set aside some years ago. In 2007, the tree species Pinus tabuliformis (Chinese pine), Quercus variabilis (Chinese cork oak), Acer truncatum (Chinese Norway maple) and Pistacia chinensis (Chinese pistachio) were planted. The aim of the study was to determine the survival rate and productivity of these tree species on formerly agricultural lands (Summa, 2013).

The tree species chosen are native to the area and therefore well adapted to the prevailing climate and soil conditions. They enable the production of valuable wood. Furthermore, seeds and plant parts are used in the pharmaceutical industry or for vegetable oil production. It was very difficult, however, to acquire suitable planting material of the above-mentioned tree species which, in regard to its genetic characteristics, meets the requirements for establishing near-natural forests. For this purpose, the seeds for growing the young forest plants should ideally come from recognized seed stocks so as to ensure large genetic diversity as well as favorable quality characteristics of the future forests. In this case, too, a globally common problem manifested itself: for larger afforestation efforts, we still need to identify a

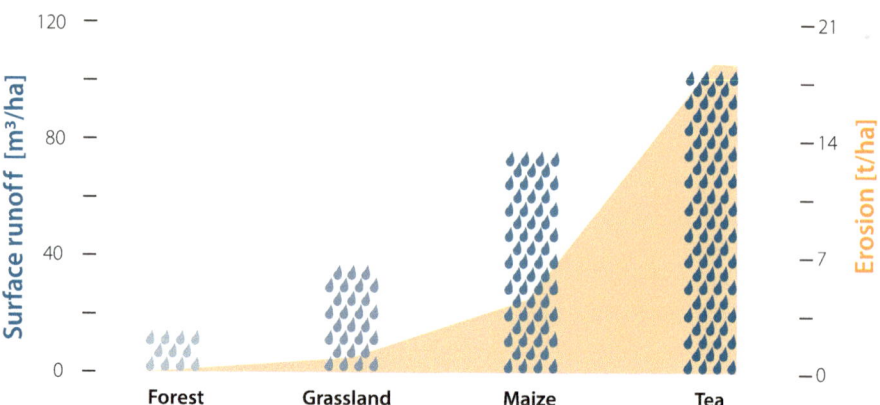

Fig. 3 Surface runoff and erosion when covered with forest, grassland, maize field, and young tea bushes in combination with root crop cultivation. Among all land uses, forests provide the best protection against erosion and increased surface runoff. (Data from El Kateb et al., 2013)

sufficient number of tested resources for forest seeds and establish a continuous testing system.

The experimental afforestation areas of 2007 have meanwhile developed into dense forest and completely replaced the previous grass vegetation. Rainwater can now seep into the soil and erosion hardly occurs anymore, even on steep slopes (Kägler, 2019). This is associated with rapid storage of CO_2 in tree biomass and soil. Compared to other forms of land use, the establishment of forests proves to be the best measure for erosion control (Fig. 3).

In an accompanying study, the local population was asked about their acceptance of afforestation. Due to the loss of agricultural land, many local residents need additional sources of income. Afforestation is therefore always of interest to the population if it can improve their economic situation. This underlines the importance of ensuring that reforestation can also contribute to securing the livelihood of the rural population (Wang et al., 2015).

Reforestation of Desert Areas in Egypt

Egypt is located in the dry subtropics. The country consists mainly of desert and is thus almost devoid of vegetation. Forests cover less than 0.1% of the country (FAO, 2020c). Remnants of natural forests exist only in the protected mountain region of Elba in the south of the country. The fertile land areas along the Nile are fully used for agriculture and increasingly as settlements for the rapidly growing population. However, Egypt has had a national forest administration since the eleventh century and has thus been able to systematically establish plantation forests since the 1970s. Within the framework of the "National Program for the Safe Use of Treated Wastewater for Afforestation," attempts have been made since the 1990s to use the large amount of urban wastewater for the cultivation of a variety of tree species in desert areas. The aim is to establish large-scale afforestation to bind CO_2, to protect cities and agricultural land from sandstorms, to protect the soil from wind erosion, to produce wood and bioenergy, and to create green spaces and new jobs for the local population.

Highly productive afforestation in the desert through irrigation with pre-cleaned (and otherwise unusable) wastewater has great potential for CO_2 sequestration. Large knowledge gaps still exist, however, in the field of silviculture and sustainable management. These gaps concern the selection of suitable tree species and seeds, matters of forest care, and technical-administrative challenges in the construction and maintenance of wastewater treatment and irrigation facilities. In a long-term practical trial, artificially irrigated afforestation areas were therefore established at three locations in Egypt with 14 native and foreign tree species from tested forest genetic resources. The aim of the studies was to resolve the following questions: what are the survival rates and productivities of the planted tree species under different irrigation regimes? And how does their wood quality develop?

The results show that the tree species Eucalyptus camadulensis (red eucalyptus), Corymbia citriodora (lemon eucalyptus), Casuariana equisetifolia (horsetail-leaved casuarina) and Khaya senegalensis (African mahogany) show exceptionally high productivity with an optimal supply of the nutrient-rich effluents and predominantly also develop good wood quality. At the age of 15 years, the stands can store 50 metric tons of CO_2 per hectare annually (El Kateb et al., 2022). Other tree species tested, however, did not cope with the conditions in the desert. Due to the social upheavals in recent years, securing irrigation was a major problem.

The experiments show that, where urban wastewater is available, desert afforestation has great potential for CO_2 sequestration. As the population continues to grow, settlements are increasingly relocated to desert areas. These newly emerging cities could form the core of new afforestation areas that store CO_2 and reduce the widespread wood scarcity. Due to the dry climate, wooden buildings in the desert have an extremely long lifespan. In addition, there are from the field of climate modelling indications that support further desert afforestation: under expected global climate conditions, the Sahara could receive more precipitation, and desert afforestation could promote cloud formation and precipitation at the regional level (Branch & Wulfmeyer, 2019).

Reforestation and Rehabilitation of Degraded Primary Forests in Ethiopia

In the tropical and subtropical areas of Africa, forests are dwindling (FAO, 2020d). The main causes of this development are poverty of the population, armed conflicts, and deficiencies in forestry legislation and supervision. Growing population density and rising world market prices for food and energy will exacerbate this situation. The increasing development of the continent enables growing access to previously untouched resources and forest areas. In view of the weaknesses of public administrations, which are expected to continue, careful management of the tropical forest can be achieved only if the sustainable use of forests offers improved income opportunities for the local population.

Ethiopia is among the longest-settled regions on earth. Despite its long history of settlement, the country has a diverse tropical flora and fauna and is one of the most important centers of biodiversity in the world. The Ethiopian flora comprises about 7000 higher plant species, 15% of which are endemic (Gebretsadik, 2016). The country has a varied topography and geology which produce a diversity of different habitats. The climate is shaped by a dry season, as is typical in the marginal tropics. At the beginning of the twentieth century, over one-third of Ethiopia was still forested (Dessie & Christiansson, 2008). Today, only 16% of the country is covered by forest (FAO, 2020d). Ethiopia's forest loss is currently above average among African countries.

One of the largest closed forest areas is the Munessa-Shashamene forest in the center of Ethiopia, covering approximately 35,000 hectares. The natural vegetation of the area is tropical mountain forest. Its structure is dominated in the canopy by the tree species Podocarpus falcatus, which is intermixed with the tree species Syzygium guineense, Prunus africana and Croton macrostachyus depending on the location (Strobl, 2011). Based on a cooperation with European research institutions, agricultural land and natural forest areas were converted into forest plantations in the past. Since the 1960s, the Shashamene Forest Industry, which is responsible for the forest area, has planted pure stands of the foreign tree species Eucalyptus saligna, Cupressus lusitanica and Pinus patula on 7000 hectares (Strobl, 2011). These monocultures have an annual timber growth of 20 to 25 cubic meters per hectare at a stand age of 20 years, with stocks averaging 300 cubic meters per hectare. Per hectare, these forest stands have thus stored about 300 metric tons of CO_2 in the wood of the trunks alone and are currently sequestering another 20 metric tons annually. However, pure stands are regularly attacked by insect pests. The local population has lost its traditional usage rights in the plantations. They practice agriculture and livestock breeding on the remaining agricultural land. Since the afforestation, remnants of natural forest have been used even more intensively for firewood and as forest pasture. Young trees are thus browsed or cleared by cattle. This gives the former primeval forest areas a park-like character. These forests are degraded over a large area and will dissolve completely over the years.

The example of the Munessa-Shashamene forest area clearly shows how problematic it is simply to transfer practices of nineteenth century European forestry to tropical Africa. The future lies more in developing sustainable utilization practices for the remaining and mostly overused natural forest remnants and in converting unnatural timber plantations back into near-natural, multifunctional forests.

Doing this requires researching the characteristics and dynamics of the natural forest. This was done in a cooperation between Ethiopian and German universities. Since 2006, a joint experimental station has been established in the Munessa-Shashamene forest area. This station has been the basis for numerous studies investigating the reaction of natural forests and plantation stands to different ecological factors and their control through silvicultural interventions. The silvicultural studies pursue two main questions: how can the overexploited natural forest be regenerated and made attractive for sustainable use by enriching it with lucrative native tree species? And how can the functions of the non-natural plantations be stabilized and improved by intensive thinning and regeneration with native tree species?

Due to overexploitation, many gaps have appeared in the canopy layer of the natural forest. At the same time, subsequent tree generations fail due to browsing and firewood use. Therefore, the native tree species Cordia africana, Juniperus procera, Prunus africana and Podocarpus falcatus were planted in these gaps as an experiment (Birru et al., 2011). The long-term aim was to determine which native tree species could be introduced into the region's overexploited natural forests under which light conditions. First results showed that especially Juniperus procera and Podocarpus falcatus can be established very successfully by planting. These two tree species were therefore recommended for the rehabilitation of degraded forests in the Ethiopian highlands.

Coniferous afforestation is highly productive but has, due to deficient forest management, developed into susceptible monocultures with poor timber quality. Currently, these stands are cut down according to plan at the age of 30 to 35 years. In a series of experiments, it was investigated how the growth of valuable wood and the stability of the stands can be improved through targeted promotion of the most vital and highest-quality trees. For this purpose, trees that impede the growth of the most promising trees were felled in thinning operations. These interventions also allow additional light to reach the soil so that native tree species can reestablish themselves and the stands can develop from pure stands to near-natural mixed stands (Strobl, 2011).

Further studies show that conversion of plantations back to more natural stands reverses adverse changes in the topsoil that accompanied the cultivation of the pure coniferous stands that were far away from close-to-nature (Ashagrie et al., 2007).

The studies in the Munessa-Shashamene forest area provided fundamental insights into the management of tropical forests. They also provided positive impulses for sustainable forest development in Ethiopia, which have been incorporated into the country's forest management guidelines. The research station and the study plots were handed over to the largely underfunded Ethiopian universities after project funding was completed. A visit 10 years later showed that the research station had fallen into disrepair, the study plots were neglected, and usable materials were found in the households of the surrounding villages. This underlines the need to involve in such projects not only scientific expertise and practical experience in afforestation and forest rehabilitation measures, but also the local population, administration, and businesses. Only in this way can such projects be successful in the long term.

The reforestation activities initiated by Prime Minister Abiy Ahmed can also be seen as an indicator of more sustainable forest development in Ethiopia. According to government figures, 350 million trees were planted in 2019. With this number of plantings, a region the size of Saarland's forest area can be covered. The government in Addis Ababa does indeed have large areas in mind: Ethiopia's participation in the African Forest Landscape Restoration Initiative (AFR100, 2021) for the restoration of 100 million hectares of land in Africa and the Great Green Wall Project (Mirzabaev et al., 2022) for the restoration of a further 100 million hectares in the Sahel are proof of this.

Rehabilitation of Degraded Forest Areas in Nigeria

Nigeria stretches from the dry forests of the Sahel to the ever-humid rainforests of the Atlantic. Nigeria is home to 220 million people. The country has one of the highest rates of deforestation. Since 1990, 20% of Nigeria's forest area has been lost. Today, forests cover 23% of the country's surface (FAO, 2020e).

In Nigeria, forest planning days for the production of timber have been observed since the beginning of the twentieth century, especially in the tropical-humid South of the country. In the Oluwa and Omo forest reserves, large-scale afforestation of degraded primary forest areas began in the 1960s. Initially, the areas were planted using the Taungya system, which originated in Asia. In this agroforestry system, the manual cultivation of tropical timber is usually carried out by contractually bound farmers or agricultural workers who are given the right to temporarily grow agricultural plants next to the still young trees. After one or two years, the tree canopy closes and the farmers cultivate a new plot of land. After a few years, the forest plantation is cleared, and the cycle begins anew. From the 1980s onwards, mechanization prevailed and replaced the taungya system.

In the beginning, mainly native tree species such as Nauclea diderrichii, Entandrophragma spp., Guarea spp., Terminalia spp., Khaya spp. and Lophira alata were used. From the 1960s onwards, however, foreign tree species dominated—especially Gmelina arborea (trade name Gmelina) and Tectona grandis (trade name Teak). In the Oluwa and Omo Forest Reserve, Gmelina arborea dominates today with a share of about 90%. The original goal was to produce pulp with a production period of 10 years. For this purpose, the soil was turned over, after which Gmelina arborea was planted and competing ground vegetation was fought with great effort. In the following years, the plantations were protected from fire. No measures were taken to increase the quality of the wood, as it was to be used exclusively as mass for the pulp industry. The only problem was that the pulp mill, which was supposed to process the wood, did not work. Therefore, the focus shifted toward producing valuable wood in longer production periods of 15 to 20 years, and the plantations were systematically thinned to increase the proportion of valuable trees and to accelerate their growth.

The sustainability of forest plantations is often questioned. Central to such criticisms are the threat of soil damage and the loss of biodiversity, which are comparable to those accompanying conversion to agricultural land. For this reason, the Technical University in Akure, in cooperation with German universities, conducted studies on Gmelina arborea plantations in the Oluwa and Omo Forest Reserves to answer the following questions (Onyekwelu et al., 2006): Does the cultivation of

Gmelina arborea lead to a loss of soil quality relative to primary forest areas? And is a loss of biodiversity to be expected with the cultivation of Gemelina arborea?

The results show that nutrient stocks in young and middle-aged plantations are somewhat lower than in the natural forest. However, there are no significant differences across all ages. On the contrary, soil conditions improve with the age of the plantations so that Gmelina forest plantations can be operated in the long term, over several generations, without loss in soil fertility. Due to the canopy closure, to the warm and moist conditions in the plantation, and to the activity of decomposers (such as earthworms), organic as well as mineral nutrients are incorporated into the soil. Organic carbon accumulates in the soil and sequesters long-term the atmospheric CO_2 that is bound by the plants. At the same time, the trees' nutrient supply and hence their productivity and CO_2-reducing effect is maintained or even improved.

The question is then not whether, but how forest plantations can be run sustainably. Much damage is caused, for example, by the cultivation of unsuitable conifer species, by soil compaction and humus losses during mechanized soil preparation, by burning off residues from clearing, and by inappropriate timber harvesting techniques. If these practices are avoided and if, in addition, the nutrient-rich leaves and twigs are left on the plantation after the timber harvest and the production period for valuable roundwood is increased to 25 years, then soil quality in the Gmelina plantations can be maintained at its original level over several tree generations. The studies show that suitable management of forest plantations can avoid the kind of a long-term loss of soil quality occurring after conversions from forest to agricultural land.

The older Gmelina plantations, which have already been intensively studied, were compared in terms of their biodiversity with degraded forests in the Oluwa Forest Reserve and with a primary forest in the neighboring Akure Conservation Area (Onyekwelu & Olabiwonnu, 2016). The results show that in regard to the older trees in the forest plantation, biodiversity is limited both in terms of the plant families and species involved. This is not surprising, as the sole management objective of the plantation was to grow Gmelina arborea. It is remarkable, however, that, since planting began 26 years ago, eight tree species from seven different families have naturally established themselves in the plantation in addition to the planted Gmelina arborea. Two of these tree species are classified as endangered.

The biodiversity of seedlings and young trees below the crown layer, however, is comparable in all three forest types. In the Gmelina plantation, 13 tree species from 17 plant families are found in the young trees. In the seedlings, there are 24 tree species from 17 plant families—including several economically important native tree species such as Cola gigantea, Celtis zenkeri, Bridelia ferruginea, Pterygota macrocarpa, Cleistopholis patens, Sterculia rhinopetala and Strombosia pustulata.

The woody layer below the canopy of the planted tree species is thus crucial for assessing the biodiversity of forest plantations. Older plantations, especially, can contribute to the preservation of biodiversity and endangered tree species. Furthermore, the natural regeneration of forests plays a crucial role in the dynamics of forest ecosystems and their long-term development. After the harvest of Gmelina arborea, the rich natural regeneration of the plantation stands would develop into a mixed stand consisting of native tree species, which would also contain various economically lucrative tree species. These stands could therefore continue to be managed in a near-natural way.

In summary, it can be concluded from these studies that, if properly managed, forest plantations can contribute to the preservation of soil fertility and biodiversity. At the same time, they produce valuable wood and sequester CO_2 long-term. In the medium term, forest plantations can also be converted into near-naturally managed commercial forests and ultimately support the restoration of natural forest areas. However, the development of the study area also showed that forest management is possible only in conjunction with adequate structures for processing and selling the wood. The development of new forest areas therefore requires cross-sectoral regional planning.

Reforestation of Tropical Mountain Rainforests in Ecuador

Ecuador is among the most biodiverse countries on Earth. Much of this biodiversity is found in Ecuador's forests, which are, however, massively threatened by one of Latin America's highest rates of deforestation (Mosandl et al., 2008).

Whereas in other South American countries, such as Chile, forest area is continuously increasing through afforestation, it is steadily decreasing in Ecuador (between 2010 and 2020 by 53,000 hectares annually, according FAO, 2020f). This trend is due to low afforestation activities and steady deforestation. As part of a large-scale research project of the German Research Foundation (DFG) in the mountain rainforest of Southern Ecuador, the processes underlying forest loss were investigated in detail (Beck et al., 2008; Mosandl & Günter, 2008; Bendix et al., 2013). The main driver of deforestation turned out to be the clearing of natural forests to obtain pastureland. As a result of the invasion of grazing areas by other vegetation (mainly bracken), which strongly impaired grazing and limited grazing periods, new grazing areas had to be created continually. This caused substantial expansion of abandoned pasture areas, which in turn spawned the thought that these degraded areas should be used in some way to reduce the clearing pressure on the still existing natural forests. Within the framework of the DFG research project, experiments were conducted to make these areas fit to be grazed again, and various reforestation experiments were established toward restoring the natural forest.

On the areas earmarked for afforestation, there were only very few naturally accrued woody plants, so that natural reforestation was not to be expected within the foreseeable future. Without afforestation, neither restoration of biodiversity nor a possibility of future use could be expected. The procurement of suitable planting material for the afforestation proved to be a major difficulty (Stimm et al., 2008). The lack of tree nurseries, seeds of indigenous trees, and knowledge about fruiting and cultivation of forest plants delayed the start of afforestation considerably. Apart from seeds of exotics such as Pinus patula and Eucalyptus saligna, there was

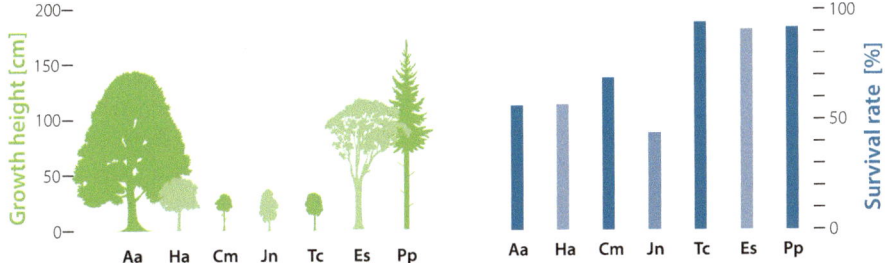

Fig. 4 Survival rate and growth heights of all trees 2 years after afforestation. (Data from Aguirre, 2007) (Aa = *Alnus cuminata*, Ha = *Heliocarpus americanus*, Cm = *Cedrela montana*, Jn = *Juglans neotropica*, Tc = *Tabebuia chrysantha*, Es = *Eucalyptus saligna*, Pp = *Pinus patula*)

practically nothing available on the market. Studies on the fructification of natural forest tree species, own seed harvests, and greenhouse trials were necessary to produce the planting material needed for the afforestation. Already 2 years after the successful planting on the afforestation plots, large differences between native and exotic tree species became apparent. The native tree species had great difficulty surviving on the strongly sunlit open areas; only Tabebuia chrysantha (current name Handroanthus chrysanthus) was able to keep up with the exotics Pinus patula and Eucalyptus saligna by achieving similarly good survival rates (Fig. 4). The growth of the native tree species was also very limited in the open areas. Only Alnus acuminata, as a pioneer tree species, could reach heights of over one meter as the two exotic tree species.

Compared to the unsatisfactory growth of the native trees in open areas, native trees planted in gaps in the natural forest (enrichment planting) or under the thinned-out canopy of pine plantations developed splendidly. This ultimately led to the recommendation for forestry practice in Ecuador to work in difficult afforestation sites (especially in open areas) with easy-to-establish pine (Pinus patula) or alder (Alnus acuminata) plantations. Under the thinned-out canopy of pines or alders, the native tree species can then later be raised without problems.

Conclusions

Forests are natural carbon sinks, removing substantial amounts of CO_2 from the atmosphere. This absorption effect of forests can be maintained or increased by protecting existing forest areas from clearing, by forest restoration, and through sustainable management. In the medium term, however, the restoration of former forest areas through reforestation is necessary for achieving a stabilizing effect on the atmosphere's CO_2 content (IPCC, 2018). Forest restoration is especially efficient in the humid tropics and subtropics, as the trees there are highly productive and hence absorb a lot of CO_2 (Strassburg et al., 2020).

The build-up of carbon stocks in newly planted forest areas takes place over many decades. However, the annual CO_2 sequestration rate of forest areas in the subtropics and tropics reaches a maximum after just a few years and then decreases steadily. The only way to sustainably maintain the high sequestration potential of these forest areas is through the targeted renewal of forests with the help of silvicultural measures. Wood obtained in this way can then permanently store the sequestered carbon in the form of durable wood products. Wood can replace raw materials with a negative CO_2 balance and thus contribute to reducing CO_2 emissions. These substitution effects must be taken into account in evaluating forest-based solutions to protect the atmosphere. Even wood from damaging events can usually still be used and then generate such substitution effects. The sustainable production and efficient use of wood should consequently be part of any decarbonization strategy.

The case studies have shown that the restoration of forests in many regions of the subtropics and tropics can contribute to stabilizing the natural balance (e.g., through protection of soil, water resources, and biodiversity) and can provide many positive impulses for rural and economic development. These significant side effects must be taken into account when evaluating forest-based measures. After all, sustainable forest management will only succeed if the local people benefit from the forest. Due to heterogeneous local and cultural conditions, implementation should therefore always involve the local population. The regionally varying climate changes also require a regionally adapted approach (IPCC, 2021).

Forest-based measures to reduce atmospheric CO_2 emissions therefore go far beyond the planting of trees (Girardin et al., 2021). They encompass nature and society. They are part of global efforts to moderate the rise of atmospheric CO_2 through "nature-based solutions" (Seddon et al., 2021). The global potential of nature-based solutions is estimated at 24 billion metric tons of CO_2 in relief for the atmosphere annually, taking into account the areas needed for agricultural production and the preservation of biodiversity (Griscom et al., 2017). However, this potential can only be realized if the costs for the metric ton of CO_2 saved are not taken into account. If costs, e.g. for planting and other forest operations are taken into account, the savings potential is reduced. At a maximum price of €100 per metric ton of CO_2, however, savings would still amount to 11 billion metric tons of CO_2 annually (Griscom et al., 2017)—equivalent to about 30% of the emissions caused by humans. "Forest-based solutions" can make a decisive contribution here (Roe et al., 2019; IPCC, 2018; Griscom et al., 2017; Felbermeier et al., 2016; Canadell & Raupach, 2008; Burschel & Fabian, 1999)—especially in the tropics and subtropics (Strassburg et al., 2020). Even if there is still high uncertainty in the estimates of CO_2 sequestration (Griscom et al., 2017) and even if it is to be feared that afforestation will often favor the wrong tree species (as in Cambodia, cf. Scheidel & Work, 2018) or the wrong sites (as in the upper reaches of the Paramo in Ecuador, cf. Quiroz Dahik et al., 2021)—there are nonetheless enough suitable areas and tree species as well as promising technologies available to restore the forests. Whether the areas estimated to be available worldwide (about 1 billion hectares, which roughly corresponds to the area of Europe) will actually be afforested and whether afforestation has the potential of binding 3–4 billion metric tons of CO_2

annually (Canadell & Raupach, 2008), are irrelevant for the afforestation efforts that are necessary now. What seems of foremost importance is that the forest-based proposals for combating the global climate crisis—formulated as early as 1999 by leading forest scientists in a Forest-Wood Manifesto—be finally taken up (Burschel & Fabian, 1999). Measures of forest/carbon management that have been shown to be carbon-ecologically effective, must be implemented on a larger scale and not be discredited from the outset as "modern trade in absolutions" or as "nature-based distractions."

Forest-based solutions offer great opportunities for successful bioeconomic development. By establishing efficient and socially balanced value chains—from ecological forest plant production and sustainable forest management to careful harvesting of wood that takes biodiversity aspects into account to the production and sale of wood products—we can achieve many positive effects that are lacking in the previous economy based on fossil resources. Taking account of the interests of the local population and making use of market mechanisms, it should be possible with appropriate economic incentives to restore large forest areas. Here the emitters of CO_2 from fossil fuels should also be involved by paying for the external costs of their economic activities and providing funds for forest restoration. However, this must not come at the expense of the priority tasks of CO_2 avoidance and the conversion to renewable energies.

References

AFR100. (2021). *The African Forest Landscape Restoration Initiative*. https://afr100.org/

Aguirre, N. (2007). *Silvicultural contributions to the reforestation with native species in the tropical mountain rainforest region of South Ecuador*. Dissertation, Technical University Munich.

Ahrends, A., et al. (2017). China's fight to halt tree cover loss. *Proceedings of the Biological Sciences, 284*(1854). https://doi.org/10.1098/rspb.2016.2559

Alkama, R., & Cescatti, A. (2016). Biophysical climate impacts of recent changes in global forest cover. *Science, 351*, 600–604.

Ashagrie, Y., Zech, W., & Guggenberger, G. (2007). Soil aggregation, and total and particulate organic matter following conversion of native forests to continuous cultivation in Ethiopia. *Soil & Tillage Research, 94*(101).

Beck, E., Bendix, J., Kottke, I., Makeschin, F., & Mosandl, R. (Eds.). (2008). *Gradients in a tropical mountain ecosystem of Ecuador* (Ecological Studies) (Vol. 198). Springer.

Bendix, J., Beck, E., Bräuning, A., Makeschin, F., Mosandl, R., Scheu, S., & Wilcke, W. (Eds.). (2013). *Ecosystem services, biodiversity and environmental change in a tropical mountain ecosystem of South Ecuador* (Ecological Studies) (Vol. 221). Springer.

Birru, G. A., El Kateb, H., & Mosandl, R. (2011). Rehabilitation of degraded natural forests by enrichment planting of four native species in Ethiopian highlands. In S. Günter, M. Weber, B. Stimm, & R. Mosandl (Eds.), *Silviculture in the tropics* (pp. 377–385). Springer. https://doi.org/10.1007/978-3-642-19986-8

Boenigk, J., & Wodniok, S. (2014). *Biodiversität und Erdgeschichte*. Springer Spektrum.

Branch, O., & Wulfmeyer, V. (2019). Deliberate enhancement of rainfall using desert plantations. *Proceedings of the National Academy of Sciences, 116*, 18841–18847. https://doi.org/10.1073/pnas.1904754116

Burschel, P., & Fabian, P. (1999). Die Wald- und Holz-Option – ein Manifest (G. Altner, Ed.). *Jahrbuch Ökologie 2000, Beck'sche Reihe, 1343*, 281–285.

Campbell, J. E., Lobell, D. B., & Genova, R. C. (2008). The global potential of bioenergy on abandoned agriculture lands. *Environmental Science & Technology, 42*(15), 5791–5794.

Canadell, J. G., & Raupach, M. R. (2008). Managing forests for climate change mitigation. *Science, 320*, 1456–1457. https://doi.org/10.1126/science.1155458

Dessie, G., & Christiansson, C. (2008). Forest decline and its causes in the south-central Rift Valley of Ethiopia: Human impact over a one hundred year perspective. *Ambio, 37*, 263–271. https://doi.org/10.1579/0044-7447(2008)37[263:FDAICI]2.0.CO;2

Duveiller, G., Hooker, J., & Cescatti, A. (2018). The mark of vegetation change on Earth's surface energy balance. *Nature Communications, 9*, 679. https://doi.org/10.1038/s41467-017-02810-8

El Kateb, H., Zhang, H., Zhang, P., & Mosandl, R. (2013). Soil erosion and surface runoff on different vegetation covers and slope gradients: A field experiment in Southern Shaanxi Province, China. *Catena, 105*, 1–10. https://doi.org/10.1016/j.catena.2012.12.012

El Kateb, H., Zhang, H., & Abdallah, Z. (2022). Volume, biomass, carbon sequestration and potential of desert lands' afforestation irrigated by wastewater on examples of three species. *Forest Ecology and Management, 504*. https://doi.org/10.1016/j.foreco.2021.119827

FAO. (2020a). *Global forest resources assessment 2020* (Main Report). Rome, Italy.

FAO. (2020b). *Global forest resources assessment 2020* (Country Reports: Report China). Rome, Italy.

FAO. (2020c). *Global forest resources assessment 2020* (Country Reports: Report Egypt). Rome, Italy.

FAO. (2020d). *Global forest resources assessment 2020* (Country Reports: Report Ethiopia). Rome, Italy.

FAO. (2020e). *Global forest resources assessment 2020* (Country Reports: Report Nigeria). Rome, Italy.

FAO. (2020f). *Global forest resources assessment 2020* (Country Reports: Informe Ecuador). Rome, Italy.

Felbermeier, B., Weber, M., & Mosandl, R. (2016). *Zur Machbarkeit eines weltweiten Aufforstungsprogramms*. Institute of Silviculture Technical University Munich.

Friedlingstein, et al. (2020). Global carbon budget 2020. *Earth System Science Data, 12*, 3269–3340. https://doi.org/10.5194/essd-12-3269-2020

Gebretsadik, T. (2016). Causes for biodiversity loss in Ethiopia: A review from conservation perspective. *Journal of Natural Sciences Research, 6*(11).

Girardin, C. A. J., et al. (2021). Nature-based solutions can help cool the planet – If we act now. *Nature, 593*, 191–194. https://doi.org/10.1038/d41586-021-01241-2

Griscom, B. W., et al. (2017). Natural climate solutions. *Proceedings of the National Academy of Sciences, 114*, 11645–11650. https://doi.org/10.1073/pnas.1710465114

Gutiérrez, J. M., Jones, R. G., & Narisma, G. T. (2021). IPCC interactive atlas. In *Climate change 2021: The physical science basis*. Contribution of Working Group I to the Sixth Assessment Report of the Intergovernmental Panel on Climate Change. Cambridge University Press.

Hanson, D. T., & Rice, S. K. (2014). *Photosynthesis in bryophytes and early land plants*. Springer.

IPCC. (2018). *Global warming of 1.5°C. An IPCC special report on the impacts of global warming of 1.5°C above pre-industrial levels and related global greenhouse gas emission pathways, in the context of strengthening the global response to the threat of climate change, sustainable development, and efforts to eradicate poverty*. Cambridge University Press. https://www.ipcc.ch/sr15/

IPCC. (2021) *Climate change 2021: The physical science basis*. Contribution of Working Group I to the Sixth Assessment Report of the Intergovernmental Panel on Climate Change. Cambridge University Press.

ITTO. (2020). *Guidelines for forest landscape restoration in the tropics*. Yokohama, Japan.

Kägler, S. (2019). *Afforestation with native tree species on degraded slopes in the Qinling Mountains, Shaanxi Province, People's Republic of China*. Master thesis, Technical University Munich.

Liu, X., Zhang, S., Zhang, X., Ding, G., & Cruse, R. M. (2011). Soil erosion control practices in Northeast China: A mini-review. *Soil & Tillage Research, 117*, 44–48. https://doi.org/10.1016/j. still.2011.08.005

Miao, L., Zhu, F., Sun, Z., Moore, J. C., & Cui, X. (2016). China's land-use changes during the past 300 years: A historical perspective. *International Journal of Environmental Research and Public Health, 13*(847).

Mirzabaev, A., Sacande, M., Motlagh, F., Shyrokaya, A., & Martucci, A. (2022). Economic efficiency and targeting of the African Great Green Wall. *Nature Sustainability, 5*, 17–25. https:// doi.org/10.1038/s41893-021-00801-8

Mosandl, R., & Günter, S. (2008). Sustainable management of tropical mountain forests in Ecuador. In S. R. Gradstein, J. Homeier, & D. Gansert (Eds.), *The tropical mountain forest: Patterns and processes in a biodiversity hotspot* (Vol. 2, pp. 177–193). Universitätsverlag.

Mosandl, R., Günter, S., Stimm, B., & Weber, M. (2008). Ecuador suffers the highest deforestation rate in South America. In E. Beck, J. Bendix, I. Kottke, F. Makeschin, & R. Mosandl (Eds.), *Gradients in a tropical mountain ecosystem of Ecuador* (Ecological Studies) (Vol. 198, pp. 37–40). Springer. https://doi.org/10.1007/978-3-540-73526-7_4

Onyekwelu, J. C., & Olabiwonnu, A. (2016). Can forest plantations harbour biodiversity similar to natural forest ecosystems over time? *International Journal of Biodiversity Science, Ecosystem Services & Management, 12*, 108–115. https://doi.org/10.1080/21513732.2016.1162199

Onyekwelu, J. C., Mosandl, R., & Stimm, B. (2006). Productivity, site evaluation and state of nutrition of Gmelina arborea plantations in Oluwa and Omo forest reserves, Nigeria. *Forest Ecology and Management, 229*, 214–227. https://doi.org/10.1016/j.foreco.2006.04.002

Paul, C., Weber, M., & Mosandl, R. (2009). *Kohlenstoffbindung junger Aufforstungsflächen. Literaturstudie.* Institute of Silviculture Technical University of Munich.

Poorter, L., et al. (2016). Biomass resilience of Neotropical secondary forests. *Nature, 211*–214. https://doi.org/10.1038/nature16512

Quiroz Dahik, C., et al. (2021). Impacts of pine plantations on carbon stocks of páramo sites in southern Ecuador. *Carbon Balance and Management, 16*(5). https://doi.org/10.1186/s13021-021-00168-5

Roe, S., et al. (2019). Contribution of the land sector to a 1.5 C world. *Nature Climate Change, 9*, 817–828.

Scheidel, A., & Work, C. (2018). Forest plantations and climate change discourses: New powers of 'green' grabbing in Cambodia. *Land Use Policy, 77*, 9–18. https://doi.org/10.1016/j. landusepol.2018.04.057

Schönwiese, C.-D. (2020). *Klimatologie.* Eugen Ulmer.

Seddon, N., et al. (2021). Getting the message right on nature-based solutions to climate change. *Global Change Biology, 27*, 1518–1546. https://doi.org/10.1111/gcb.15513

Stimm, B., et al. (2008). Reforestation of abandoned pastures: Seed ecology of native species and production of indigenous plant material. In E. Beck, J. Bendix, I. Kottke, F. Makeschin, & R. Mosandl (Eds.), *Gradients in a tropical mountain ecosystem of Ecuador* (Ecological Studies) (Vol. 198, pp. 417–429). Springer. https://doi.org/10.1007/978-3-540-73526-7_40

Strassburg, B. B., et al. (2020). Global priority areas for ecosystem restoration. *Nature, 586*, 724–729.

Strobl, S. (2011). *Analysis of the "nurse-tree effect" of exotic shelter trees on the growth of the indigenous Podocarpus falcatus in an Ethiopian montane forest.* Dissertation, University Bayreuth.

Summa, J. (2013). *Rehabilitation of degraded land ecosystems by afforestation with native tree species in the Qinling mountains of the Southern Shaanxi Province, China.* Dissertation, Technical University Munich.

Wang, X., Felbermeier, B., Kateb, H. E., & Mosandl, R. (2015). Household forests and their role in rural livelihood: A case study in Shangnan County, Northern China. *Small-Scale Forestry, 14*, 287–300. https://doi.org/10.1007/s11842-015-9288-8

Wenhua, L. (2004). Degradation and restoration of forest ecosystems in China. *Forest Ecology and Management, 201*, 33–41. https://doi.org/10.1016/j.foreco.2004.06.010

Bauhaus Earth

Sustainable Use of Wood in the Construction Sector

Hans Joachim Schellnhuber

Even if, from spring 2020 onwards, the Corona pandemic has driven man-made global warming from the headlines, all indications show that our civilization has maneuvered itself into a global predicament through unchecked fossil fuel consumption. Many imponderables notwithstanding, science today firmly concludes that the global environment begins to feel ill at 1.5 degrees of "Earth fever" and that the natural foundations of human life are threatened if this fever exceeds the 2-degree mark for an extended period. No one likes to imagine a world 4 or 5 °C warmer than today, even though quite realistic scenarios present global society as staggering toward this very world. There is even a non-negligible danger that anthropogenic climate change could develop its own fatal dynamic through powerful feedback loops (such as the self-reinforcing unlocking of carbon reservoirs in the Arctic and the tropics) (Steffen et al., 2018).

Ahead of Us the Hothouse Earth?

In the summer of 2021, the drumbeat from Lytton, a village in the Canadian province of British Columbia, was heard around the world. At the end of June, temperatures there rose to nearly 50 °C, values never recorded in Canada or indeed north of our planet's 50th latitude, since the beginning of instrumental weather recording. On the evening of 30 June, a forest fire broke out near the village, destroying Lytton within hours. There can hardly be a more macabre illustration of the connection between climate change, extreme weather events, and tragedy. In Germany, the terrible July floods in Rhineland-Palatinate and North Rhine-Westphalia have finally

H. J. Schellnhuber (✉)
Potsdam, Germany

International Institute for Applied Systems Analysis (IIASA), Laxenburg, Austria
e-mail: schellnhuber@iiasa.ac.at

© The Author(s) 2024
K. Wiegandt (ed.), *3 Degrees More*,
https://doi.org/10.1007/978-3-031-58144-1_8

roused people from their sweet sleep of climate ignorance. Like a monster crawling out from under our bed after a short sleep, climate change is now beginning to attract attention and dismay once again.

In fact, the average temperature at the Earth's surface in 2020 was already 1.25 degrees above the pre-industrial level. So we are moving unchecked towards the guard rails set by the 2015 Paris Climate Agreement and will break through them in a few decades. The faint planetary fever that the industrial revolution has now produced is already inflicting great suffering on the planet's individual creatures in ever more rapid succession: the Australian bushfires of the 2019/20 fire season destroyed over 20% of the continent's forested area and killed about a billion(!) animals of higher species. On 20 September 2017, Hurricane Maria ripped through the island nation of Puerto Rico with winds of around 250 km per hour. Almost 3000 people fell victim to the extreme event and its aftermath; economists estimate that the region's economic development has been set back by 20 years.

In 2020, a record number of 29 tropical cyclones were recorded over the Atlantic Ocean, which is undoubtedly related to the exceptionally high ocean surface temperatures. One more current example was the Mediterranean storm Daniel in September 2023, which caused havoc in Libya in form of heavy rainfall, flooding and devastation. This storm was referred to as a 'medicane', a Mediterranean hurricane, which seems to be a new phenomenon caused by climate change. Due to the rapidly increasing CO_2 content in the atmosphere, the energy balance of the Earth system is severely disturbed, so that our planet absorbs substantially more energy via solar radiation than it can radiate back into space. The excess is stored as heat in all system components, mainly in the oceans (90%) down to depths of 2000 m and more (von Schuckmann et al., 2020). Growing almost unnoticed, this heat giant will accompany humankind for many centuries to come.

Only 2% of the excess energy is currently warming the Earth's atmosphere, while as much as 4% of it is responsible for melting glaciers and sea ice. This is especially noticeable in the Arctic: in 2019, for example, the Greenland Ice Sheet lost mass at a rate of about one million metric tons per minute! And in two decades, the Arctic Ocean might already be completely ice-free during late summer.

"So what?" is what people in Central Europe might think, without reckoning with the jet stream, the band of strong winds that, at an altitude of 10–12 km, separates the cold polar air from the moderately warm air from the tropics around the Equator.

Since the Arctic is warming about three times faster than most other regions of the Earth and the jet stream is ultimately driven by the temperature difference between regional air masses of the Arctic and the Tropics, the westerly wind band is weakening in the meantime and is experiencing huge bulges ("Rossby waves") ever more frequently. If, by chance, these wave bulges become stationary (because they get stuck on the continental shelves, for example), we will have an extreme weather situation that can manifest itself as a biblical heat wave or even a deluge in Europe, North America, or North Asia (the physics behind this critical phenomenon was already discussed in 2013) (Petoukhov et al., 2013).

The Elephant in the Climate Shop

This brings us to Germany, which endured three consecutive years of drought from 2018 onwards. In 2021, by contrast, we experienced a wet episode with tropical-like rainfall, which in western Germany and Belgium led to one of the worst flood disasters in living memory. These direct experiences reflect in a startling way the above-outlined theoretical insights from the physics of climate change. They make evident that the dry heat will return and cause incalculable damage, especially to our forests. In the 1980s, there was talk of forests dying because of acid rain which, thanks to modern filtration technology in fossil-fuel power plants and refineries, has become insignificant today. The current climate-related forest dieback, in which droughts, fires, storms, and pests fatally interact, is more devastating and very difficult to control in the short term.

The information pages of the German Federal Ministry of Food and Agriculture (BMEL) display a sad statistic: some 285,000 hectares of land currently need reforestation in Germany. Mostly spruce stands are affected, but deciduous trees such as the copper beech also show serious to fatal damage. German forests, "built so high up there," as Joseph von Eichendorff wrote, must be fundamentally transformed if they are to survive in a changed climate. How can this transformation succeed in terms of forest economy, in a way that achieves profitability? The answer lies in the construction sector in Europe and worldwide—the "elephant in the climate shop" which has long been ignored or at least trivialized in environmental debates.

In fact, nearly 40% of the greenhouse gases currently emitted worldwide (primarily CO_2) are produced during the construction, operation, and demolition of buildings and infrastructure. In western industrialized countries, more than half of all waste mass is generated during construction and demolition. And in Germany, about 45 hectares of near-natural land are converted into settlement and transport infrastructure each day, with serious consequences for flora and fauna. Surprisingly, these facts are not part of society's basic knowledge, although the sight of construction cranes or cement trucks and the noise of jackhammers or concrete mixers have long been part of everyday life in the modern age. In Bertolt Brecht's words, this could be explained as stupidity eluding perception once it has reached a certain size.

But we can no longer afford climate stupidity or ignorance today. This is why the built environment must play a central role in society's sustainability strategies—and not only in Germany: if all housing projects planned in Asia, Africa, and Latin America are implemented with conventional methods and materials, with such "modern" building materials as concrete, steel, aluminum, glass, and plastic, then the Paris Agreement will degenerate into multilateral propaganda (WBGU, 2016). The battle to preserve a civilization-bearing Earth climate will be won or lost outside Europe.

Due to its technological, financial, institutional, and diverse cultural capacities, the EU is probably capable of the fastest transformation of its construction sector, thus able to lead the way toward a circular economy in this critical sector. This can be achieved only with some appropriate humility, because the transformation requires harnessing the best the world has to offer by way of traditional and contemporary building styles.

The Geological Journey of C

The climate history of the Earth is primarily a history of carbon (Schellnhuber, 2015). This is why solving the human-made climate problem largely depends on whether and how we get a grip on "God's element" with the chemical symbol C. Figuratively speaking, we need to coax the most powerful of all genies in our planetary environment back into a safe bottle.

For this, it is worth taking a brief but deep look into the Earth's past. In the late Proterozoic, more precisely in the period between 750 and 580 million years ago, there were probably four large-scale or total glaciations of our planet ("Snowball Earth"). The plate tectonic causes of this do not need to interest us here, but the mechanism by which the Earth was able to free itself from the ice cover does: countless volcanoes, especially at the edges of the continental plates, persistently emitted CO_2 whose steady accumulation in the atmosphere was helped by the fact that the snow and ice cover on the Earth's surface significantly reduced CO_2 sequestration through weathering processes. This continuously intensified the greenhouse effect, the atmosphere heated up, and the great melting finally began.

The subsequent Palaeozoic Era (about 540–250 million years ago) was followed by climatic phases of varying warmth. The most significant for the development of our evolved technical civilization was the Carboniferous period (about 360–300 million years before today). With an atmospheric CO_2 concentration about twice as high as today's (i.e., about 800 parts per million or ppm) and a comparable mean temperature of the Earth's surface (about 14–15 degrees), huge forest and swamp landscapes developed. The flora was dominated by various ferns, especially horsetails 20–40 m tall as well as scale and seal trees. This enormous biomass decayed, partly in the absence of oxygen (anaerobic), in the wetlands and formed the basic substance for thick coal seams, which formed biogeochemically over the course of millions of years. Thus accumulated the fuel for the Industrial Revolution, which began in the eighteenth century in England's north-west and in turn, through various waves of innovation, landed us somewhat casually in today's climate emergency.

Ironically, however, the environmental events in the Carboniferous show an ideal solution out of this predicament, which could lead not only to climate stabilization, but even to partial climate restoration. The lush plant life of the Carboniferous extracted more and more CO_2 from the atmosphere through photosynthesis, which, however, was not completely returned to the air due to wet decay but rather accumulated in fossil deposits (see above). In addition, the deep-rooted plants loosened the soils, creating additional weathering surfaces through which atmospheric CO_2 was also extracted. As a result, the CO_2 concentration in the atmosphere dropped to near 100 ppm, which almost led to another major glaciation (Feulner, 2017). These findings of palaeo research make clear how massively the biosphere can influence the global climate system.

Today, it is our civilization that has thrown this system out of kilter within a geologically extremely short period of a few hundred years. Until recently, the global economy was firmly on course to industrially burn, by 2200 at the latest, almost all the fossil energy resources (coal, oil, gas) that had accumulated naturally over hundreds of millions of years. This artificial oxidation event, unprecedented in the Earth's history, would double or even quadruple atmospheric CO_2, depending on the reactions of the individual elements and processes in the planetary system.

In the meantime, however, in nearly all countries the realization is growing that this unwanted environmental experiment is likely to be punished with devastating climate consequences and must therefore be stopped as fast as possible. On the one hand, this means that global greenhouse gas emissions must be reduced to almost zero by 2050 at the latest; and in highly developed industrialized countries like Germany the phase-out of fossil fuels should be achieved even by 2035. For this we already have coherent roadmaps that take account of the relevant sectors and technologies (e.g., Rockström et al., 2017).

Unfortunately, this is no longer sufficient. Due to human activities, too much CO_2 has already accumulated in the atmosphere, with today's concentration at 420 ppm or about 50% above the pre-industrial level. The associated increase the Earth's mean surface temperature, as already mentioned, is about 1.2 degrees—with the damaging effects described in the first section. A few years ago, the Intergovernmental Panel on Climate Change (IPCC) estimated how these consequences would come to a head beyond the 1.5-degree level and how it might still be possible to hold this forward defense line against global environmental chaos (IPCC, 2018). The result is clear: in addition to the rapid decarbonization of all sectors of the economy we need extensive "negative emissions," that is, processes in which CO_2 is actively removed from the atmosphere.

These processes should be initiated as soon as possible and may have to be sustained for many decades. However, the largely technical approaches discussed so far are rather naive to the point of being scary: the IPCC itself is backing the so-called BECCS process (e.g., IPCC, 2014). In this process, biomass is to be produced on a large scale, the energy stored in it is to be extracted, the remaining carbon is to be captured and deposited somewhere (!) for the long term. A convincing potential-cost analysis for this proposal is still lacking. Even more absurd to my mind are certain geo-engineering processes (Lawrence et al., 2018) in which huge amounts of CO_2 are to be directly removed from the atmosphere by physical-chemical processes. For this, however, a global infrastructure would have to be financed and built from scratch, which would not create any significant value apart from "air washing" and would ultimately degenerate into the biggest technological ruin of all time.

An elaborate synopsis and evaluation of all the existing 1.5-degree scenarios has recently shown that the technical extraction of CO_2 is often sold as an illusory solution to the climate dilemma and that there is generally no single robust strategy for the full implementation of the Paris Agreement (Warszawski et al., 2021). This is precisely where the idea of *turning the built environment into a powerful carbon sink* comes in—making a virtue of necessity! For if organic materials were to be used as much as possible as construction materials for future buildings and infrastructure, then one could not merely avert enormous amounts of greenhouse gas emissions that are generated in the production of concrete, steel, etc., but could also additionally capture historic CO_2 emissions from the atmosphere via photosynthesis. In addition, the carbon bound through photosynthesis during plant growth would be safely stored long-term in buildings and products.

The unique advantage over other ways of generating negative emissions lies in the fact that climate protection would here come as a welcome side effect of meaningful and attractive value creation—making this a genuine win-win option.

I discuss the most important aspects of this potentially decisive weapon in the fight against global warming in the next sections. The first systematic assessment of this novel approach has recently been published by an international researcher group that I initiated (Churkina et al., 2020). The almost frighteningly large vision behind it is based on carbon's above-outlined journey through the Earth's ages: a climatically meaningful amount of carbon—the constituent of life on our planet, after all—would be laid to rest for centuries in "built forests" and similar civilizational constructs after passing through many stages (volcanoes, palaeo-atmosphere, biosphere of the Carboniferous Age, fossil deposits, recent atmosphere). In this way, the cooling effect of natural carbon extraction some 300 million years ago could be replayed, as it were, as a large-scale cultural project in time lapse. Figure 1 summarizes the central insights of this section as a planetary infographic.

This could "initially" return the climate to the range of the Holocene—which began about 11,700 years ago and, due to its stable environmental conditions, enabled the explosive development of human culture (Neolithic Revolution). This would probably mean a centuries-long slide, with tenths-of-a-degree steps out of the risk zone. Apart from the climate issue, the move away from fossil-intensive construction also offers a great opportunity to completely rethink modernity in one of its core aspects. More on this in the last two sections of this article.

Bio-Based Architecture

It is clear from the above that a royal road to climate protection should lead directly through the construction sites of the world. As I recently noted in an essay, there has long been a signpost to this effect in my immediate neighborhood, namely in idyllic Caputh near the Schwielowsee (Schellnhuber, 2021a). We are talking about Einstein's summer house, which the architect Wachsmann built out of wood in 1929 at the express wish of the physicist of the century. The main materials used were timber of Californian redwood and Galician fir; peat panels were used to insulate the walls. This place has a magical aura and once again exemplifies Einstein's far-sightedness outside the natural sciences: barely 100 years before Greta Thunberg's climate strike, he chose the building materials with which we can contain global warming.

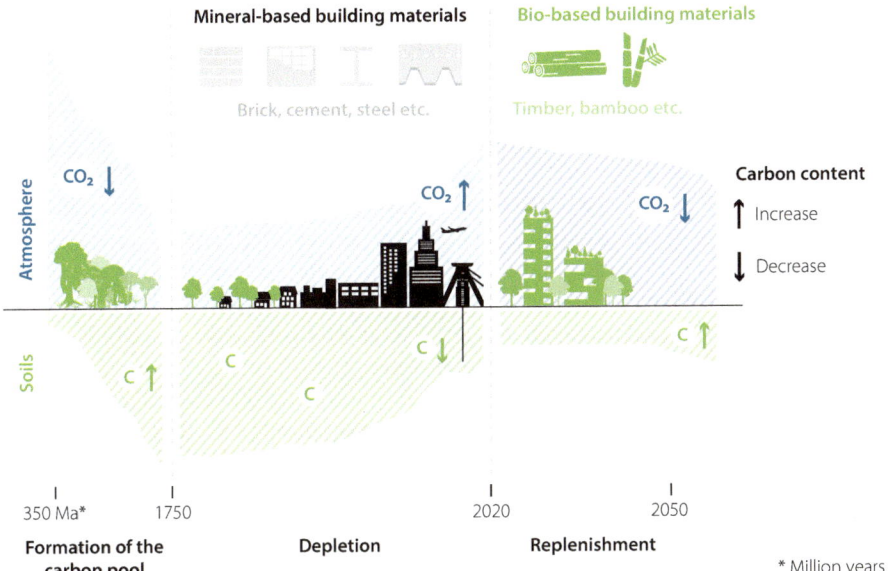

Fig. 1 Over millions of years, the carbon pool formed on land (left). Due to this process and other mechanisms (such as the weathering of rocks), the CO_2 content of the atmosphere slowly decreased. Middle: The urban and industrial growth triggered by the Industrial Revolution gradually depleted the carbon stock on land and increased the CO_2 concentration in the atmosphere again. The towering and load-bearing urban buildings of concrete and steel, made with raw materials and fuels from ever deeper layers of the Earth's crust, consume a lot of energy and cause high greenhouse gas emissions. Right: Settlements built from bio-based materials such as wood and bamboo can serve as artificial carbon sinks. The storage and conservation of carbon in the built environment will help replenish terrestrial C stores, thus reducing current atmospheric CO_2 concentrations or offsetting future emissions (Churkina et al., 2020)

Konrad Ludwig Wachsmann, the architect of Einstein's house, was chief architect of a company specializing in wooden buildings in Upper Lusatia from 1926 onwards. Because of his Jewish origins, he emigrated to the U.S. in 1941, where he met a certain Gropius. Together, the two developed a prefabricated wooden house system (General Panel System) that supposedly enabled five unskilled workers to completely erect a house in just under 9 hours. But Walter Gropius, born in Berlin in 1883, left a broader mark in architectural history. In 1919, he founded the Staatliches Bauhaus in Weimar as a school of art that, through its concept, aspirations, and impact, revolutionized the construction sector of the twentieth century as the New Bauhaus Movement.

Today, the name Bauhaus is usually identified with brutalist reinforced concrete constructions of the post-war period or even with a cheap hardware store chain that has been spreading across Europe's industrial areas since 1960. Wachsmann's masterful use of organic materials has been all but forgotten, and timber construction itself has largely disappeared from the modern cityscape. Even agricultural barns are now hardly ever carpentered but cast and assembled from mineral materials.

This systemic change, especially in commercial and private multi-story construction, took place around 1900 for many different reasons, of which at least three had a decisive joint effect.

First, with the trend in the construction sector toward serial industrial production of elements that are as uniform and homogeneous as possible (see, for example, the famous GDR type WBS 70 board), the particular substance wood quickly fell into disuse. This is because organic substances reflect the evolutionary complexity of life itself: they are depth-structured, anisotropic, and environmentally dynamic—to name just a few characteristic properties. However, these "competitive disadvantages" can be transformed into significant advantages in a highly intelligent building industry, the contours of which are already becoming visible today (see below).

Second, the development of enormous fossil energy resources (especially oil from the Middle East) for the world market from around 1960 onward enabled the comparatively cheap production of concrete, steel, aluminum, glass, plastic, and other non-sustainable building materials. Associated "externalities"—such as the destabilization of the global climate through massive greenhouse gas emissions—were of course not included in the individual cost-benefit calculations of the relevant companies.

Third, the same energy glut forced the historically unprecedented use of heavy machinery in civil engineering. Growing up, I often had to help my father (a master glazier with a penchant for creativity) on building sites after school and experienced how individual physical strength and dexterity were increasingly displaced by diesel-smelling equipment with rough bearings.

And this is how, in the second half of the twentieth century, construction became reliant on rather crude but ubiquitously available and always compliant materials. Expert or even artistic design has retreated almost exclusively to spectacular niche projects for luxury and prestige. To be fair, this has created affordable and hygienic housing for the masses (at least in the Western economies), but should this really be the end of architectural history? Apart from preserving the natural foundations of life, doesn't humanity also need built culture in its environment, that is, a milieu that comforts the soul even in everyday life?

But let us first return to the scientific analysis of the climate problem. As far as building materials are concerned, the all-important difference is expressed in the following three process equations:

$$\text{Iron ore smelting (Davis et al., 2018)} > 2Fe_2O_3 + 3C \rightarrow 4Fe + 3CO_2 \uparrow \qquad (1)$$

$$\text{Lime burning } 1000\,^{\circ}C > CaCO \rightarrow CaO + CO \uparrow Q = 181\,kJ * mol^{-1} \qquad (2)$$

$$\text{Photosynthesis} > 6CO_2 + 6H_2O \rightarrow C_6H_{12}O_6 + 6O_2 \qquad (3)$$

The laws of chemistry themselves thus dictate CO_2 emissions, which are inextricably linked to the conventional production of reinforced concrete; the laws of biophysics themselves force CO_2 uptake, which is inextricably linked to the growth of biomass! This fundamental asymmetry in greenhouse effect alone justifies the transition from mineral to organic building materials (such as wood, bamboo, jute, flax, hemp, lichen, algae, etc.). Or, to put it casually, if you had the choice between a climate-destroying material and a climate-healing one that tends to have even better properties (see below), which would you choose?

Incidentally, the energy required for the production, transport, processing, and assembly of building materials does not play a fundamental role in such a comparison: we graciously assume that in this respect the mineral-based economy will also quickly switch to the use of renewable energies without direct greenhouse gas emissions. The largely climate-neutral production of steel with "green hydrogen" would play a more systemic role in this context; but thus far this approach is more hype than based on evidence. The promises of the concrete industry and other relevant sectors to offset their CO_2 emissions through reforestation projects in exotic parts of the world are a dead end: why not use sustainably produced wood directly in construction? In general, the conventional construction sector's rhetoric of sustainability is unconvincing (e.g., The Concrete Imitative, 2021), but the economic and political significance of this well-organized industry is enormous.

Not least, stale myths about organic architecture in general and timber construction in particular play into its hand: that timber is much more vulnerable to fire, earthquakes, storms, and insects than reinforced concrete; that timber does not allow large constructions due to its lack of load-bearing capacity and dimensional stability; that wooden houses pose insuperable problems to sound insulation; that the whole thing would be too expensive anyway; and that wood will soon be in short supply worldwide. Many of these bluntly parroted prejudices are as groundless— and as effective—as the slogans that were used for many years to discredit the sustainable transformation of the energy and transport system. As late as the 1990s, people in Germany were still claiming, without good reason, that the local electricity supply from wind and sun would never exceed 4% of total electricity demand. In 2022, this share was already between 37% and 38% (Enerdata, 2023).

The built environment, too, is blighted by a twisted vision of progress, exemplified when ingenious architects and engineers, contracted by autocrats or oligarchs, erect futuristic buildings of concrete, steel, and glass reaching up into the sky. But when bourgeois medium-sized companies dare build an 8-story administration building made of solid wood on behalf of a Rhineland municipality, skepticism abounds: visionary work should be left to California charismatics.

Fortunately, the enlightenment has now arrived even in the world of architecture. Through a mutually accelerating interplay of practical experience and research-led innovation, the myths listed above have been largely debunked. Unfortunately, the relevant information is mostly buried in specialist journals and project reports; but the volume of literature comprehensible to the interested layperson is also growing. Exemplary here are the current books by Pablo van der Lugt, in which almost all critical aspects of bio-based construction are addressed with numerous references to pertinent original works (van der Lugt, 2020).

The greatest reservations about timber construction are related to the latent fear of urban fires, which has become a collective trauma over the centuries, especially in Europe. A case in point is the "Great Fire of London," which destroyed about four fifths of the still largely medieval city within a few days in September 1666. The fire broke out at night in a bakery in Pudding Lane and, due to incompetence on the part of the authorities, quickly developed into an inferno. It doesn't take divine wrath, just a little human stupidity and arrogance to unleash hell on Earth.

With appropriate knowledge and care, on the other hand, bio-based materials can nowadays be installed in a fire-safe manner. Important criteria such as flammability, structural fire stability, fire propagation dynamics, etc., must be assessed and considered in the design. Essential in this context is a holistic examination of the building's core and its casing because modern fire disasters often involve the cladding (the 2017 "Grenfell Tower Fire," also in London, is an example). This is not a technical paper on fire safety, so I refer to the technical literature (such as the website of "Informationsdienst Holz" with articles on "Fire protection in timber construction" etc. (von Winter, 2013)). Here it suffices to point out that, when exposed to fire, solid wood forms a protective surface layer of charcoal which usually slows down or prevents the fire's further advance. This response can be understood as an evolutionary adaptation of tree species to the ubiquitous oxidation risk associated with the equally evolutionary accumulation of oxygen in the atmosphere (Lenton & Watson, 2011).

Of course, active fire protection systems (such as sensor-supported sprinkler systems) can and will also be used in organic construction. Perhaps more important, however, is "passive" fire prevention through appropriate building safety design, adapted to the building's particular purposes and the actually chosen bio-based materials (see below). Similarly for sound insulation: design and composition must consider and exploit the physical building conditions, sometimes down to subtle and seemingly unimportant aspects. At first, this seems to be a disadvantage compared to conventional building, but with advanced digital methods it may well be turned into an advantage (there is more to be said about this, too). Special attention should be paid to the construction of floors and ceilings.

While wood conducts sound well, as can be experienced during a violin concert, its thermal conductivity is very low. The relevant technical parameter in this context is the so-called thermal resistance R which, for building components, should be as low as possible. For comparison: R is 50 for steel, 1.9 for concrete, and only 0.1–0.2 for wood, depending on tree type. This comparative advantage is cleverly used in the design of timber constructions, especially when these are to meet especially demanding requirements. A spectacular example is the Filmarchiv Austria, which was realized in 2004 with organic architecture and whose inside temperature must be kept within a 2-degree range year-round (Thoma, 2016). In everyday construction, wood is used for windows and exterior doors because of their good insulating properties.

But now to the question of all questions: what about gravity? How big and, above all, how high can one safely build with organic materials? Well, certainly bigger and higher than is commonly assumed—and this applies not only to the present and the future. Probably the most impressive historical example is the Sakyamuni wooden pagoda in Shuozhou (China). It was built in 1056, has 9 stories and towers 67 m high (including the base and top)! The European counterpart is the beautiful Heddal

Stave Church (Norway), often referred to as a "Gothic pine cathedral." This church was probably consecrated in 1242 and measures 26 m.

For the reasons outlined above, timber construction has for long been shunned in the modern industrial age. But today we witness a dramatic development to which the invention and spread of new types of wood materials (engineered and modified wood) have contributed decisively. This progress in a rather traditional industry stands in striking contrast to the aversion to innovation that, according to experts and insiders, has pervaded the success-saturated mineral construction industry for decades. The key innovation is simple but compelling: instead of, like in the past, searching for the optimal tree that can be sawn to size and installed directly, today's construction involves the composition of suitable wood elements. Solid wood materials, whose structural elements are joined and fixed together by glues, dowels, nails, etc., are becoming increasingly important. This approach is certainly reminiscent of the "additive manufacturing" of consumer goods (vulgo: 3D printing), in which the desired object is not created by milling off a blank, etc., but by skillfully assembling the necessary ingredients in a material-saving way.

It all began quite humbly in the mid nineteenth century in Thuringia with the serial production of plywood from wood veneer panels that were glued and pressed together perpendicular to the grain direction. This made it possible to largely "lock" the material dynamics through swelling and shrinkage under the influence of environmental fluctuations (in temperature, humidity, etc.) and to achieve a high dimensional stability. The process was simple, and the corresponding products were considered functional but inferior compared to objects made of grown solid wood. Since then, the approach has gained dramatically in importance and reputation, especially with the post-1990 introduction and perfection of cross-laminated timber (Fig. 2) in Germany and Austria (e.g., Schickhofer, 2013). These are suitably dimensioned solid wood panels made of at least three crossed board layers, which can be prefabricated in the factory and used in construction for almost all components

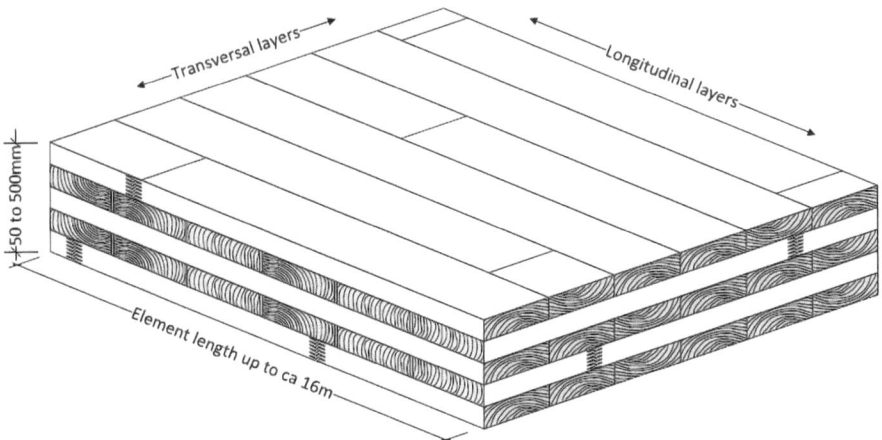

Fig. 2 Schematic representation of cross laminated timber (Schmid et al., 2018)

(exterior and interior walls, roof trusses, ceilings, railings, balconies, façades, etc.). The construction principle, which is as simple as it is ingenious, largely compensates for the natural heterogeneity and anisotropy of raw wood—almost any desired property (stiffness, strength, elasticity, fire resistance, etc.) can be realized through design and choice of material.

In addition to cross-laminated timber, there is today a whole arsenal of solid wood, veneer, chipboard, and wood fiber materials to choose from for almost every application. There is also a whole range of composite materials in which, for example, wood fragments are mixed with concrete or plastic. However, these are practically non-recyclable and therefore unacceptable for sustainable construction.

Ever since the creation of the New York skyline of skyscrapers in 1908, "modern" construction has been associated with high-rise buildings. Today, projects around the world that make use of the approaches just outlined prove that such architectural feats can also be achieved with wood. In Heilbronn, Germany, there is already the "Skaio", a 34-m-tall 10-story residential building made of organic material. In Hamburg's Hafencity, an eco-skyscraper is now rising to a height of 65 m. For Berlin-Schöneberg, the "WoHo" is planned, a 98-m hybrid timber building that will offer 29 floors of commercial and living space for all income levels. In Vienna-Donaustadt, the namesake "HoHo" has been in existence since 2019 and already towers 84 m. This makes it the second tallest wooden building in the world after "Mjostarnet" in Brumunddal, Norway (85.4 metres, 18 stories), which was completed in the same year.

These buildings will soon be overshadowed by a wooden skyscraper (strengthened with reinforced concrete) that, commissioned by software company Atlassian, will rise to 180 m near Sydney's central railway station. In Chicago, the true home of skyscrapers, 228 m is the target for a similar construction. And in Japan, the "PlyscraperW350" is being planned as the 350-m-tall new headquarters of a traditional timber construction company. Apparently, one of the most earthquake-prone countries in the world trusts the fabulous stability properties of cross-laminated timber.

If we look back at the less spectacular, but much more climate-relevant mass residential construction in Germany, we observe a rapid and encouraging development, especially in the area of detached and semi-detached houses. Of the total number of new buildings approved in this category in Germany in 2020 (105,962), 22.2% were prefabricated timber houses—with even much higher proportions in Baden-Württemberg (38.4%) and Hesse (32.4%)! In multi-story buildings, on the other hand, the relative market volume of timber construction is still an order of magnitude smaller. A similar picture emerges in commercial construction, where timber as a material is somewhat on the retreat in agriculture (e.g., barns), but is increasingly used for other commercial buildings.

Timber construction is also continuing to advance in other European countries (especially Austria, Switzerland, and Scandinavia). In North America, the corresponding building materials (mainly sawn timber from conifers) are traditionally popular in family housing. However, solid wood is hardly ever used, which has a negative to catastrophic effect on comfort, safety, durability, and environmental compatibility. Such lightweight construction is evidently due to the desire to keep the costs of a suburban home with a double garage as low as possible. Yet it is already possible to build high-quality timber houses at very competitive prices, as is to be demonstrated on a large scale in the so-called Schumacher Quartier on the former Berlin-Tegel airfield. It is planned to build 5000 flats for 10,000 people there in timber construction over the next 10–12 years—and about 20% cheaper than conventional construction (Tegel Projekt GmbH, 2020, 2021).

In general, the killer argument, that sustainable business is always more expensive than working with fossil fuels and extracted minerals and therefore a doomed elite project, will not save the concrete industry from extinction. In contrast to building with organic materials today, photovoltaic power generation, for example, was still hopelessly uncompetitive with its climate-damaging rivals (especially coal-fired power) 30 years ago. However, targeted political support (such as the legendary German Renewable Energy Sources Act of 2000) and a vibrant innovation dynamic triggered by it have pushed the production costs for solar power at favorable locations into a range that is unattainable for fossil energy sources. And this does not even take account of the "externalities": the enormous losses to the common good (climate damage, illnesses, ecosystem destruction, etc.) caused by electricity generation with coal, oil, or gas! If the learning curve of the sustainable building industry is now steeply bent upwards through improved framework conditions, increased research and development (R&D) activities, and economies of scale, there will no longer be any socio-economic reason to continue building in the dismal post-war style.

This sets in motion a self-reinforcing innovation spiral that will also help overcome the other challenges to "alternative" construction. In addition to resistance to earthquakes, hurricanes, and vibrations, this also involves the less spectacular but enormously important protection against fungal attack, insects, moisture, etc. Whereas before the turn of the millennium, aggressive chemical interventions were still heavily relied on here, ever-improving thermal processes are now being developed which remove, even before the wood is installed, many wood constituents that could later attract pests (such as woodworms). Furthermore, such treatment ensures that the wood moisture content remains permanently below the critical level of 20%, which stops the advance of fungi and microbes. Comparable results with concrete buildings are often only achieved through problematic coatings.

It is noteworthy in this context that innovations in organic construction also aim to avoid as far as possible substances harmful to climate or health. In the past, such substances were used on a massive scale in the gluing of wooden elements. Research, development, and practice have not yet been able to completely overcome this substance problem, but several solutions are already emerging (see, for example, the publications of the Fraunhofer Institute for Wood Research WKI).

Forestry's Transition Toward Sustainability

As explained above, large-scale bio-based architecture is not only technically and operationally possible but could well dominate building in the twenty-first century. However, we must now face an objection that is raised in every relevant discussion right after the allegedly uncontrollable fire problem: there would not be enough organic building material—in Germany, in Europe, worldwide—for the desired transformation. Or rather, the necessary quantities could only be procured through a brutal industrial plantation system. The latter would not only counteract climate protection itself, but also trample on other key sustainability goals such as the preservation of biodiversity.

In 2021, this skeptical attitude has stiffened even further. On the one hand, this is due to price volatility on timber markets, which is related to some mainly temporary special factors: massive shifts in demand due to the COVID-19 pandemic, disruptions in the global supply chain system, reduction in sawing capacity in the wake of the 2008 financial crisis, forest fires and bark beetle infestations in North America, etc.

On the other hand, some activists and nature conservation associations have recently intensified their campaigns for a purist forest policy. Two ideas seem to be central to this. First, that "the forest," in contrast to farmland, is a piece of "pure nature"; and second, that forest ecosystems undisturbed by humans store the most carbon.

Both ideas are romantic misconceptions. If we look at Western Europe, for instance, we find only tiny remnants of post-glacial (or even pre-glacial) primary forests ("primeval forests"). And if we look at the whole continent, it hardly looks any different. Only Finland and the Carpathian arc still have significant original forest areas, dominated by conifers in the former case and deciduous trees (mainly beech) in the latter. According to the WWF, 6000 years ago, 80% of Europe was covered with forest. Today, about 40% of our continent is still forested, and less than 0.2% of this area is primeval forest. Germany has no primary forests left at all; of its 11.4 million hectares of forest area 94% are even actively managed.

As far as the storage capacity of forests is concerned, a fundamental error is often made—consciously or unconsciously—by not considering the entire system cycle. Of course, a tree can only absorb atmospheric CO_2 so long as it forms additional photosynthetic biomass: grows in height, width, or depth. After the plant has matured, the carbon remains stored for decades to millennia, depending on the species, until decay finally sets in. Then the tree falls and rots, the CO_2 returns to the atmosphere, and space is made for the regrowth of the same species or competing species. Without human influence, forest ecosystems mature after 600–2000 years (!) in a dynamic equilibrium, with a natural distribution of species and ages according to site conditions. The carbon content of these systems remains largely constant over long periods through balanced supply and removal of CO_2.

By contrast, through the targeted extraction of biomass by means of sensible management, an autonomous ecosystem can be transformed into a driven one (in the sense of control theory) that can organize a net flow of carbon from the atmosphere into a stock of long-lasting material assets (wooden houses, roof trusses, wood-based materials, furniture, etc.). With the right planting and felling strategy, maximum growth (and thus maximum CO_2 extraction) is ideally achieved on a permanent basis. And the conversion of harvested biomass into useful objects takes the place of rotting—with the crucial difference that the re-release of CO_2 does not occur until several centuries later. The terrestrial dwell time of atmospheric carbon is thus artificially stretched, buying humanity valuable time in the fight against global warming! In the next section, I will explain this approach further with the help of a system diagram.

But of course, the size of the climate protection effect outlined above depends on the extent and nature of the forest area used. After many years in which the challenge of a transformative global land use strategy was often trivialized to a simple call to individual actors ("Let's plant a tree!"), a serious debate on deforestation and reforestation has finally begun. The research group led by Jean-Francois Bastin of ETH Zurich made an important contribution to this with their assessment of the "Earth's tree restoration potential" (Bastin et al., 2019)—although (or perhaps because) the publication attracted massive criticism from specialist colleagues. Based on extensive empirical data sets and with the help of machine algorithms, the scientists developed a model that identifies the potential geographical forestation cells worldwide with relatively high resolution (30 arc seconds). The model even calculates the respective percentage area of tree canopy, which increases from 0% in dry deserts to 100% in dense equatorial forests.

Result of the study: under today's climate conditions, 4.4 billion hectares of the Earth's land surface could be covered by trees. If one subtracts from this the existing canopy area (2.8 billion hectares) as well as the land currently occupied by agriculture and settlement, then an estimated 0.9 billion hectares remain. On this globally scattered area (see Fig. 3), afforestation or reforestation could in principle take place in the sense of a global repair measure. More than half of the potential is concentrated in just six countries (Russia, USA, Canada, Australia, Brazil, and China).

According to the study, if this potential were fully exploited, the added forest areas, when fully grown, would store additional 205 gigatonnes of carbon. If this vegetation were to continue to exist (without the biomass decoupling described above), this alone could compensate for more than two thirds of the CO_2 budget that would still be available to our civilization if the 2-degree guard rail were to be observed. If the newly created forests were mainly used for the regenerative production of raw materials for organic construction, demand could be met on a large scale (see below).

Desertified areas could prove to be a wild card in the great reforestation game, especially dry subtropical areas that once were more or less densely forested. Australian agronomist Tony Rinaudo received the "Alternative Nobel Prize" in 2018 for his conceptual and practical contributions to the revitalization of deserts (especially in West Africa). His work is driven by two critical insights. One is his rediscovery of the "subterranean forest" during the terrible Sahel drought of the 1970s. During an excursion in Niger, he noticed the scattered low bushes sprouting

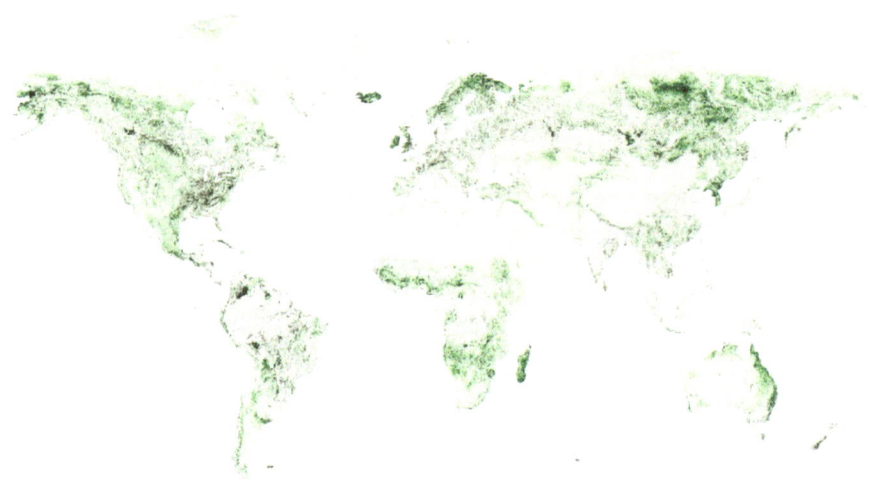

Fig. 3 The global reforestation potential. Today's existing tree cover as well as agricultural and urban areas according to GlobCover were subtracted from the worldwide potential tree cover (Bastin et al., 2019)

from the seemingly sterile desert floor. On closer inspection, these turned out to be branches of living tree roots hidden by the sand. With a simple knife, these roots can be pruned so that young trees sprout from them again. And indeed, there are millions and millions of hidden tree trunks in the Sahel and also in India, Brazil, Australia, etc.

Rinaudo's other critical insight is that reforestation of suitable areas should be realized by and for local farmers (FMNR—*Farmer Managed Natural Regeneration*). Often, large-scale governmental or international reforestation campaigns fail in the Global South because imported tree nursery plants are introduced into the dry soil and then irrigated at enormous expense in people and material. Behind the planting front, which continues to move, a high percentage of the young trees usually die. Rinaudo's double-local approach, however, has been crowned with great success. FMNR has already reclaimed over 20 million hectares of forest in several African countries!

More spectacular, but rather utopian by contrast, is the project of the "Weather Makers," a Dutch company of engineers of a special kind (Rose, 2020). This group around co-founder Van der Hoeven wants to reforest about half of Egypt's Sinai Peninsula, an area of about 3 million hectares. What seems insane at first glance becomes possible on closer inspection. For the now desolate peninsula was still green until a few millennia ago, and satellite images reveal ancient river networks that apparently collected abundant precipitation and transported it to the Mediterranean. Presumably due to unsustainable human practices (excessive use of biomass, overgrazing, etc.), the vegetation disappeared, the soils eroded, and fertile humus was washed away—partly into the large mud lake Bardawil on the north coast. The sediments of this lake, now dozens of meters thick, will be the starting point of the restoration campaign by Van der Hoeven and his fellow campaigners.

Their bold undertaking is inspired by China's remarkably successful attempt to re-vegetate the legendary loess plateau in the center of the country and thereby to stop the soil erosion that has prevailed there for a millennium, giving the mighty Huang He ("Yellow River") its name. With massive support from the World Bank, which provided 300 million US dollars as part of a 12-year project, a breathtaking restoration campaign took place in the former granary of the Middle Kingdom. With relatively simple measures (tree planting, terracing, soil enrichment with organic carbon, water retention, and pasture management), the conversion of almost 1 million hectares of desert into a vital ecosystem was achieved by the end of the campaign in 2009. This gives hope for many degraded areas around the world, which may be transformed back into fertile and site-appropriate forested landscapes sooner than had long been assumed (e.g., Blaustein, 2018).

This is precisely why the United Nations has called for the *Decade of Ecosystem Restoration,* which aims to revegetate around 350 million hectares of degraded land by 2030, thereby removing up to 26 billion tonnes of carbon from the atmosphere.

However, forestry calculations should not be made without the climate economist. Bastin et al. estimate that in the most pessimistic of the common emission scenarios (RCP 8.5), the Earth's tree carrying capacity would shrink by over 200 million hectares compared to today. In this case, large increases in boreal forests (e.g., in Siberia) due to climate change would be offset by even larger losses of tree crowns in the tropical rainforests (e.g., in the Amazon region). However, the authors themselves emphasize that the corresponding model calculations involve enormous uncertainties and do not satisfactorily model a whole slew of processes, disturbances, and feedbacks. Nonetheless, based on this analysis, there does not seem to be a fundamental shortage of forest development options.

On the other hand, in view of the existential importance of the topic, one can by no means be satisfied with the current state of scientific knowledge. As already indicated, the study has provoked numerous responses from colleagues in the field, who point out various shortcomings in terms of content and methodology. Whoever is ultimately right on the individual points, the research has been set in motion. At the Potsdam Institute for Climate Impact Research, we are currently trying to accelerate it by asking how much forest area would actually be needed for timber harvesting in the various scenarios of climate-positive transformation of the building sector. The starting point is the work by Churkina et al., already cited above, which estimates the global carbon sink effect of organic architecture. In the most recent follow-up study (Mishra et al., 2022), a scenario is developed in which 90% of the new residential and commercial buildings constructed by the end of the century are made largely of wood. The background is a plausible socio-economic development of the world society, in which humanity's degree of urbanization rises to 80%.

The quantitative analysis is performed with the integrated model complex MAgPIE which simulates the cost-optimal global production of food, feed, bioenergy, and timber in the twenty-first century. Special attention is paid to the terrestrial biosphere, where the dynamics of managed forest systems and the behavior of natural vegetation (primary and secondary forests, non-forested vegetation areas) must be modelled. The calculations are based on extensive data sets from international organizations (FAO, IUCN, UNESCO, etc.). The latter indicate, among other things, the areas that are highly productive for agriculture and those that should be off-limits to forestry for various reasons (intact natural forests, hotspots of biodiversity, retreat areas of indigenous cultures, etc.).

The most important result of the model calculations is that the renewable raw materials needed for the "wooden cities" can be provided without grossly violating other sustainability goals. However, the productivity of the cultivated areas would have to be increased through appropriate measures and investments, and a considerable expansion of global forest plantations would be needed. Concretely, up to 143 million hectares of such plantations would have to be newly established by 2100 (see Fig. 4 for the optimal geographical distribution).

The necessary areas could be taken primarily from unprotected or degraded forests and non-forested areas with low biodiversity. In addition, our model simulation envisions a shrinking of pasture areas, which would have a twofold emission reduction effect, especially in the beef industry. Timber harvesting in the 90% scenario (90 pc in Fig. 4) could save 122 billion tonnes of CO_2 by the end of the century—this alone would relieve the carbon budget agreed with the 2-degree guard rail by about 11%! Obviously, the share in relation to the 1.5-degree budget would be much

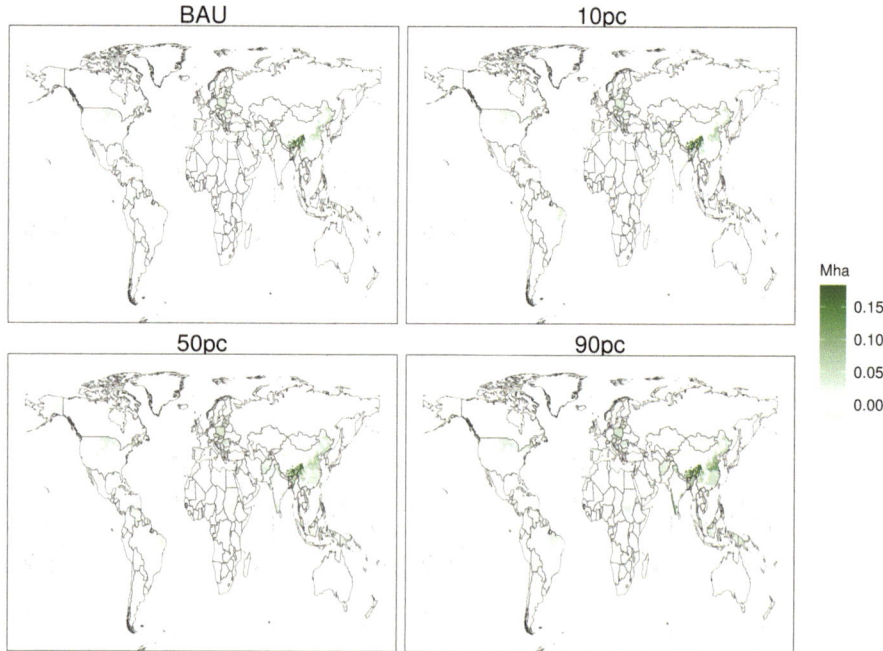

Fig. 4 Difference in forest area in million hectares (Mha) between 2100 and 2020. Shades of green represent the increase in plantation area in 2100 compared to 2020 (Mishra et al., 2022)

larger still. Thus, on a global scale, it becomes clear that bio-based settlements can make a substantial and feasible contribution to climate stabilization and restoration.

Accordingly, more and more studies have recently been devoted to the question of the basic availability of organic building materials, whereby very different aspects are being considered. The relationship between production and suitability of the plant types in question is of special interest. In Germany, for example, under the current climate conditions, 75% of the forest areas would naturally be covered with beech and 17% with oak. Spruce would only occur at altitudes above 700 m. Contemporary forestry turns this distribution upside down, because currently 25% of our forest areas are planted with spruce and 23% with pine! (Richter, 2021) In spite of all the romanticization, the "German forest" is a highly artificial thing whose construction began in early eighteenth century (see the legendary *Sylvicultura oeconomica* by Hans Carl von Carlowitz (2013/22)). The main criteria for management were rapid growth to make up for the forest losses of the previous centuries and good material properties for building and furniture production. The clear winner in this competition was the spruce, whose suitability is still raved about by modern timber architects, but the existence of this tree in the Central European lowlands is seriously threatened by human-made environmental changes.

Is this why the age of the beech is dawning, at least in Germany? This is by no means certain, because climate change seems to be affecting this deciduous tree—in contrast to the oak—more than predicted. Nevertheless, beech will probably play an important role in forestry in the future. So far, however, the construction industry has been reluctant to make friends with this wood, but that could change: since 2014, the Thuringian company Pollmeier has been offering the wood material "BauBuche" (building beech), which is produced by completely peeling open the tree trunks and skillfully smoothing, pressing, cutting, and gluing the resulting thin strips. This technique produces a very homogeneous material with a load-bearing capacity close to that of reinforced concrete. In addition, the approach is extremely resource-efficient, which in turn drives down costs. However, "BauBuche" and its undoubtedly numerous successors in the hardwood industry must be used optimally, which in turn is only possible if architects and engineers learn to handle the new material, and the state building codes no longer block the corresponding innovations.

Besides beech as a renewable raw material, we need to think about other tree species that can withstand global warming, serve the relevant value chains, and provide vital ecosystem services. Oak wood, for instance, is well suited for construction, but difficult to work because of its hardness. New processes and production facilities will probably also have to be created for the increased use of pine. The elastic all-rounder ash has recently been threatened by various pests that have migrated to Central Europe (such as the fungus *Hymenoscyphus pseudoalbidus*, which originated in Japan). Contemporary strategies for the conservation and utilization of native species are therefore needed, but the carefully considered introduction of tree types of non-European origin must not be taboo either if forestry is to keep pace with the enormous global dynamics. In this context, experts often mention the robinia (origin: North America; preferred climate: warm, moderately humid; properties: flexible, resilient, and rot-resistant wood) and the bluebell tree (origin: China; preferred climate: warm-dry; properties: fast-growing, light, stable, and fire-resistant wood).

Twenty-first century forestry faces the unprecedented challenge of composing location-appropriate forest ecosystems from many species and age classes to meet a variety of objectives. This is a veritable paradigm shift: less than one third of European forested land has trees of diverse ages; 30% are plantations with only one tree species, 51% still have two or three species, and only 5% are home to six or more tree species (Science for Environment Policy, 2021)!

But even the fate of the spruce, the conventional bread tree of German forestry, is not sealed. A recently published research paper runs through different variants of a national forestry strategy for Great Britain, which would allow 30,000 hectares of land to be replanted annually in the period from 2020 to 2050 (Foster et al., 2021). This study provides a wealth of interesting insights, but above all shows that newly planted commercial coniferous forests have by far the greatest potential as carbon sinks. This is so because the most important factor in realizing "natural" negative emissions is identified as the growth rate of the ecosystem under consideration. But this takes a long time: the cumulative climate protection effect of stands with continuous logging exceeds that of unused stands only after 120 years. The study is hopefully the prelude to a whole series of analyses that will also map Europe's biogeographical diversity.

So, while there is much to suggest that carefully managed forests in temperate climate zones can contribute significantly to climate restoration, there is a heated scientific-political debate about the tropics and subtropics. The "naturalist" position, which vehemently advocates the permanent non-use of reforested areas, is represented by environmental researchers such as Simon Lewis (Lewis et al., 2019). But the arguments against plantations do not consider the important option of biomass processing into durable products such as building materials. At the same time, this option must evidently be assessed differently in the Global South than, for example, in Europe: tropical hardwood cannot play the same value-added role as softwood from temperate climates due to its ecological properties and economic conditions, as must be discussed in more detail elsewhere.

By contrast, the latitudes near the equator offer a unique plant species that played a formative role in traditional architecture there and is likely to regain similarly prominent importance in future construction: bamboo. This is a family of sweet grasses (more than 12,000 species) whose culms are hollow on the inside and become increasingly stiff on the outside. The stability of the walls derives from the numerous fibers and the high silica content (e.g., Schönauer, 2021). The giant bamboo is the fastest growing plant in the world, gaining half a meter a day and eventually reaching a height of over 40 m. Interestingly, the culms push their final cross-section out of the ground right from the start; so there is no thickness growth as with woody plants. In summary, bamboo is undemanding, highly regenerative, light, and yet resistant to pressure and tension, thus in principle an ideal climate-positive material.

To assess bamboo's potential for modern sustainable construction on a global scale, the questions of geographical availability and further processing in line with demand must be answered with priority. First, it should be said that the species spectrum of this miracle grass covers an altitude range from 0 to 4000 m and a temperature range from −28 to +50 °C. Bamboo species are native to all continents except Europe and Antarctica; some of them can also be planted and cultivated almost anywhere in temperate latitudes. As far as its use as a building material is concerned, several innovative processes and tools are currently being developed that should finally bring bamboo out of its South Seas kitsch niche and directly into the dynamic new world of architecture.

This involves a wide range of aspects, such as the dynamic bonding of various tube geometries, the production of flat components with wood-panel-like properties, or the use of advanced digital methods in individual building design. An up-to-date overview is given, for example, in Chapter 5, "Bamboo Technology," in the aforementioned book by Pablo van der Lugt. In my view, we are witnessing the beginning of a development that could lead to flourishing mid-sized economies based on regional resources in countries such as Ecuador, Ghana, Indonesia, or Vietnam.

At the end of this section, to preempt two potential allegations, it is important to state two self-evident facts. First, intact primary forests absolutely must be protected, if necessary with appropriately designed trade agreements and new types of financial instruments. But, for the sake of the climate, secondary or degraded forests should also be protected from further human intervention and damage. In the tropics especially, such ecosystems have proven to be highly resilient. They can largely regain important forestry properties such as the original growth rate just 20 years after a massive disturbance—if nature is left in charge (Poorter et al., 2021).

Second, there are critical elements of the terrestrial biosphere that have stored enormous amounts of carbon but could lose this storage function in a practically irreversible way due to civilizational interventions (*Irrecoverable Carbon*: IC). A recent paper maps these ecosystems, explains their special importance, and presents remarkable figures: half of the IC is located on just 3.3% of the planet's land surface, 23% of the total is in protected areas, and 33.6% is currently used in a sustainable way by indigenous communities (Noon et al., 2021). This highlights the opportunities but also the risks that must be considered in preserving this important component of the Earth system. The further displacement of indigenous peoples from their traditional habitats alone could cause fatal reservoir losses.

Conclusion: The "reforestation of the world" for the benefit of our biosphere, climate, culture, and economy is possible in principle. But due to the multiple target functions and the heterogeneity of the fields of action, it is probably the most complex challenge in the history of our civilization. It is high time to jettison the simplistic and short-sighted strategies of the post-war era and steer a far-sighted, evidence-based course between brutal forestry and radical environmental protection.

The Forestry Construction Pump

We can now combine the insights and facts of the three previous sections into a "Technical Guide to Climate Restoration." The inverted commas are meant to signal self-irony in view of the fearsome dimension of the claim and to emphasize the provisional nature of the approach: the operationalization of the following basic considerations still requires enormous analytical, empirical, experimental, and social-science efforts. System theories, data collection, simulation models, demonstration projects, instrument discourses, etc. of unprecedented breadth and depth must be carried out on the topic in the coming years. "Horizon Europe," the EU's current research and innovation program, could play an essential role.

The concept itself, however, can be outlined: the built environment must be transformed from a monstrous CO_2 source into a powerful CO_2 sink. This requires (1) bio-based material (especially in the construction phase), (2) sustainable energy (mainly in the operational phase), and (3) cycle-oriented component use (mainly in the demolition phase). While the problem associated with point 2 could be quickly overcome with the brilliant triumph of renewables and dramatically improved energy system thinking, we are still in the early stages with points 1 and 3. However, if land use and settlements are consistently thought and designed together, we can be successful on these two fronts as well. In essence, it is about bringing supply and demand for climate-positive building materials into a sustainable balance at a significantly higher level. How the demand side could be greatly improved was outlined in Section "The Geological Journey of C"; how the supply side could be shaped in harmony with other sustainability goals, in Section "Forestry's Transition Toward Sustainability".

The concrete "TA Climate Restoration" is shown in the following graphic (Fig. 5):

In this illustration, the migration of carbon through different compartments of the Earth system is traced as part of an active surplus strategy. In the final analysis, this strategy transforms atmospheric CO_2 from the burning of fossil fuels into wooden buildings and infrastructure of high utility; in the best case, exhaust gases are transformed into kindergartens made of wood. A better "value proposition" will be hard to find in the sustainability market.

In natural equilibrium, there is a stable atmospheric CO_2 concentration maintained by several biogeochemical processes. In one of the most important of these processes, terrestrial vegetation through photosynthesis removes CO_2 from the atmosphere that is, however, largely returned to the atmosphere via metabolic (respiration, etc.) and transformative (decay, etc.) processes. A small proportion is stored in the soil as organic carbon; the deficit in the air is made up, for example, by CO_2 emissions from volcanoes.

Due to the enormous technical oxidation of fossil fuels, this equilibrium has been greatly disturbed for about 200 years. This has not only resulted in an increase in the global mean temperature, but has also substantially altered the productivity of the biosphere. The latter absorbs a good quarter of anthropogenic CO_2 emissions and

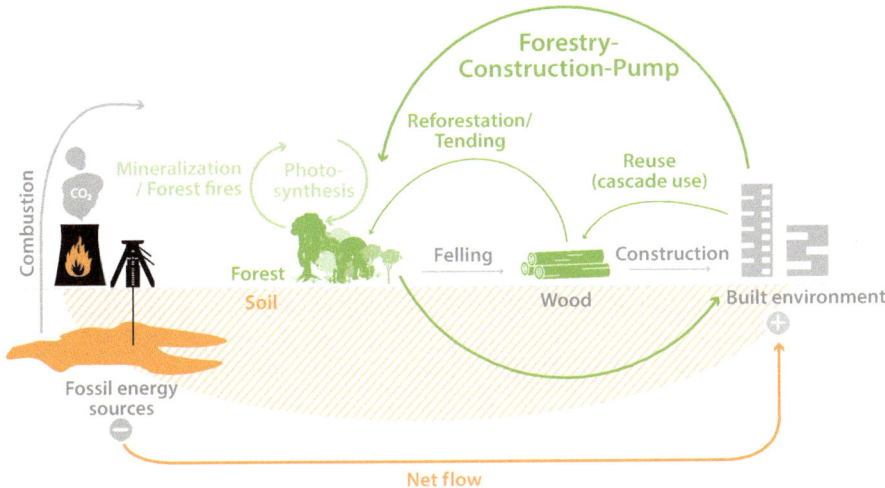

Fig. 5 The forestry construction pump. A cyclical carbon dynamic among terrestrial vegetation, atmosphere, soil, timber, and built environment (Schellnhuber & Köllner, unpublished)

thus tends to increase in mass. Through targeted forestry and construction measures, the atmospheric CO_2 surplus can now be siphoned off again and stored at least in the medium term (centuries to millennia). This is essentially done via two cycle-like processes: via the harvest-plant-tend-harvest cycle for living biomass and via the composition-decomposition-recomposition cycle for processed biomass. In both cases, these are not true closed circular processes, but the repetitive nature is clearly evident. In cycle 1, permanently stimulated growth through decoupling of mature plants is essential, in cycle 2, the most value-preserving and reversible is assembling of the construction components. Through the interaction of the various natural and civilizational processes, a kind of pump is created for the targeted cleaning of the atmosphere from unwanted pollution that was a by-product of the fossil-industrial economy. This organizes a (largely after-care) flow of carbon from the ancient deposits of the Carboniferous into present and future human settlements.

To preempt the undoubtedly numerous objections to this utopian-seeming but, if successful, existentially important project, I would like to state explicitly here that each individual element of the diagram requires strict scientific scrutiny, that the approach must be explored with great care geographically and explicitly for all continents and economic areas, that a dynamic view taking into account different time scales (e.g. how often per century can the "pump" be operated?) is essential and that the envisioned system can ultimately only be developed and planned with the help of quantitative simulation modelling. This also outlines a highly ambitious research agenda.

Finally, whether the political-economic preconditions for the realization of this project can be created for purposeful planetary design is almost impossible to predict. The special strength of this approach—apart from its reliance on largely nature-based solutions and practices—lies in its *technical and social scalability:*

similar to photovoltaics, anyone and everyone in almost any place can make an almost arbitrarily large contribution to the overall solution. On the supply side, this may be planting a single tree or preserving a primary rainforest the size of Germany. On the demand side, this may be the climate-positive addition of a two-family house or the construction of an entire new metropolis from bio-based building materials. Therefore, this humanitarian project can become a project of all humanity.

As a physicist, I am used to subjecting every task first to an elementary order-of-magnitude analysis which, in this case as well, provides interesting insights. Recently, the scientific journal *Nature* published an article by an Israeli research group that was both illuminating and shocking (Elhacham et al., 2020). It demonstrates in a simple and convincing way that mankind has become a force that is massively reshaping the Earth. Following Paul Crutzen, this development is often described with the term "Anthropocene," mostly referring to the change in energy flows on our planet and the resulting amplification of the natural greenhouse effect (Schellnhuber, 2021b). In fact, the interventions of our civilization in the material flows in the Earth system are rather more dramatic (e.g., Smil, 2013). The above-mentioned Nature publication explicitly shows that in 2020, for the first time, the "Anthropomass"—i.e., the total mass of man-made materials (concrete, bricks, asphalt, metals, plastics, etc.) on Earth—exceeds the sum of all living biomasses!

What quantities are we talking about? Well, if we really only want to compare sheer orders of magnitude, then the planetary living biomass before the beginning of human civilization in the Holocene weighed roughly 2 trillion tonnes. About half of this has already been destroyed by economic activities; in its place, an ensemble of artificial masses weighing about 1 trillion tonnes has been created.

Now we can do a lumberjack (rather than milkmaid) calculation: in the last ten millennia, humankind has increased the atmospheric CO_2 concentration from about 280 to about 420 ppm, which corresponds to a weight increase of about 1 trillion tonnes. In the long term, this mass must be removed from the atmosphere to complete the desired climate restoration. Disregarding all natural processes that are taking place anyway, in which, by the way, soils play a critical role, this would be realized mainly through reforestation and organic architecture for settlements and infrastructures. An average wooden house weighs about 100 tonnes and thus binds nearly 200 tonnes of CO_2. If about 2 billion houses were to be constructed from bio-based materials over the next centuries, then some 400 billion tonnes of atmospheric CO_2 would be extracted enduringly removed from the climate system. Another 100 billion tonnes could be stored in infrastructure, so that about half of the restoration rent (500 billion tonnes of cumulative negative CO_2 emissions) would be raised via the built environment. The other half would have to be taken care of by an intelligent sustainable forest strategy (see also Section "Forestry's Transition Toward Sustainability").

The bottom line is that the anthroposphere would also undergo a fundamental material transformation in addition to the energetic one: *The corresponding demineralization processes would largely replace today's trillion tonnes of artificial masses with harvested or living biomass and thus successfully complete a renaturation on a planetary scale.* Whether such a breathtaking plan can really succeed is another matter. But the path and goal have now been relatively clearly outlined. And, as I explained above, implementation can begin at any time with local contributions. Even "realpolitikers" and cynics should find it difficult to refuse here.

The New Bauhaus Movement

The transformation of the built environment is a decisive prerequisite for correcting a climate development that seriously endangers the foundations of human life. But today there are also myriad other reasons to look at the housing system as a whole and to rethink it thoroughly.

For "what if CO_2 did not influence the planetary radiation balance?"—as some confused minds stubbornly claim? Would we then—freed from climate fears—find our existing built environment socially and aesthetically appealing? Must we really surround ourselves with dysfunctional ugliness that literally cements social differences?

Structurally, built modernity has failed in the concentric logic of megacities, which in daily, weekly, and seasonal shuttle movements suck in enormous quantities of living beings, goods, energies, and information, digesting and then discharging them back "into the hinterland." This principle of operation has a multitude of fatal consequences, of which I mention just a few:

- There are the sprawling scattered settlements on the periphery of the metropolises that have no identity beyond their geometric relationship to the center. The country house kitsch runs riot there in detached houses on tiny plots. You can't push social atomization much further, but the poisonous current "home ownership debate" shows where the vanishing point of middle-class longing still lies many decades after the destruction of World War II.
- There is the cultural emptying of the so-called provinces, where the meeting places in the villages and small and medium-sized towns are becoming deserted, the traditional places of *joie de vivre* are falling into disrepair and even the most elementary supply facilities are disappearing in the direction of the upper center. If young people must turn the town's only bus stop into an inhospitable substitute pub, we need not be surprised if, at the next election, resentment and envy of the urban elites steer the pen to the right.
- Then there are the 1–2 billion people (no one knows exactly how many) who live in informal settlements—mostly in slums on the outskirts of big cities, but also under motorway bridges, in disused docks, even on smoldering rubbish tips. Often only a few kilometers away from the glamorous city centers where, for example, the hip penthouse with a view of New York's Central Park or the mostly empty luxury apartment in one of the "Dark Towers of London costs tens of millions of dollars or pounds (Schellnhuber, 2021a)."

I quote here from the original manuscript of my essay published in the FAZ in 2021, in which I argue that building culture, urban development, and spatial planning have been in the wrong movie for a long time—at least not one set in the twenty-first century. In the right movie, we are at a turning point reminiscent of the great social ruptures that caused the traditional social orders to disintegrate at the end of the First World War. From the ruins, something new quickly emerged, not least the *Staatliches Bauhaus* (State Bauhaus), which Walter Gropius founded as an art school in Weimar in 1919. Through its concept, aspirations, and impact this school revolutionized the twentieth century world of architecture.

"The *Bauhaus* was decidedly contemporary at the time because it chose a holistic approach that brought together all trades from carpenters to painters. It studied and enthusiastically employed the possibilities of its technical epoch (such as modular construction) and saw itself as a social progress project that wanted finally to create decent living space for the masses 'below' the elites. Organic materials were definitely part of the school's fabric canon, but the actual idea of sustainability was understandably as alien to it as the dark side of the moon.

But what if Gropius and his comrades-in-arms (from the architect Mies van der Rohe to the graphic artist Feininger) were suddenly to find themselves in today's world and were to direct their programmatic and creative energies to its dramatically transformed challenges? They would certainly place the planetary ecological crisis at the center of their work (ibid)."

Since these historical personalities are unlikely to be reborn, people of the present must take appropriate action. In this context, the Association of German Architects (BDA) has been playing a lead role in Germany for quite some time, especially with its magnificent manifesto on climate-friendly housing development (BDA, 2019). For some years now, I have been exchanging ideas with leading BDA members.

Independently of this, however, I had long entertained the idea of reviving the basic approach of the Weimar Bauhaus—*striving for a holistic building culture mindful of social responsibility*—in view of the climate emergency. The anniversary year 2019, 100 years after the Bauhaus founding, was an appropriate time to take action.

I succeeded in bringing the *Bauhaus Earth* initiative circle to life, which includes some remarkable personalities and has met for the first time in December 2019 in Caputh near Potsdam, close to Einstein's summer house. Among other things, this resulted in the "Caputh Declaration," which calls for an early transformation of the built environment within planetary boundaries and consciously embraces the holistic approach of the historic Bauhaus. Quote: "If civilization is not merely to survive, but to develop in diversity and solidarity, we must take a new holistic view of the built environment."

But the declaration also clearly expresses that the Bauhaus approach must be transported into the twenty-first century and given a new direction that corresponds to the requirements and possibilities of our time: "*A main goal of all building culture must be the good life of people in harmony with nature.* The guiding concept of the sustainable modernity that is now to be created through architecture, art, design, manufacture, infra-structure, urban development, landscape design and spatial planning could therefore be the organic (or the natural)—in contrast to the mechanical which was the lodestar of the declining industrial modernity."

I will expand on the last sentence in the next and final section. It is worth reporting here that *Bauhaus Earth* is developing excellently, taking on its first institutional forms and setting in motion interesting projects. The initiative is now spreading far and wide. First and foremost in this context is the *New European Bauhaus* (NEB), a major European project that has enormous transformative potential. The NEB was launched by Ursula von der Leyen, President of the European Commission, in her *State of the Union Speech* of 16 September 2020. According to the NEB website, she said: "The new *European Bauhaus* is a creative interdisciplinary initiative to create a meeting place where future ways of living are shaped at the intersection of art, culture, social inclusion, science, and technology. It brings the Green Deal to the center of our lives and is a call to develop and realize together ideas of a sustainable, inclusive, intellectually and emotionally engaging future."

This is necessarily "Commission jargon," which strings together as many positively connoted words as possible to underpin a certain program. Also in my capacity as official advisor to the NEB, I should therefore emphasize the most important terms as they ultimately give direction to the whole: this is about the future of building/the building of the future, in which *sustainability, participation, and beauty* are more important standards than profit, competition, and prestige. This can be dismissed as a noble but unrealistic claim. However, the debates on the NEB that are already being organized across Europe suggest that our civilization is finally ready to break out of the post-war global paradigm to seriously discuss better narratives of modernity.

In order not to go beyond the scope of this article, I would like to highlight only one aspect here. Can there be such a thing as a "mass movement towards beauty"? Is aesthetics and especially building culture not the privilege of the wealthy or educated elites, while the lower strata of society are (must be) politically fobbed off and mentally anaesthetized with a "class movement towards kitsch"? These questions touch on an explosive socio-cultural problematic and must be discussed or answered only with great seriousness.

But I do want to bring one thought into play here: well-preserved small historic towns and villages in Europe can often boast a remarkably valuable functional aesthetic, while in the contemporary provinces, architectural horror usually dwells. The latter settlements have lost almost all connection to the (topographical, climatic, resource-economic, regional-political, art-historical, etc.) conditions of their location, while the former have grown out of a long collective confrontation with these very conditions. Umbrian hill towns such as Spello or Spoleto, for example, are supremely beautiful architectural implementations of the principle of defense against raids by Saracens or Vikings. The "people" can build if they are given sufficient time and resources to do so. In the exhausting battles over reconstruction after the Second World War, the concepts and practices of collective-evolutionary settlement development were unfortunately largely lost. At best, the eccentric constructs of the "star architects" who serve the whims of a wafer-thin global upper class stand out from the desolation of today's commercial and residential ensembles.

Hi-Tech Meets No-Tech: Entering the Cyborganic Age

To conclude this chapter, let me venture a perspective that looks beyond the built environment and attempts to glimpse what might constitute the essence of a post-industrial civilization. I believe this essence will be decisively shaped by the *reconciliation of humans with nature on a cultural plateau that can still be reached in this century*. The essay justifying this in more detail has yet to be written, but I would like to place a few signposts here already.

I quote once again from the FAZ manuscript repeatedly mentioned: We must finally muster the will to "comprehend the exuberant inventiveness of evolution. This requires courageous entry into the comprehensive school of nature. Unfortunately, we have set fires to their most precious libraries (the Amazon, the Congo, Borneo, and Sumatra), and the first thing to do is to put them out. If we are then willing to learn, both the gain in knowledge and the practical benefits will be enormous. Construction and operation in the twentieth century were mainly oriented towards mechanics and thermodynamics, toward the directly calculable, under the dictate of optimization that maintains its stranglehold on economic thinking even today. The built environment in the twenty-first century should instead take inspiration from elements in the Earth's ecosystem that grow, bend, circulate, flourish late or by accident.

Some may regard this as an excessive romanticism about nature. But after some decades of scientific study of non-linear processes and complex systems, I must conclude that the evolutionary creation of a rose bush far outshines the engineering design of a powerful diesel engine. Incidentally, the engineer's brain is itself an outgrowth of nature: for example, millions of years of environmentally controlled selection processes have led to the mounds of certain termite species in South Asia being so cleverly designed in terms of shape and material that the structures achieve optimal ventilation in the diurnal cycle of the outside temperature (King et al., 2015).

The living world is full of such success stories, but they are usually discussed only in the exclusive circles of bionics experts. For the disciples of modernity, progress is still synonymous with detachment from the nature that created us. In *Bauhaus Earth*, this matricidal obsession can be ended by combining high-tech with no-tech to create innovations the likes of which the building industry has never seen (Schellnhuber, 2021a)."

In this context, Marc Weissgerber and I introduced—in a contribution to the catalogue of the exhibition "urbainable—stadthaltig: positions on the European city for the twenty-first century" (Schellnhuber & Weissgerber, 2020)—the concept of cyborganics in settlement design. This term anticipates a possible new architectural epoch oriented towards the following principles:

Cybernetic principle. Buildings, empowered by appropriate artificial intelligence (AI) tools, are self-regulating and learning systems that adapt optimally to their respective environmental conditions and the dynamic needs of their users. "Cybernetic" is essentially understood here as "digitally controlled."

Organic principle. Buildings are developed like ecosystems, using climate-friendly materials, employing and offering important ecosystem services. This principle could be boldly implemented by, for example, creating "living" houses that integrate vital trees or other plants. The constructions would then mature over time and could also be rejuvenated.

This may sound like hyperbolic sci-fi babble, but just a few decades ago even gene therapy would have been dismissed as hype. And really no one would have dreamed that a microbiological method with the cryptic abbreviation CRISPR would play a heroic role. The scientific age has only just begun ...

This is especially true for the construction industry where, for half a century, barely any innovations have been stimulated or productivity increases realized. The industry is overripe for transformation. Interestingly, this is already advancing from the margins into the established center, where the least imagination still promises the greatest profit. This may change soon as many small and medium-sized companies, particularly ones driving wooden architecture, are now beginning to conquer the markets—initially in Central Europe and the Far East, but surely soon in other parts of the world as well. These companies perfectly embody the hi-tech-meets-no-tech paradigm with their value creation: they work with renewable raw materials but use customized AI methods for this purpose, i.e., the most advanced cognitive tools in human history to date. *Forward to nature!*

References

Bastin, et al. (2019). The global tree restoration potential. *Science, 365*(6448), 76–79.

BDA. (2019). Das Haus der Erde. Positionen für eine klimagerechte Architektur in Stadt und Land. Berlin. Bund Deutscher Architektinnen und Architekten, www.bda-bund.de/wp-content/uploads/2020/06/2020_BDA_DasHausDerErde_Monitor.pdf

Blaustein, R. (2018). Turning desert to fertile farmland on the Loess Plateau, Rethink; https://rethink.earth/turning-desert-to-fertile-farmland-onthe-loess-plateau/

Churkina, G., et al. (2020). Buildings as a global carbon sink. *Nature Sustainability, 3*(4), 269–276.

Davis, S. J., et al. (2018). Net-zero emissions energy systems. *Science, 360*(6396), eaas9793.

Elhacham, E., et al. (2020). Global humanmade mass exceeds all living biomass. *Nature, 588*(7838), 442–444.

Enerdata. (2023). Accessed 11/07/2023. https://www.enerdata.net/publications/daily-energy-news/germanys-power-consumption-falls-2022-generation-renewables-rises.html

Feulner, G. (2017). Formation of most of our coal brought earth close to global glaciation. *Proceedings of the National Academy of Sciences, 114*(43), 11333–11337.

Forster, et al. (2021). Commercial afforestation can deliver effective climate change mitigation under multiple decarbonization pathways. *Nature Communications, 12*(1), 3831.

Fraunhofer Institute for Wood Research WKI. www.wki.fraunhofer.de/de/fachbereiche/hofzet/profil/publikationen.html

IPCC. (2014). *Climate change 2014: Synthesis report*. Contribution of Working Groups I, II and III to the Fifth Assessment Report of the Intergovernmental Panel on Climate Change. Genf/Schweiz.

IPCC. (2018). *Global warming of 1.5°: An IPCC special report*. Genf/Schweiz.

King, H., Ocko, S., & Mahadevan, L. (2015). Termite mounds harness diurnal temperature oscillations for ventilation. *Proceedings of the National Academy of Sciences, 112*(37), 11589–11593.

Lawrence, M. G., et al. (2018). Evaluating climate geoengineering proposals in the context of the Paris Agreement temperature goals. *Nature Communications, 9*(1), 3734.

Lenton, T., & Watson, A. (2011). *Revolutions that made the earth*. Oxford University Press, Oxford, UK.

Lewis, S. L., et al. (2019). Restoring natural forests is the best way to remove atmospheric carbon. *Nature, 568*(7750), 25–28.

Mishra, A., Humpenöder, F., Churkina, G., et al. (2022). Supplementary figure 15; Land use change and carbon emissions of a transformation to timber cities. *Nature Communications, 13*, 4889. https://doi.org/10.1038/s41467-022-32244-w

Noon, M. L., et al. (2021). Mapping the irrecoverable carbon in Earth's ecosystems. *Nature Sustainability*. https://doi.org/10.1038/s41893-021-00803-6

Petoukhov, V., et al. (2013). Quasiresonant amplification of planetary waves and recent Northern Hemisphere weather extremes. *Proceedings of the National Academy of Sciences, 110*(14), 5336–5341.

Poorter, L., et al. (2021). Multidimensional tropical forest recovery. *Science, 374*(6573), 1370–1376.

Richter, R. (2021). *So kann's gehen: Holzbau*. Die Zeit.

Rockström, J., et al. (2017). A roadmap for rapid decarbonization. *Science, 355*(6331), 1269–1271.

Rose, S. (2020). ›Our biggest challenge? Lack of imagination‹: The scientists turning the desert green. The Guardian, www.theguardian.com/environment/2021/mar/20/our-biggestchallenge-lack-of-imagination-thescientists-turning-the-desert-green; The Guardian (2021): ›China floods death toll rises to 302 with 50 people still missing‹, www.theguardian.com/world/2021/aug/02/china-floods-death-toll-rises-people-stillmissing-henan-province.

Schellnhuber, H. J. (2015). Selbstverbrennung:Die fatale Dreiecksbeziehung zwischen Klima, Mensch und Kohlenstoff. München.

Schellnhuber, H. J. (2021a). Bauhaus für die Erde. Frankfurter Allgemeine Zeitung, www.faz.net/aktuell/feuilleton/debatten/vorschlag-zur-rettung-der-weltschellnhuber-ueber-holzbau-17305173.html

Schellnhuber, H. J. (2021b). Paul Josef Crutzen: Ingeniousness and innocence. *Proceedings of the National Academy of Sciences, 118*(17), e2104891118.

Schellnhuber, H. J., & Weissgerber, M. (2020). Bauen im Anthropozän. In T. Rieniets, M. Sauerbruch, J. Walter, & (Hrsg.) (Eds.), *urbainable/stadthaltig: Positionen zur europäischen Stadt für das 21.* Jahrhundert.

Schickhofer, G. (2013). Starrer und nachgiebiger Verbund bei geschichteten, flächenhaften Holzstruk-turen. Monographic Series TU Graz, www.tugraz-verlag.at/gesamtverzeichnis/bauingenieurwissenschaften/starrer-und-nachgiebigerverbund-bei-geschichteten-flaechenhaftenholzstrukturen-ebook/

Schmid, J., et al. (2018). Simulation of the fire resistance of Cross-laminated Timber (CLT). *Fire Technology, 54*(5), 1113.

Schönauer, M. (2021). *Bambus:Das Supergras*. Die Zeit.

Science for Environment Policy. (2021). *European Forests for biodiversity, climate change mitigation and adaptation* (Future Brief No. 25). Bristol/UK, https://ec.europa.eu/environment/integration/research/newsalert/pdf/issue-25-2021-11-europeanforests-for-biodiversity-climate-changemitigation-and-adaptation.pdf

Smil, V. (2013). *Making the modern world: Materials and dematerialization*. Wiley.

Steffen, W., et al. (2018). Trajectories of the earth system in the Anthropocene. *Proceedings of the National Academy of Sciences, 115*(33), 8252–8259.

Tegel Projekt GmbH. (2020). Bauhütte 4.0. In Berlin TXL soll ein Prototyp für nachhaltige Stadtentwicklung entstehen, www.tegelprojekt.de/pressematerial/detail/bauhuette-40-in-berlin-txl-soll-ein-prototypfuer-nachhaltige-stadtentwicklungentstehen.html

Tegel Projekt GmbH. (2021). Berlin TXL—Schumacher Quartier, www.tegelprojekt.de/fileadmin/10.0_Presse/Basistexte_Facts_Figures/2021-08-05_Basistext_SQ.pdf

The Concrete Initative. (2021). A New European Bauhaus: The Concrete Initative Manifesto.

Thoma, E. (2016). *Holzwunder: Die Rückkehr der Bäume in unser Leben*. Elsbethen /Österreich.

van der Lugt, P. (2020). *Tomorrow's Timber: Towards the next building revolution* (Detrix Edition); van der Lugt, P. (2017): *Booming Bamboo: The (re)discovery of a sustainable material with endless possibilities* (Mosco Edition).

von Carlowitz, H. C. (2013/22). *Sylvicultura oeconomica*. München.

von Schuckmann, K., et al. (2020). Heat stored in the earth system: Where does the energy go? *Earth System Science Data, 12*(3), 2013–2041.

von Winter, S. (2013) Brandschutz im Holzbau. In: Cheret, P.; K. Schwaner; A. Seidel (Hrsg.): Urbaner Holzbau: Chancen und Potenziale für die Stadt. Handbuch und Planungshilfe, , https://informationsdienst-holz.de/urbaner-holzbau/kapitel-4-der-zeitgenoessische-holzbau/brandschutzim-holzbau

Warszawski, L., et al. (2021). All options, not silver bullets, needed to limit global warming to 1.5 °C: A scenario appraisal. *Environmental Research Letters, 16*(6), 064037.

WBGU. (2016). Der Umzug der Menschheit: Die transformative Kraft der Städte, https://www.wbgu.de/de/publikationen/publikation/der-umzug-der-menschheitdie-transformative-kraft-der-staedte#sektion-downloads

Peatland Must Be Wet

Advance Rewetting, Stop Peat Extraction

Hans Joosten

Peatlands are in everyone's mouth these days. This is quite literally true, because—as few people know—nowadays nearly all fruits and vegetables ending up on our plates have at some point grown on peat, as have most of the ornamental plants or shrubs we plant in our gardens. Frequently mentioned are the so-called ecosystem services of peatlands, especially their potential to mitigate climate change. This chapter presents an overview of the relevant facts, challenges, and solutions.

What Peatlands Are and What Makes Them So Special

Intact (natural, living, growing) peatlands are ecosystems without a closed cycling of matter. Due to permanent water saturation of the substrate, the remains of dead plants decompose more slowly than new plant material is produced. Over time these remains—in various stages of decomposition—accumulate as thick layers of carbon-rich organic material, which we call "peat". Worldwide, peatlands—by which we conventionally understand areas covered with at least 30 cm of peat—cover about 4 million km^2 or 3% of the Earth's land area, almost as much as the European Union. Because of the enormous carbon density of peat, a threshold value of 10 cm (instead of 30) would be better from a climate perspective and would substantially "increase" the global peatland area.

H. Joosten (✉)
Greifswald Mire Centre, Greifswald, Germany
e-mail: joosten@uni-greifswald.de

© The Author(s) 2024
K. Wiegandt (ed.), *3 Degrees More*,
https://doi.org/10.1007/978-3-031-58144-1_9

179

Fig. 1 Known occurrences of peatlands, as of 2021. In the red-brown colored areas, peatlands occupy more than 50% of the mapping unit, in the light-brown colored areas 20–50% (Greifswald Mire Centre, 2022)

Peatlands exist in 169 of the UN's 193 member states; on the European continent they cover an area of about 600,000 km², 12,800 of these in Germany (Tanneberger et al., 2017a, b). Peatlands are predominant in three humid climate zones: in the northern subarctic/boreal zones, in the equatorial tropics, and in the subantarctic regions (Fig. 1, Joosten, 2016). In the latter zone, at the southern end of the inhabited world, peatlands are insignificant in terms of area because there is hardly any land at those latitudes. Relatively speaking, however, they loom large there; Tierra del Fuego, the Falkland Islands, Tasmania, and the South Island of New Zealand all have substantial peatland cover. As peatlands exist all over the world, it is inevitable that, despite all their commonalities, they differ widely in their biodiversity.

In the long term, growing peatlands cool the climate because they act as CO_2 sinks. But their direct, short-term importance should not be overestimated. On the one hand, the net carbon sequestration due to peat formation in the still growing peatlands (about 100 million metric tons of carbon per year) compensates for only 1% of anthropogenic CO_2 emissions. On the other hand, living peatlands also produce and emit methane (CH_4) due to the prevailing oxygen-free conditions (which enable peat formation and preservation in the first place). Worldwide CH_4 emissions from peatlands are "only" 30 million metric tons per year; but because methane is a much more potent greenhouse gas (GHG), the positive climate effect of CO_2 sequestration is arithmetically more than offset by the methane emissions. Seen from a short term perspective, living peatlands are therefore, on the whole, no help against climate change. In the longer term, however, they have a cooling effect because methane oxidizes quickly in the atmosphere, thus losing its strong climate effect. As a result, the permanent methane production of living peatlands does not lead to a permanently increasing concentration of methane in the atmosphere, while the ongoing sink effect continuously reduces the CO_2 concentration in the atmosphere.

In this way, peatlands have cooled the global climate over the last 10,000 years by about 0.6 °C (Frolking & Roulet, 2007; Joosten et al., 2016; Günther et al., 2020).

Typical Features of Peatlands
- high content of organic matter and carbon in the soil;
- permanent water saturation and a slow but continuous rise of the ground-water level as well as the surface elevation;
- relative lack of nutrients and often high acidity;
- a cooler and more humid mesoclimate compared to the surrounding area;
- occurrence of harmful organic substances, toxic, chemically reduced elements, and black water.

All these factors shape the habitats of the often very specialized biota typical of peatlands.

Furthermore, peatlands are characterized by their capacity for:

- long-term carbon sequestration and storage;
- water purification and retention and regulation of run-off;
- accumulation and preservation of palaeoecological information and archaeological artefacts.

The sophisticated interplay of plants, peat, and water enables the long-term development of self-regulation and self-organization, making peatlands resistant, long-lasting ecosystems with often fascinating surface structures and unique ecosystem biodiversity (Convention on Wetlands, 2021a; Couwenberg et al., 2022).

Of much greater, direct climatic importance is the role of peatlands in storing carbon—i.e. as possible sources of CO_2. Compared to other ecosystems, peatlands contain a disproportionately large amount of carbon: in the boreal zone on average seven times as much per unit area, in the tropics ten times as much as ecosystems growing on mineral soils. And although they cover only 3% of the world's land, peatlands store 600 billion metric tons of carbon, almost twice as much as the biomass of all the world's forests, which account for about one third of the land area (see also Chapter "Humus Enrichment of Soils") (Joosten & Couwenberg, 2008; Joosten et al., 2016; Kirpotin et al., 2021; Temmink et al., 2022).

The importance of peatlands in terms of their biodiversity and ecosystem services has long been overlooked. Richard Lindsay called this at the 6th Conference of the Parties to the Ramsar Convention on Wetlands in Brisbane in 1996 "the Cinderella Syndrome" (Lindsay, 1996). It took until 2002 for the Convention to recognize the importance of peatlands and to explicitly call for their protection and restoration (Barthelmes et al., 2015). The Convention on Biological Diversity (CBD) has likewise thus far barely taken notice of peatland biodiversity, even though, due to their special peat-forming properties, peatlands are home to many species that are found nowhere else and thereby contribute disproportionately to regional biodiversity. Furthermore, as already mentioned, peatlands possess an eco-system biodiversity—largely independent of genetic diversity—with a great rich-ness of peatland forms and surface patterns (Joosten et al., 2017; UNEP, 2022). Instead of appreciating this characteristic, international decision-making processes often merge peatlands with other wetlands that lack a peat layer and have com-pletely different functional characteristics (Minayeva et al., 2017).

For a long time, the Climate Convention (UNFCCC) treated peatlands in a simi-lar manner as the Ramsar and Biodiversity Conventions. It was only in 2012 that it recognized the importance of peatlands for climate protection and included the rewetting of peatlands as an eligible measure in the Kyoto Protocol. At the same time, within the framework of the REDD+ mechanism (Reducing Emissions from Deforestation and Forest Degradation), the peat soils of the swamp forests were recognized as an inseparable part of "forest carbon" (Joosten, 2011).

The Condition of Peatlands—Globally, in Europe, and in Germany

In relative terms, the world's peatlands seem to be doing not so badly: Their area is currently larger than it has been in most of the last 100,000 years. The global loss of intact peatlands amounts to about 15%, while a third of all former forest areas on earth have disappeared (Ritchie, 2021). Globally, peatland area is decreasing by 0.1% per year versus 0.3% for tropical primary forests; the global peat volume is also being reduced by 0.1% annually versus 2% for crude oil reserves (Joosten & Clarke, 2002; World Resources Institute, 2022; Worldometer, 2023).

The global peatland problem has two aspects. (1) Peatland losses are concen-trated in regions where the climate is suitable for agriculture, especially arable farming. As populations keep growing, and warming increases, the pressures to exploit these peatlands will continue to rise. (2) Peatland degradation immediately leads to enormous impacts on the climate because very large amounts of rapidly releasable carbon are present in small areas.

The State of the Peatlands

According to the most recent figures, as mentioned already, about 4 million km^2 of peatlands are currently inventoried. But unfortunately, the inventory is inadequate; so the figures in circulation (including ours) are not as precise as they appear and should rather be regarded as suggesting an order of magnitude. In fact, so little is known about some areas that huge "new" peatlands are still being discovered, in large parts of Africa, for example, and in South and Central America (Kirpotin et al.,

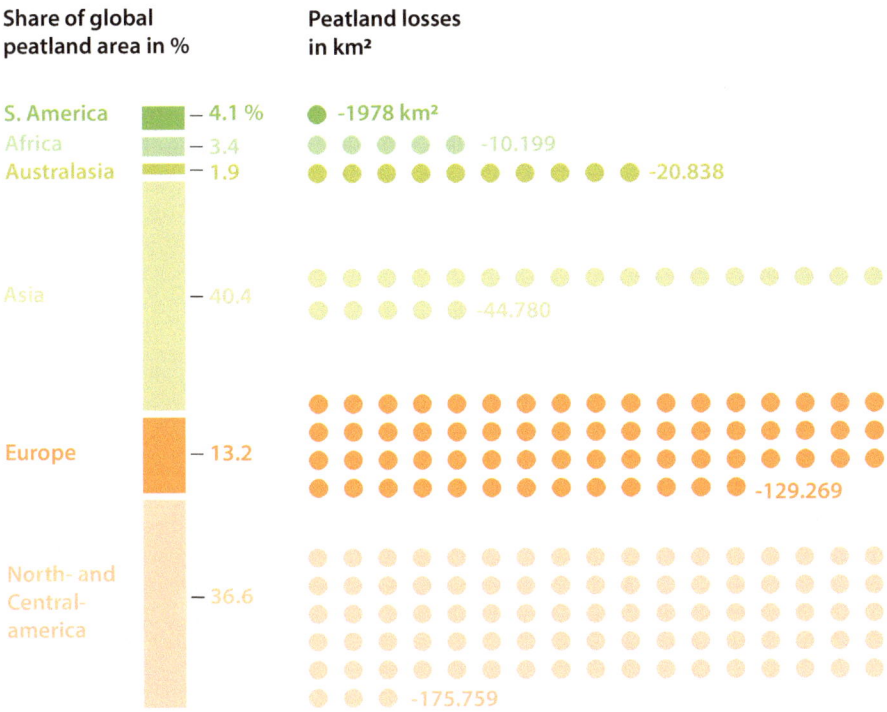

Fig. 2 Current global proportional distribution of peatlands (with over 30 cm of peat) and peatland losses per continent (in km²) relative to the maximum extent reached during the Holocene. (After Joosten, 2009 and Joosten, unpublished)

2021; Dargie et al., 2017; Draper et al., 2014; Elshehawi et al., 2019; Peters & Tegetmeyer, 2019).

Worldwide, about 85% of today's peatlands are (still) in a largely natural state; in parts of Canada, Alaska, and Siberia, they can still be found over very large areas (UNEP, 2022). However, over an area of 500,000 km², peatlands have been disturbed to such an extent that peat no longer forms; on the contrary, the organic matter that accumulated over thousands of years is gradually disappearing (Fig. 2). In short, pristine peatlands are concentrated in the (sub)arctic and boreal zones, while drained and degraded peatlands are found in the temperate and increasingly also in the (sub)tropical climate zones.

Europe's long cultural history, high population density, and climatic suitability for agriculture have made it the continent with the globally largest peatland losses. So much peat has disappeared from 20% of the original peatland area that they no longer qualify as peatlands (Fig. 2), more than half no longer accumulates peat, outside Russia even two thirds of the peatlands are "dead", in many European countries more than 90%.

Every year, a further 5000 km^2 (about 0.1%) of natural peatlands are destroyed worldwide by human activities. This is a fast rate of destruction, considering that over the last 10,000 years peatlands have expanded ten times more slowly. The main causes of peatland loss were and are drainage for agriculture and forestry, peat extraction (formerly for energy production, nowadays mainly for horticultural substrates), and the expansion of built infrastructure, including reservoirs, or urbanization (Joosten, 2016).

Climate Impact of Drained Peatlands

When peatlands are drained, penetration of oxygen into the peat soil leads to a steady depletion of the peat body. The accumulated organic matter oxidizes and disappears into the atmosphere in the form of the GHGs carbon dioxide (CO_2) and nitrous oxide (N_2O). These emissions correlate strongly with the average level of the groundwater table in the peatlands. In Central Europe, each lowering by 10 cm leads to additional emissions of 5 metric tons of CO_2e per hectare and year; in the tropics the figure can be as high as 9 metric tons (Jurasinski et al., 2016; Hooijer et al., 2006; Couwenberg et al. 2010; Carlson et al., 2015). The concrete consequences of this are, exemplarily, as follows:

An arable field on drained peatland soil in Germany emits 37 metric tons of CO_2e per hectare per year—the same amount of GHGs that a medium-sized car with a gasoline engine releases when it is driven 185,000 km per year (over four times around the Earth) (Joosten, 2017).

Effusively advertised peatland potatoes, carrots, and maize should thus be labelled and treated as fossil, not as renewable raw materials, crops, and fuels because more carbon disappears from the peat soil where they grow than they fix in their biomass.

Each liter of milk from a cow fed mainly from peatland grass- or farmland is almost as CO_2-intensive as burning 2 L of gasoline; the CO_2 footprint of 1 kg of peatland cheese is 45 kg of CO_2e, that of a kilogram of peatland butter almost 100 kg of CO_2e (Chemnitz & Becheva, 2021).

"Bioenergy" produced from plants grown on drained peatlands (such as maize, the Chinese reed Miscanthus, sugar cane, palm oil, or wood) releases more fossil carbon per unit of energy produced than fossil fuels do. "Biogas" from "peatland maize" is eight times more harmful to the climate than burning lignite (Couwenberg, 2007).

One liter of edible palm oil produced on peatland emits 15 kg of CO_2 (Hiraishi et al., 2014; Couwenberg & Hooijer, 2013; Schleicher et al., 2019), six times as much as burning 1 L of gasoline. With the CO_2 that one hectare of palm oil plantation on peatland produces each year, one could fly 50 times per year, i.e. nearly once a week, from Berlin to Jakarta and back, economy class (Hiraischi et al., 2014).

The Climate Impact of Degraded Peatlands

While natural peatlands have been cooling the climate for over 10,000 years, drained and degraded peatlands are significant sources of GHGs and thus contribute to global warming. These GHGs are produced mainly by the microbial oxidation of organic matter once air enters the formerly water-saturated peat.

The drier conditions also increase the risk of fires. In addition to GHG emissions, smoldering peat fires cause haze that can extend far beyond its region of origin and be life-threatening. As a result of the 2015 peat fires in Indonesia, over 100,000 people have died, half a million people had to be hospitalized, and economic damage ran into tens of billions of euros (Koplitz et al., 2016; Marlier et al., 2019; Kiely et al. 2021). Emissions from peatland drainage, degradation, and fires currently amount to over 2 billion metric tons of CO_2e annually and thus for almost 5% of all anthropogenic GHG emissions.

Without countermeasures, emissions from drained peatlands might consume between 12 and 41% of the GHG emissions budget we have left to keep global warming below 2 or 1.5 degrees by 2100 (Leifeld et al., 2019). Another projection shows that the entire land area of the Earth would by 2100 just be a net carbon sink if all currently intact peatlands are preserved and at least 60% of degraded peatlands are rewetted (Humpenöder et al., 2020). This means that the entire terrestrial carbon sink capacity would then be required to compensate for the carbon losses from the remaining 40% of degraded peatlands and would contribute nothing to the net carbon sink capacity required to meet the Paris climate targets (Convention on Wetlands, 2021a). For peatlands to be carbon neutral globally, 80% to 85% of drained peatlands need to be rewetted. Only if we rewet all drained peatlands worldwide, peatlands can once again fully serve their natural function as a global carbon sink (Convention on Wetlands, 2021b).

The German peatlands that have been drained for agriculture, forestry, or peat farming emit 53 million metric tons of GHGs annually (BMU, 2021), nearly 40% more than the Bełchatów lignite-fired power plant in Poland, which is considered the most climate-damaging thermal power plant in the world. The vast majority (83%) of these emissions come from agricultural land. The German Federal Environmental Agency currently assesses the damage caused by emitting one metric ton of CO_2 at 195 euros (Umweltbundesamt, 2020). This means that drained peat soils in Germany cause climate damage of 10 billion euros annually—a sum almost as high as the net value added of the entire German agricultural economy (Statista, 2023). These figures make abundantly clear that drained peatlands cause economic damage that far exceeds the value of the agricultural and forestry products produced on them.

In 2020, drained peatlands worldwide emitted 1.56 billion metric tons of CO_2e (Fig. 3). Just under 1.4 billion metric tons were due to CO_2 from microbial peat oxidation and dissolved organic carbon (DOC) discharge; the rest came from the

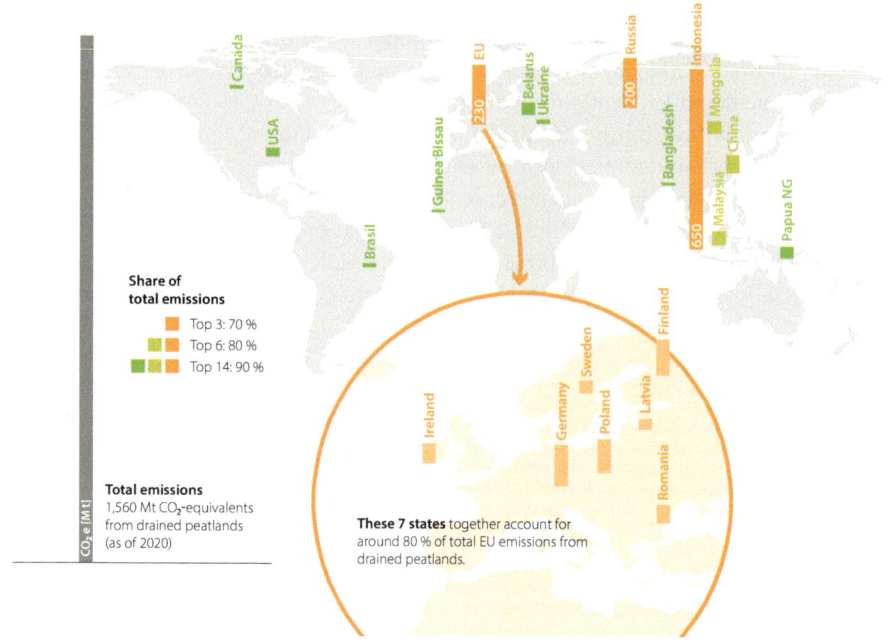

Fig. 3 GHG emissions (in CO₂e as of 2020) from degraded peatlands of the members of the United Nations Framework Convention on Climate Change (UNFCCC), excluding peatland fires (using IPCC default values (Hiraischi et al., 2014; Greifswald Mire Centre, 2022)

2.14 million metric tons of methane (from drainage ditches) and the 440,000 metric tons of N_2O (also from peat oxidation) (Greifswald Mire Centre, 2022).

In addition, there are, varying from year to year (and difficult to quantify), GHG emissions from open and smoldering peat fires in the order of 0.5 to 1 billion metric tons of CO_2e per year (Joosten, 2009; Rossi et al., 2016).

This means that 0.4% of the Earth's land area accounts for nearly 5% of all global anthropogenic GHG emissions (cf. Ritchie et al., 2020). Major emission sources are Indonesia, the European Union, and Russia, whereas half of the EU share is taken up by Germany, Finland, and Poland (Fig. 3).

Other (Societal) Harms from Peatland Drainage

Beyond climate damage, there are many other problems associated with peatland drainage (Joosten, 2017). A lower water level leads directly to a reduction in evapo-transpiration cooling of the landscape and to a loss of typical peatland biodiversity. The nitrogen mineralized by peat oxidation leads to the flushing out of nitrate and subsequently to the eutrophication of water bodies, including the oceans. Furthermore, because peat is primarily composed of water, drainage causes compaction of the peat body. This changes the peat's hydraulic properties, reducing the peatland's capacity to store water and to regulate runoff (Zeitz, 2016).

Mephistopheles Would Have Known Better ...
While Faust emphasized:

> *A swamp lies there below the hill,*
> *Infecting everything I've done;*
> *My last and greatest act of will*
> *Succeeds when that foul pool is gone.*

. and issued the order:

> *Report on progress every day,*
> *The length of ditch earth dug away.*

Mephistopheles said half aloud:

> *They speak, as was the word they gave,*
> *Not quite of ditch, but more of – grave.*

(FROM GOETHE, FAUST II).

A serious problem that has received too little attention so far is peatland subsidence. Drained peatlands lose—depending on climate and use—between a few millimeters and several centimeters of thickness per year due to microbial oxidation and compaction. The resulting subsidence of the peatland surface requires deepening of the drainage ditches if the peatland continues to be used conventionally. This in turn promotes further subsidence and requires further ditch deepening. This spiral is called the "vicious circle of peatland use." Gravity-based drainage is thus becoming increasingly difficult and may finally, to keep the peatland dry, necessitate construction of polders with dikes and pumps. In the peatland-rich Netherlands, large areas have, after some 1000 years of peatland drainage, demonstrably sunk by over 8 m, and much of the country now lies below sea level (Erkens et al., 2016). Financially, the damage to roads and sewage infrastructure due to peatland subsidence there amounts to around 200 million euros per year, and it is estimated that by 2050 subsidence damage to buildings will amount to 80 billion euros (Van den Born et al., 2016; Tieleman, 2020).

Many peatlands in temperate latitudes, the subtropics, and the tropics are located near the ocean. Especially in often densely populated coastal areas, peatland subsidence increases the risks of flooding and the intrusion of salty seawater. While sea levels rise due to global warming, the adjacent peatland is literally bogged down. Considerable parts of the Malaysian and Indonesian peat swamps near the ocean, which have recently been drained for the cultivation of oil palm and pulpwood, will in the next decades be flooded by the sea due to intensive subsidence (Hooijer et al., 2015). Diking, poldering, and pumping – the interim solution tried in the Netherlands, northern Germany, England, California, and Florida to keep drained peatlands dry — will not work in Indonesia and Malaysia because of the huge extent of the areas and the enormous amounts of rainfall. It also would not stop subsidence and would achieve only an insignificant delay in the inevitable abandonment of this drainage-based land use (Dommain et al., 2016).

In more continental climates, frequent water level fluctuations lead to the formation of cracks in the drained peat, which prevent capillary water supply and thus lead to an even more frequent and deeper drying out of the soil. Due to the action of soil organisms, a loose, fine-grained, and water-repellent topsoil then develops, which finally can sustain at best only dry grassland species (Joosten et al., 2016; Zeitz, 2016). In this way, millions of hectares of former peat wetlands in Eastern Europe have, within a few decades, been transformed into deserts (Joosten et al., 2012).

Drainage-based peatland use is thus not only a disaster for the climate but also a threat to the productive potential of peatland soil.

Solutions and Challenges

Drainage-based peatland use is harmful to the climate, has no long-term prospects, and often entails a net social loss. Instead of continuing with outdated and unsustainable production methods, we should ask how we can manage peatlands without causing such grave environmental problems.

The answer is clear: peatlands must be wet! Wet peatlands must remain wet, drained peatlands must become wet again, and if we absolutely must use peatlands, we must use them wet (Joosten et al., 2012).

Threats to, and Protection of, the Remaining Natural Peatlands

Most of the world's peatlands are still largely unused and undrained. But threats and actual damage are on the rise, even in areas that received little attention a few decades ago. One example is the rapid expansion of agriculture and forestry in the tropics, often accompanied by deforestation: plantations of oil palm, fiber wood, aloe, pineapple, and banana are encroaching on forests and peatlands, as are smallholder farms, also in significant quantities. While until recently only peatlands in Southeast Asia, especially Malaysia and western Indonesia (Sumatra and Kalimantan), fell victim to rapidly expanding drainage-based peatland use, recent years have seen increasing encroachment of exploitation on natural peatland areas on the island of New Guinea, in western Amazonia, and in West Africa. Something similar is happening in East African countries, such as Uganda, Rwanda, and Madagascar, where increasing population pressure is driving people to seek a livelihood in the last still unused areas, the peatlands (Joosten et al., 2012).

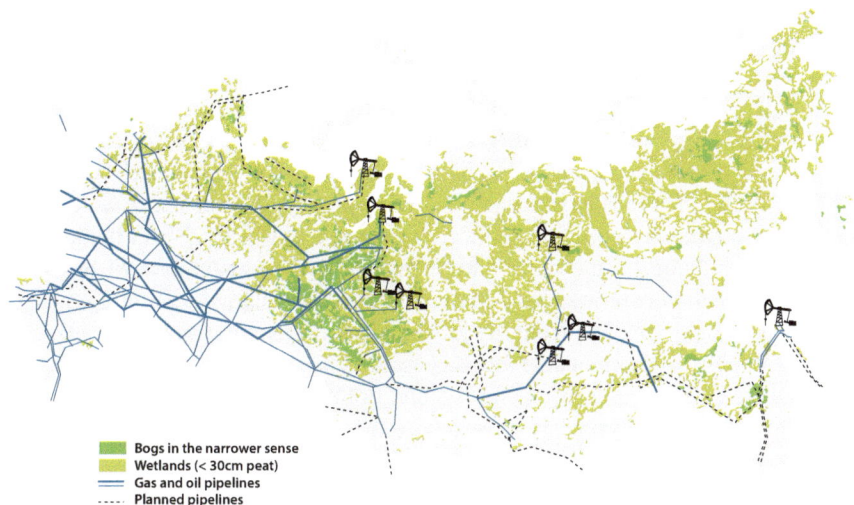

Fig. 4 Peatland distribution and location of existing and planned oil and gas pipelines in the Russian Federation. (Data from Timashev, 2019; Kirpotin et al., 2021)

The damage from new infrastructure projects, such as roads or oil and gas exploration and extraction, is more local (but radiates out from there). Important sources of liquid energy are found in peatlands such as Prudhoe Bay (Alaska), the Pastaza-Marañón Foreland Basin (Peru), the Niger Delta (Nigeria), Siberia, and Kamchatka (Russia). The pipelines to the metropolises largely run through forest and peatland areas (Fig. 4).

The huge forest peatlands in the Congo Basin, only recently surveyed and barely explored, have now been almost entirely licensed out to oil and gas explorers (Dargie et al., 2019; Global Witness, 2020). Oil exploration in the Alberta oil sands (Canada) is currently destroying peatlands on a large scale (Rooney et al., 2012). In the Hudson Bay Lowlands (Ontario, Canada), the second largest peatland area on Earth, there are currently 13,296 active mining claims for diamonds, chrome, nickel, copper, gold, and platinum, covering approximately 2127 km^2 within Ontario's "Ring of Fire" (Harris et al., 2022; McIntosh, 2021; Ministry of Mines 2022). For peatlands in Mongolia and mountainous areas like the Andes, mining is also a strong threat, as is overgrazing (Assessment Report, 2017; Maldonado Fonkén, 2014/2015).

In 2017, Rwanda opened its first peat-fired power plant in Gishoma, on the border to the Democratic Republic of Congo. Another and much larger peat-fired power plant in Gisagara, in the east of the country, is soon to feed 80 megawatts (MW) into the national grid. Currently, only 25% of households have access to the 190 MW of electricity generated in Rwanda, but capacity is set to increase to 563 MW. This increase will be achieved partly with peat extracted from the large valley peatland that runs along the border with Burundi (Cole, 2017; Mugerwa et al., 2019).

Stopping such dangerous developments is of course anything but easy. Arlette Soudan-Nonault, Minister for the Environment in the Republic of Congo, recently put it in a nutshell: peatlands are an important ecosystem not only for the two Congos "but also for the entire planet." The Republic of Congo, she said, is committed to protecting the peatlands. However, "we are not naïve, and we do not intend to stop our development just so the planet can breathe easier. The invaluable ecosystem services that our peatlands and forests render to the planet cannot remain free forever, to the detriment of our population's aspiration for well-being" (Cannon, 2021). The situation is thus as clear as it is precarious: on the one hand, we cannot expect (poor) countries to provide services for the whole world without being adequately paid for them; on the other hand, they are sawing at the branch they are sitting on, because short-term gains will quickly turn into major disadvantages, also for their own populations. We from the global North, especially we Europeans, are probably least able to communicate this successfully: we have no credibility because we have cleared a large part of our forests and drained our peatlands—and apparently become rich in the process. This is why it is so important that with the *UN Global Peatlands Initiative* (2023), the *Brazzaville* Declaration (UNEP, 2018), as well as the Peatlands Resolution of the United Nations Environmental Assembly (UNEA, 2019), a South-South link has been established in which the tropical peatland giants exchange views and consult with one another. Thus, to share lessons learned, Indonesia has sent many of its specialists to Africa and invited representatives from the two Congos and Peru to Indonesia.

Rewetting of Drained Peatlands

The rewetting of drained peatlands is the ultimate solution to the problems outlined above. Rewetting, that is, raising the groundwater table to or above the peatland surface, significantly reduces GHG emissions and can reactivate carbon sequestration (Wilson et al., 2016; Nugent et al., 2018; Mrotzek et al., 2020). While drained peatlands release nitrate through peat mineralization and agricultural use, rewetted peatlands cost-effectively remove this fertilizer through the process of denitrification (which produces molecular nitrogen; Trepel, 2010). However, because rewetting of overfertilized areas can also lead to the release of phosphate, it makes sense to assess this risk in advance (Emsens et al., 2017).

Rewetted peatlands also protect against flooding. This is so because peatlands suffer less damage from floods (there is no risk of crop failure) and also can absorb and retard flood waters, thereby protecting sensitive areas downstream (Joosten et al., 2015). Since rewetted peatlands delay the runoff of water from the landscape, the groundwater reserves in the catchment area increase—and this, together with flood retention, is an important adaptation to climate change which, apart from general warming, also leads to redistributed precipitation patterns, to more frequent and more intensive heavy rainfalls, and to longer dry spells (Quante & Colijn, 2016).

Rewetting changes the heat balance towards more cooling because *more* of the incident solar energy is used for evaporation, leaving *less* for significant heating (see also Chapter "Strengthen Terrestrial Water Cycles"). Over the peatland area of the Kieve polder in Mecklenburg-Western Pomerania, Germany, rewetting has produced a cooling effect of about 3 watts per square meter and has thus more than compensated for the GHG-induced warming of approximately 2.6 W/m² since 1750 (Joosten et al., 2015). With large-scale peatland rewetting, we can thus expect regional cooling effects. Last but not least, the increase in peatland biodiversity must be mentioned. In Mecklenburg-Western Pomerania, a few years after the rewetting of formerly heavily drained river valley peatlands, many peatland-typical and some highly endangered bird species have reestablished themselves (Herold, 2012).

In recent decades, several hundred thousand hectares of peatland have been rewetted in Europe (Tanneberger et al., 2017a). This sounds good at first, but in view of some 30 million hectares of drained peatlands in Europe, it must be considered a "drop in the ocean." Many rewetted peatlands are nature reserves, or abandoned, more or less exhausted peat extraction areas, or areas in very extensive use, with correspondingly relatively low emissions.

In all areas dedicated to "peatland nature conservation," rewetting is in any case the method of choice. Only a small proportion of peatlands (such as peat meadows and peat heaths) need minor drainage (and active management) to preserve their typical, anthropogenic, semi-natural biodiversity (Joosten, 2017). The centuries-long focus on drainage has fostered in Europe (even among conservationists!) a far too "dry" idea of how wet a living peatland is supposed to be. Since 1992, many hundreds of peatlands have been rewetted and revitalized through the European Union's LIFE programme (Camarsa et al., 2015).

Former peat extraction areas have been rewetted on a large scale in Germany, Belarus, and Russia – in Russia especially after the huge peat fires of 2011 (Tanneberger & Wichtmann, 2011; Barthelmes et al., 2021; Sirin et al., 2021). In Ireland, the parastatal peat company *Bord na Móna* has completely stopped peat extraction in 2021 and begun to restore over 30,000 hectares of peat extraction areas—as a core element of the Irish climate strategy (Bord na Móna, 2021; Kelleher, 2021). The largest such projects in the UK are mainly located on blanket bogs, which are used (very) extensively for sheep farming and grouse and deer hunting. In part, their rewetting is carried out as compensation for the construction of wind farms (Grouse Moor Management Review Group, 2019).

Rewetting of Peatlands in Intensive Use: Paludiculture

Most urgent for the climate and most challenging politically is the rewetting of heavily drained peatlands that are intensively farmed. This includes vegetable cultivation in California (to supply San Francisco), carrot cultivation in Norway, lettuce and potato cultivation in the east-anglian fens (UK), potato cultivation in the Teufelsmoor and Donaumoos (Germany) as well as cultivation of maize, Miscanthus, and reed canary grass (for "bio energy" in Germany and the Nordic countries), of sugar cane (Florida), of pulpwood and rubber (Indonesia), and of oil palm (mainly in Southeast Asia and more recently in New Guinea, Africa, and Amazonia). Dairy farming is also very common on drained peatlands (as in the Netherlands, Germany, and Poland).

Until now, peatlands have mostly been taken out of use after rewetting and converted into new "wet wilderness areas" or kept in low intensity management for biodiversity purposes. But we will no longer be able to afford doing this everywhere. All these peatlands have been drained to produce biomass, of which we need more and more as the human population continues to increase and we need to reduce the abject poverty of many. Moreover, in Paris 2015, humanity has agreed to largely decarbonize the world in the coming decades. Most fossil raw materials and fuels − coal, oil, gas, ores, and minerals − are to be replaced by biomass, which must be supplied by sustainable agriculture and forestry. We can no longer afford to degrade valuable production areas through (additional) drainage or to take degraded and then rewetted peatlands out of production on a large scale, unless we reduce the consumption of animal protein substantially. So long as a different consumption pattern has not become widely established, we must urgently develop new land use options that avoid the environmental damage of conventional peatland use while enabling us to use peatlands productively (Joosten et al., 2012; Joosten, 2017). Such uses exist, they are known as "paludicultures" (Wichtmann et al., 2016).

Paludicultures
are agricultural and forestry systems that aim to produce crop or livestock products on peat soils, conserving the carbon stock of the peat body and minimizing GHG emissions from the peat soil. Whether these goals are achieved depends not only on which crops are grown, but mainly on whether the growing conditions preserve the peat soil and keep it permanently wet (Wichtmann et al. 2016; Convention on Wetlands, 2021a).

Paludicultures make it possible to maintain productive agriculture and forestry while also restoring important ecosystem services of wet peatlands: reduction of GHG emissions, improvement of water quality, flood protection, evaporative cooling, maintenance and increase of peatland biodiversity. Paludiculture, by definition, stops peatland subsidence and has the potential to preserve employment in rural areas, to enable regional value creation, and to regionalize the supply of raw materials and energy (Wichtmann et al., 2016).

Paludiculture methods are currently being tested worldwide (Ziegler et al., 2021). In Europe, on nutrient-rich fen peatland sites, the focus is on reeds (for high-quality building materials), cattails (for building panels, insulation, animal feed, peat and plastic substitutes, etc.), unspecified biomass (for heat and energy) and alder (for furniture and veneers); the raising of water buffalo (for meat) is also interesting. On nutrient-poorer bog soils (cut-over bogs and bog grasslands), the focus is on peat mosses (as a substitute for fossil peat in horticulture) and sundew (for medicinal purposes) (Wichtmann et al., 2016).

The Southeast Asian lowlands provide more options. There are 1376 species of higher plants growing in the swamp forests there, of which 534 (39%) are already being used: 222 species produce useful wood, 221 have medicinal uses, 165 species provide food (such as fruits, nuts, and oils), another 165 are suitable for "other uses" (such as latex, fuels, and dyes). Many species have multiple uses, 81 non-timber products are "economically important" (Giesen, 2013, 2015). As rural communities are essentially farming communities, and paludiculture enables sustainable farming (albeit with modified techniques and alternative crops), paludiculture has great potential to preserve and revitalize local livelihoods while re-wetting drained peatlands (Convention on Wetlands, 2021a).

Paludiculture can also be a sensible way of protecting species and habitats (Närmann et al., 2021; Martens et al., 2023). Successful examples of the combination of species conservation and paludiculture are the occurrence of the globally threatened sedge warbler *(Acrocephalus paludicola)* in commercially used reed areas in western Poland (Tanneberger et al., 2009) and the spontaneous mass occurrence of sundew *(Drosera rotundifolia* and *D. intermedia)*, cranberry *(Vaccinium oxycoccos)*, and beak rush *(Rhynchospora alba)* as well as rare arthropods (such as spiders and dragonflies) on peatmoss farms in Lower Saxony, Germany (Muster et al., 2015, 2020; Gaudig & Krebs, 2016).

The economic benefits are so promising that one wonders why paludiculture has not yet been implemented on a large scale. The following are some obstacles that should urgently be eliminated:

- Paludiculture is fighting against the historical legacy of 10,000 years of dry agriculture. Our society does not have a "wet mentality," has never learned to farm in the wet and always striven to drain water as quickly as possible.
- Current rules and laws are not adapted to wet agriculture. In many places, landowners are obliged to maintain their ditches and drain their land—old regulations that still have an impact today. Farmers who convert to paludiculture receive lower EU subsidies (direct payments), as many paludiculture crops are not recognized as agricultural crops. This is changing in 2023 with the new EU agricultural rules. The ban on grassland conversion, introduced to protect soil carbon and biodiversity, is blocking the implementation of some paludiculture practices, even though they are more valuable than drained peatland grassland in terms of carbon and biodiversity protection.
- Incentives still exist to continue drainage-based management of peatlands, notably direct payments under the EU's Common Agricultural Policy (CAP).
- Paludiculture may create biotope types (e.g., reeds) that are protected by law and whose use is subject to legal restrictions (Czybulka & Kölsch, 2016).

Improving this situation requires creativity—and goodwill on the part of the authorities—in addition to the thoughtful and swift adaptation of rules and laws.

Paludiculture is more than just switching from one crop to another. It usually requires a redesign of the entire production chain, starting with acceptance and training, followed by selection and breeding of suitable plant species. New techniques must be developed, new infrastructure and logistics planned and implemented. It takes time to develop a fully integrated value chain. Those embarking on paludiculture need curiosity and patience as many crops are still in the pilot stage and in need of further research and development (Wichtmann et al., 2016).

A Future Strategy for Our Peatlands

Peatland rewetting is one of the most efficient land-based climate protection measures. It is actually peerless. But it requires either a complete cessation of productive land use or a comprehensive transformation to new, wet forms of cultivation. Under the Paris Climate Agreement, Germany must re-wet 50,000 hectares of drained peatlands by 2050—annually. For the European Union the commitment is 500,000 hectares, for Europe 1000,000, and worldwide 2,000,000 hectares per year. To meet these huge targets, the peatland nations need strategies (Fig. 5).

Just as lignite miners were not asked to organize the German coal phase-out, the phase-out of drainage-based peatland use must be seen as a task for society as a whole, and one of similar dimensions. If one takes the German coal phase-out budget (50–57 billion euros) as indicative of what society is willing to pay to reduce GHG emissions in a socially acceptable way, a budget of 36–41 billion euros for the climate protection measure "rewetting of arable land and grassland on peat soils" should be politically justifiable (Sommer & Lakner, 2021).

Such a cross-sectoral peatland protection strategy requires ambitious goals and policies, a time frame with defined intermediate steps, and action recommendations

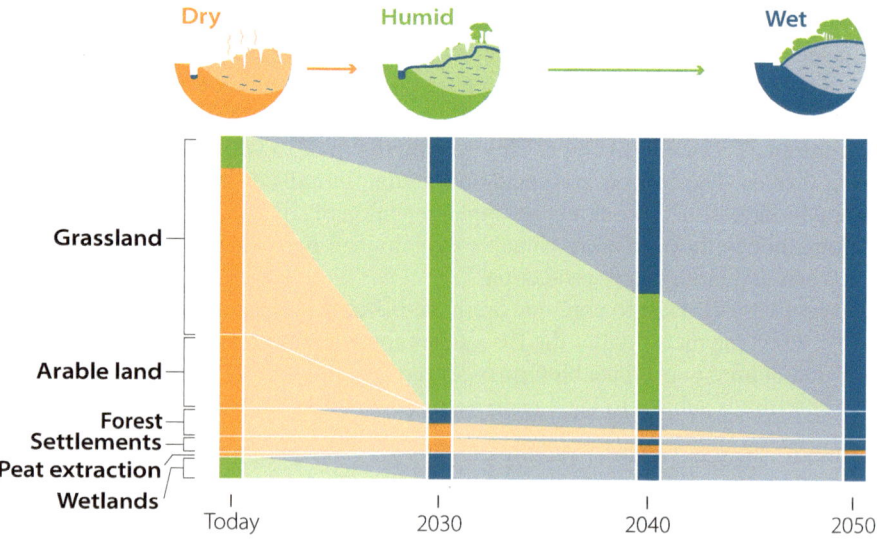

Fig. 5 Development paths and milestones for the various land use categories on peat soils in Germany pursuant to the overall emissions path recommended by the IPCC (2018). (After Tanneberger et al., 2021). Dry = deeply drained (strongly peat-consuming); wet = slightly drained (weakly peat-consuming); wet = under water (peat-preserving)

for specific actors. A broad-based peatland protection commission (similar to the German coal commission, Kommission „Wachstum, Strukturwandel und Beschäftigung", 2019) should prepare and accompany the phase-out of drainage-based peatland use. To this end, measures to strengthen structures and promote innovation are essential for the affected peatland regions. Peatland rewetting should here be integrated (with the involvement of water resource management) into the reorganization of landscape hydrology, which in any case is needed to adapt to climate change.

Because of the "species crisis," part of today's agriculturally used peatland must be restored to protect biodiversity. In this way, climate and biodiversity goals can be efficiently combined. As most paludiculture is still in the experimental phase, research and development must be more strongly promoted, focusing here on the unique structural and physiological properties of wetland-adapted plants so as to avoid competition with "dry" agriculture and thereby to enable wet agriculture to become self-sustaining in the medium term and to assert itself in the economic mainstream.

All these tasks can and should be state responsibilities. But their necessary, massively expansive implementation cannot be realized in a fully centralized manner. The purchase of all drained areas is too expensive: land prices are largely determined by rights, tax benefits, and subsidies (such as direct state payments)—ultimately, we taxpayers would be paying twice. The size of the relevant area as well as the necessary time flexibility and creativity require conditions that can be better supplied by the "free market."

A Fast-Track Paludiculture Programme Should Aim for
- simplification of the legal rules relating to implementation (laws governing water management, planning, agriculture, public procurement, land consolidation, etc.) and expansion of planning and technical capacities;
- identification of promising crops;
- selection and breeding to increase productivity and quality;
- improvement of farming and processing technologies;
- cross-company standardization of processing procedures;
- creation by the EU of a Life Cycle Assessment (LCA)-based *Conformité Européenne* (CE) certification to give preference to paludiculture products;
- market incentive programs for paludiculture products (e.g., in the sectors construction, horticultural substrates, energy, etc.).

Rewarding climate protection services should facilitate these measures. Successful peatland rewetting can avoid 20–25 metric tons of CO_2e per hectare per year in Central Europe (Joosten et al., 2016). More and more businesses, institutions, and cities are setting themselves ever more ambitious goals on the path to climate neutrality, goals they often cannot achieve on their own within their self-chosen schedules. Consequently, there is rapid increase in the demand for, and price of carbon "offsets." The price of a metric ton of CO_2 in the leading EU Emission Trading

System (ETS) market has increased 20-fold in the last 5 years and has in 2023 temporarily exceeded 100 euros per metric ton of CO_2 (https://tradingeconomics.com/commodity/carbon). How prices of carbon certificates from peatland rewetting will develop depends on the quality, that is, the reliability of the product; but similar increases might happen here, as is suggested by the experience with the regional MoorFutures® carbon certificates (https://www.moorfutures.de). Performance-based remuneration from the sale of carbon certificates can facilitate a rapid start-up of peatland rewetting, making it economically competitive with drainage-based, conventional peatland use until paludiculture will be technically fully developed and economically self-sustaining. The state should promote such carbon markets or at least not impede initiatives to create them.

Such a peatland emission trading system requires the development of:

efficient and transparent Measuring, Reporting, Verification (MRV) approaches (through adaptation/ elaboration of existing standards and methods);

similar processes for CO_2 sequestration (sinks) in paludi soils and products (*cf. Harvested Wood Products*);

regional agencies and specialist bodies for emission assessment and certification, and

reliable (national or regional) CO_2 exchanges.

A Peatland Emission Trading System Might Function as Follows (Isermeyer et al., 2019)

- The state issues landowners cost-free emission rights in the amount of their current emissions each year until 2045. These rights can be used to continue drainage-based agricultural production or, after rewetting, be sold as CO_2 certificates on the exchange.
- The state guarantees a minimum price for the CO_2 certificates (perhaps 60 euros per metric ton of CO_2 in 2025, rising to 100 euros in 2040) and, if the market price does not reach this threshold, pays the difference.
- The landowner may use the rewetted areas for paludiculture.
- The issuance of free emission rights ends by 2045. Landowners wanting to continue conventional use of drained peat soils will then have to buy the necessary emission rights on the exchange.
- Politics determines how to proceed if, among the landowners of a hydrologically connected area, most favor but some reject rewetting; the law is to be adjusted accordingly.[1]

The advantages of such a system are:

- high attractiveness for users: the land remains in private hands, and the owners decide how to manage it;
- rapid rewetting, because it is more profitable the sooner it is initiated;
- a fair balancing of interests: legitimate expectations are protected in the medium term and the polluter-pays principle prevails long-term.

---

[1] The red-green-yellow government in Germany envisions in its coalition agreement possible expropriation ("peatland protection is in the public interest").

After 2045/50—because everyone must be at "net zero"—it will no longer be possible to generate carbon certificates by avoiding CO_2 emissions. Income from agricultural peatlands must then be secured differently: through (1) competitive paludiculture, (2) CO_2 sequestration, and/or (3) rewards for other social benefits (Joosten et al., 2015).

Conclusion

Although peatlands contain more carbon than the entire global forest biomass, their importance has long been overlooked. Drained peatlands (0.4% of the world's land area) cause disproportionately high anthropogenic GHG emissions amounting to almost 5% of the worldwide total. To meet the Paris climate goals, all intact peatlands must remain wet, and drained peatlands must be rewetted; peatland use should take place only under wet conditions. The greatest challenges for rewetting lie with the peatlands now used intensively for agriculture. There is an urgent need to develop and implement (wet) production methods (paludiculture) that allow productive use of peatlands while avoiding the environmental harm from conventional peatland use. This requires the state to implement a cross-sectoral peatland protection strategy with measures to strengthen structures and promote innovation. The massive implementation on the ground, however, is best left to the self-directed momentum of enterprises and markets steered by climate protection rewards to facilitate a rapid start-up of peatland rewetting. In this way, companies can compete with drainage-based peatland use until paludiculture is technically fully mature and economically self-sustaining.

Peatland must be wet: for the peat, for the land, for the climate, forever!

References

Assessment Report. (2017). *Strategic planning for peatlands in Mongolia.* Asian Development Bank Technical Assistance TA-8802. Wetlands International, Ede.

Barthelmes, A., Couwenberg, J., Risager, M., Tegetmeyer, C., & Joosten, H. (2015). *Peatlands and climate in a Ramsar context - A Nordic-Baltic perspective.* TemaNord 2015:544. Nordic Council of Ministers, Copenhagen, 244 p.

Barthelmes, A., Abel, S., Barthelmes, K.-D., Couwenberg, J., Kaiser, M., Reichelt, F., Tanneberger, F., & Joosten, H. (2021). Evaluierung von Moor-Wiedervernässung in Deutschland. *Naturschutz und Biologische Vielfalt, 171,* 121–148.

BMU. (2021). Nationale Moorschutzstrategie. Bundesministerium für Umwelt, Naturschutz und Nukleare Sicherheit, Berlin.

Bord na Móna. (2021). Bord na Móna announce formal end to all peat harvesting on its lands. https://www.bordnamona.ie/bord-na-mona-announce-formal-end-to-all-peat-harvesting-on-its-lands/ Accessed 3 Nov 2023.

Camarsa, G., Toland, J., Hudson, T., Nottingham, S., Jones, W., Eldridge, J., Severon, M., Rose, C., Sliva, J., Joosten, H., & Thévignot, C. (2015). *LIFE and climate change mitigation* (p. 90). Publications Office of the European Union.

Cannon, J. (2021). *Carbon and communities: The future of the Congo Basin peatlands*. https://news.mongabay.com/2021/12/carbon-and-communities-the-future-of-the-congo-basin-peatlands/. Accessed 3 Nov 2023.

Carlson, K., Goodman, L. K., & May-Tobin, C. C. (2015). Modeling relationships between water table depth and peat soil carbon loss in Southeast Asian plantations. *Environmental Research Letters, 10*, 074006.

Chemnitz, C., & Becheva, S. (Eds.). (2021). Meat atlas. Facts and figures about the animals we eat. Heinrich Böll Stiftung, Berlin, Friends of the Earth Europe, Brussels, Bund für Umwelt und Naturschutz, Berlin.

Cole, L. (2017). Rwanda adds to energy mix with first peat-fired power plant in Africa. https://theconversation.com/rwanda-adds-to-energy-mix-with-first-peat-fired-power-plant-in-africa-74380. Accessed 3 Nov 2023.

Convention on Wetlands. (2021a). *Global guidelines for peatland rewetting and restoration*. Ramsar Technical Report No. 11. Gland, Switzerland: Secretariat of the Convention on Wetlands.

Convention on Wetlands. (2021b). *Restoring drained peatlands: A necessary step to achieve global climate goals*. Ramsar Policy Brief No. 5. Gland, Switzerland: Secretariat of the Convention on Wetlands.

Couwenberg, J. (2007). Biomass energy crops on peatlands: On emissions and perversions. *IMCG Newsletter, 2007-3*, 12–14.

Couwenberg, J., & Hooijer, A. (2013). Towards robust subsidence-based soil carbon emission factors for peat soils in south-east Asia, with special reference to oil palm plantations. *Mires & Peat, 12*, Article 01, 1–13.

Couwenberg, J., Dommain, R., & Joosten, H. (2010). Greenhouse gas fluxes from tropical peatlands in south-east Asia. *Global Change Biology, 16*, 1715–1732.

Couwenberg, J., Baumann, M., Lamkowski, P., & Joosten, H. (2022). From genes to landscapes: Pattern formation and self-regulation in raised bogs with an example from Tierra del Fuego. *Ecosphere, 13*, e4031.

Czybulka, D., & Kölsch, L. (2016). The legal framework. In W. Wichtmann, C. Schröder, & H. Joosten (Eds.), *Paludiculture – productive use of wet peatlands* (pp. 143–149). Schweizerbart Science Publishers.

Dargie, G. C., Lewis, S. L., Lawson, I. T., Mitchard, E. T. A., Page, S. E., Bocko, Y. E., & Ifo, S. A. (2017). Age, extent and carbon storage of the Central Congo Basin Peatland Complex. *Nature, 542*, 86–90.

Dargie, G. C., Lawson, I. T., Rayden, T. J., Miles, L., Mitchard, E. T. A., Page, S. E., Bocko, Y. E., Ifo, S. A., & Lewis, S. L. (2019). Congo Basin peatlands: Threats and conservation priorities. *Mitigation and Adaptation Strategies for Global Change, 24*, 669–686.

Dommain, R., Dittrich, I., Giesen, W., Joosten, H., Rais, D. S., Silvius, M., & Wibisono, I. T. C. (2016). Ecosystem services, degradation and restoration of peat swamps in the South East Asian tropics. In A. Bonn, T. Allott, M. Evans, H. Joosten, & R. Stoneman (Eds.), *Peatland restoration and ecosystem services: Science, policy and practice* (pp. 253–288). Cambridge University Press.

Draper, F. C. H., Roucoux, K. H., Lawson, I. T., Mitchard, E. T. A., Honorio Coronado, E. N., Lähteenoja, O., Montenegro, L. T., Valderrama Sandoval, E., Zaráate, R., & Baker, T. R. (2014). The distribution and amount of carbon in the largest peatland complex in Amazonia. *Environmental Research Letters, 9*, 124017.

Elshehawi, S., Barthelmes, A., Beer, F., & Joosten, H. (2019). *Assessment of Carbon (CO_2) emissions avoidance potential from the Nile Basin peatlands*. NBI Technical Reports- WRM 2019–13, 75 p.

Emsens, W.-J., Aggenbach, C. J. S., Smolders, A. J. P., Zak, D., & van Diggelen, R. (2017). Restoration of endangered fen communities: The ambiguity of iron-phosphorus binding and phosphorus limitation. *Journal of Applied Ecology, 54*, 1755–1764.

Erkens, G., van der Meulen, M., & Middelkoop, H. (2016). Double trouble: Subsidence and CO2 respiration due to 1,000 years of Dutch coastal peatlands cultivation. *Hydrogeology Journal, 24*, 551–568.

Frolking, S., & Roulet, N. T. (2007). Holocene radiative forcing impact of northern peatland carbon accumulation and methane emissions. *Global Change Biology, 13*, 1079–1088.

Gaudig, G., & Krebs, M. (2016). Nachhaltige Moornutzung trägt zum Artenschutz bei: Torfmooskulturen als Ersatzlebensraum. *Biologie in unserer Zeit, 46*, 251–257.

Giesen, W. (2013). *Paludiculture: Sustainable alternatives on degraded peat land in Indonesia.* Euroconsult Mott MacDonald.

Giesen, W. (2015). Utilising non-timber forest products to conserve Indonesia's peat swamp forests and reduce carbon emissions. *Journal of Indonesian Natural History, 3*(2), 10–19.

Global Witness. (2020). *What lies beneath.* https://www.globalwitness.org/en/campaigns/forests/what-lies-beneath/. Accessed 3 Nov 2023.

Greifswald Mire Centre. (2022). *Global Peatland Database/Global Peatland Map.* https://greifswaldmoor.de/files/dokumente/Global%20Peatland%20Database/GloPeMap_update_20210912_COP21_low%20res.png. Accessed 3 Nov 2023.

Grouse Moor Management Review Group. (2019). *Report to the Scottish Government.* The Scottish Government, Edinburgh.

Günther, A., Barthelmes, A., Huth, V., Joosten, H., Jurasinski, G., Koebsch, F., & Couwenberg, J. (2020). Prompt rewetting of drained peatlands reduces climate warming despite methane emissions. *Nature Communications, 11*, 1644.

Harris, L. I., Richardson, K., Bona, K. A., Davidson, S. J., Finkelstein, S. A., Garneau, M., McLaughlin, J., Nwaishi, F., Olefeldt, D., Packalen, M., Roulet, N. T., Southee, F. M., Strack, M., Webster, K. L., Wilkinson, S. L., & Ray, J. C. (2022). The essential carbon service provided by northern peatlands. *Frontiers in Ecology and the Environment, 20*, 201–268.

Herold, B. (2012). Neues Leben in alten Mooren. Brutvögel wiedervernässter Flusstalmoore. Bristol-Schriftenreihe 34. Haupt, Bern.

Hiraishi, T., Krug, T., Tanabe, K., Srivastava, N., Baasansuren, J., Fukuda, M., & Troxler, T. G. (Eds). (2014). 2013 Supplement to the 2006 IPCC Guidelines for National Greenhouse Gas Inventories: Wetlands. Intergovernmental Panel on Climate Change, Switzerland.

Hooijer, A., Silvius, M., Wösten, H., & Page, S. (2006). *PEAT–CO$_2$ – Assessment of CO$_2$ emissions from drained peatlands in SE Asia.* Delft Hydraulics report Q3943.

Hooijer, A., Vernimmen, R., Visser, M. & Mawdsley, N. (2015). *Flooding projections from elevation and subsidence models for oil palm plantations in the Rajang Delta peatlands, Sarawak, Malaysia.* Deltares report 1207384.

Humpenöder, F., Karstens, K., Lotze-Campen, H., Leifeld, J., Menichetti, L., Barthelmes, A., & Popp, A. (2020). Peatland protection and restoration are key for climate change mitigation. *Environmental Research Letters, 15*, 104093.

IPCC. (2018). Global warming of 1.5 °C. An IPCC special report on the impacts of global warming of 1.5 °C above pre-industrial levels and related global greenhouse gas emission pathways, in the context of strengthening the global response to the threat of climate change, sustainable development, and efforts to eradicate poverty. Intergovernmental Panel on Climate Change, Switzerland.

Isermeyer, F., Heidecke, C., & Osterburg, B. (2019). Einbeziehung des Agrarsektors in die CO2-Bepreisung, www.thuenen.de/media/publikationen/thuenen-workingpaper/ThuenenWorkingPaper_136.pdf. Accessed 3 Nov 2023.

Joosten, H. (2009). *The Global Peatland CO2 Picture. Peatland status and drainage associated emissions in all countries of the World.* Wetlands International, Ede..

Joosten, H. (2011). Recent achievements on the peatland/climate front. *IMCG Newsletter, 2011-2/3*, 4–10.

Joosten, H. (2016). Peatlands across the globe. In A. Bonn, T. Allott, M. Evans, H. Joosten, & R. Stoneman (Eds.), *Peatland restoration and ecosystem services: Science, policy and practice* (pp. 17–43). Cambridge University Press.

Joosten, H. (2017). Ökosystemdienstleistungen und Biodiversität der Moore. *Ber. d. Reinh.-Tüxen-Ges., 29*, 27–37.

Joosten, H., & Clarke, D. (2002). *Wise use of mires and peatlands – Background and principles including a framework for decision-making.* International Mire Conservation Group / International Peat Society.

Joosten, H., & Couwenberg, J. (2008). Peatlands and carbon. In F. Parish, A. Sirin, D. Charman, H. Joosten, T. Minaeva, & M. Silvius (Eds.), *Assessment on peatlands, biodiversity and climate change* (pp. 99–117). Global Environment Centre/Wetlands International.

Joosten, H., Tapio-Biström, M. -L., & Tol, S. (Eds.). (2012). *Peatlands – Guidance for climate change mitigation by conservation, rehabilitation and sustainable use.* Mitigation of Climate Change in Agriculture Series 5 (FAO). Rome ².

Joosten, H., Brust, K., Couwenberg, J., Gerner, A., Holsten, B., Permien, T., Schäfer, A., Tanneberger, F., Trepel, M., & Wahren, A. (2015). MoorFutures® Integration of additional ecosystem services (including biodiversity) into carbon credits – standard, methodology and transferability to other regions. BfN Skripten 407, Bundesamt für Naturschutz, Bonn.

Joosten, H., Sirin, A., Couwenberg, J., Laine, J., & Smith, P. (2016). The role of peatlands in climate regulation. In A. Bonn, T. Allott, M. Evans, H. Joosten, & R. Stoneman (Eds.), *Peatland restoration and ecosystem services: Science, policy and practice* (pp. 63–76). Cambridge University Press/British Ecological Society.

Joosten, H., Moen, A., Couwenberg, J., & Tanneberger, F. (2017). Mire diversity in Europe: Mire and peatland types. In H. Joosten, F. Tanneberger, & A. Moen (Eds.), *Mires and peatlands of Europe – Status, distribution and conservation* (pp. 5–64). Schweizerbart Science Publishers.

Jurasinski, G., Günther, A., Huth, V., Couwenberg, J., & Glatzel, S. (2016). Greenhouse gas emissions. In W. Wichtmann, C. Schröder, & H. Joosten (Eds.), *Paludiculture – productive use of wet peatlands* (pp. 79–93). Schweizerbart Science Publishers.

Kelleher, O. (2021). Midlands bogs to be restored as part of climate strategy, www.irishtimes.com/news/ireland/irish-news/midlands-bogs-to-be-restored-as-part-ofclimate-strategy-1.4719713. Accessed 3 Nov 2023.

Kiely, L., Spracklen, D. V., Arnold, S. R., Papargyropoulou, E., Conibear, L., Wiedinmyer, C., Knote, C., & Adrianto, H. A. (2021). Assessing costs of Indonesian fires and the benefits of restoring peatland. *Nature Communications, 12,* 7044.

Kirpotin, S. N., Antoshkina, O. A., Berezin, A. E., Elshehawi, S., Feurdean, A., Lapshina, E. D., Pokrovsky, O. S., Peregon, A. M., Semenova, N. M., Tanneberger, F., Volkov, I. V., Volkova, I. I., & Joosten, H. (2021). Great Vasyugan Mire: How the world's largest peatland helps addressing the world's largest problems. *Ambio, 50,* 2038–2049.

Kommission „Wachstum, Strukturwandel und Beschäftigung". (2019). Abschlussbericht. Bundesministerium für Wirtschaft und Energie. https://www.bmwi.de/Redaktion/DE/Downloads/A/abschlussbericht-kommission-wachstum-strukturwandel-und-beschaeftigung.pdf. Accessed 3 Nov 2023.

Koplitz, S. N., Mickley, L. J., Marlier, M. E., Buonocore, J. J., Kim, P. S., Liu, T., Sulprizio, M. P., DeFries, R. S., Jacob, D. J., Schwartz, J., Pongsiri, M., & Myers, S. S. (2016). Public health impacts of the severe haze in Equatorial Asia in September–October 2015: Demonstration of a new framework for informing fire management strategies to reduce downwind smoke exposure. *Environmental Research Letters, 11,* 094023.

Leifeld, J., Wüst-Galley, C., & Page, S. (2019). Intact and managed peatland soils as a source and sink of GHGs from 1850 to 2100. *Nature Climate Change, 9,* 945–947.

Lindsay, R. (1996). Themes for the future: Peatlands – a key role for Ramsar. In C. D. A. Rubec (Ed.), *Global mire and peatland conservation. Proceedings of an international workshop. North American Wetlands Conservation Council Report 96–1* (pp. 7–9).

Maldonado Fonkén, M. (2014/2015). An introduction to the bofedales of the Peruvian High Andes. *Mires and Peat, 15,* Article 05, 1–13.

Marlier, M. E., Liu, T., Yu, K., Buonocore, J. J., Koplitz, S. N., DeFries, R. S., Mickley, L. J., Jacob, D. J., Schwartz, J., Wardhana, B. S., & Myers, S. S. (2019). Fires, smoke exposure, and public health: An integrative framework to maximize health benefits from peatland restoration. *GeoHealth, 3,* 178–189.

Martens, H. R., Laage, K., Eickmanns, M., Drexler, A., Heinsohn, V., Wegner, N., Muster, C., Diekmann, M., Seeber, E., Kreyling, J., Michalik, P., & Tanneberger, F. (2023). Paludiculture can support biodiversity conservation in rewetted fen peatlands. *Scientific Reports.* https://doi.org/10.1038/s41598-023-44481-0

McIntosh, E. (2021). *Miners competing over Ontario's Ring of Fire have contentious relationships with Indigenous communities in Australia.* https://thenarwhal.ca/ring-of-fire-noront-bhp-wyloo/. Accessed 3 Nov 2023.

Minayeva, T. Y., Bragg, O. M., & Sirin, A. A. (2017). Towards ecosystem-based restoration of peatland biodiversity. *Mires and Peat, 19,* Article 01, 1–36.

Ministry of Mines. (2022). *Ontario's Ring of Fire.* https://www.ontario.ca/page/ontarios-ring-fire. Accessed 3 Nov 2023.

Mrotzek, A., Michaelis, D., Günther, A., Wrage-Mönnig, N., & Couwenberg, J. (2020). Mass balances of a drained and a rewetted peatland: On former losses and recent gains. *Soil System, 4*(16), 1–14.

Mugerwa, T., Rwabuhungu, D. E., Ehinola, O. A., Uwanyirigira, J., & Muyizere, D. (2019). Rwanda peat deposits: An alternative to energy sources. *Energy Reports, 5,* 1151–1155.

Muster, C., Gaudig, G., Krebs, M., & Joosten, H. (2015). Sphagnum farming: The promised land for peat bog species? *Biodiversity and Conservation, 24,* 1989–2009.

Muster, C., Krebs, M., & Joosten, H. (2020). Seven years of spider community succession in a Sphagnum farm. *The Journal of Arachnology, 48,* 119–131.

Närmann, F., Birr, F., Kaiser, M., Nerger, M., Luthardt, V., Zeitz, J., & Tanneberger, F. (2021). Klimaschonende, biodiversitätsfördernde Bewirtschaftung von Niedermoorböden. BfN Schriften 616. Bundesamt für Naturschutz, Bonn.

Nugent, K. A., Strachan, I. B., Strack, M., Roulet, N. T., & Rochefort, L. (2018). Multi-year net ecosystem carbon balance of a restored peatland reveals a return to a carbon sink. *Global Change Biology, 24,* 5751–5768.

Peters, J., & Tegetmeyer, C. (2019). *Inventory of peatlands in the Caribbean and first description of priority areas.* Proceedings of the Greifswald Mire Centre 05/2019.

Quante, M., & Colijn, F. (Eds.). (2016). *North Sea region climate change assessment. Regional climate studies.* Springer Verlag.

Ritchie, H. (2021). *Deforestation and forest loss. Explore long-term changes in deforestation, and deforestation rates across the world today.* https://ourworldindata.org/deforestation. Accessed 3 Nov 2023.

Ritchie, H., Roser, M., & Rosado, P. (2020). *CO_2 and greenhouse gas emissions.* https://ourworldindata.org/co2-and-greenhouse-gas-emissions. Accessed 3 Nov 2023.

Rooney, R., Bayley, S., & Schindler, D. W. (2012). Oil sands mining and reclamation cause massive loss of peatland and stored carbon. *Proceedings of the National Academy of Sciences, 109,* 4933–4937.

Rossi, S., Tubiello, F. N., Prosperi, P., Salvatore, M., Jacobs, H., Biancalani, R., House, J. I., & Boschetti, L. (2016). FAOSTAT estimates of greenhouse gas emissions from biomass and peat fires. *Climate Change, 135,* 699–711.

Schleicher, T., Hilbert, I., Manhart, A., Hennenberg, K., Ernah, Vidya S., & Fakriya, I. (2019). Production of palm oil in Indonesia. Öko-Institut e.V. / Universitas Padjadjaran (UNPAD) Indonesia.

Sirin, A., Medvedeva, M., Korotkov, V., Itkin, V., Minayeva, T., Ilyasov, D., Suvorov, G., & Joosten, H. (2021). Addressing peatland rewetting in Russian Federation climate reporting. *Land, 10,* 1200.

Sommer, O., & Lakner, S. (2021). Der Kohleausstieg als Modell zur Förderung der Wiedervernässung von Mooren? In: Strategien für den Agrar- und Ernährungssektor und den ländlichen Raum in Zeiten multipler Krisen. 31. Jahrestagung der Österreichischen Gesellschaft für Agrarökonomie, Tagungsband 2021: 63–68.

Statista. (2023). Brutto- und Nettowertschöpfung der Landwirtschaft in Deutschland in den Jahren 2011 bis 2022. https://de.statista.com/statistik/daten/studie/242840/umfrage/bruttowertschoepfung-der-landwirtschaft-in-deutschland/. Accessed 3 Nov 2023.

Tanneberger, F., & Wichtmann, W. (Eds.). (2011). *Carbon credits from peatland rewetting. Climate, biodiversity, land use.* Schweizerbart Science Publishers.

Tanneberger, F., Tegetmeyer, C., Dylawerski, M., Flade, M., & Joosten, H. (2009). Commercially cut reed as a new and sustainable habitat for the globally threatened Aquatic Warbler Acrocephalus paludicola. *Biodiversity and Conservation, 18*, 1475–1489.

Tanneberger, F., Joosten, H., Moen, A., & Whinam, J. (2017a). Mire and peatland conservation in Europe. In H. Joosten, F. Tanneberger, & A. Moen (Eds.), *Mires and peatlands of Europe: Status, distribution, and nature conservation* (pp. 173–196). Schweizerbart Science Publishers.

Tanneberger, F., Tegetmeyer, C., Busse, S., Barthelmes, A., Shumka, S., Moles Mariné, A., Jenderedjian, K., Steiner, G. M., Essl, F., Etzold, J., Mendes, C., Kozulin, A., Frankard, P., Milanović, Đ., Ganeva, A., Apostolova, I., Alegro, A., Delipetrou, P., Navrátilová, J., Risager, M., Leivits, A., Fosaa, A. M., Tuominen, S., Muller, F., Bakuradze, T., Sommer, M., Christanis, K., Szurdoki, E., Oskarsson, H., Brink, S. H., Connolly, J., Bragazza, L., Martinelli, G., Aleksāns, O., Priede, A., Sungaila, D., Melovski, L., Belous, T., Saveljić, D., de Vries, F., Moen, A., Dembek, W., Mateus, J., Hanganu, J., Sirin, A., Markina, A., Napreenko, M., Lazarević, P., Šefferová Stanová, V., Skoberne, P., Heras Pérez, P., Pontevedra-Pombal, X., Lonnstad, J., Küchler, M., Wüst-Galley, C., Kirca, S., Mykytiuk, O., Lindsay, R., & Joosten, H. (2017b). The peatland map of Europe. *Mires and Peat, 19*, Article 22, 1–17.

Tanneberger, F., Abel, S., Couwenberg, J., Dahms, T., Gaudig, G., Günther, A., Kreyling, J., Peters, J., Pongratz, J., & Joosten, H. (2021). Towards net zero CO_2 in 2050: An emission reduction pathway for organic soils in Germany. *Mires and Peat, 27*, Article 5, 1–17.

Temmink, R. J. M., Lamers, L. P. M., Angelini, C., Bouma, T. J., Fritz, C., van de Koppel, J., Lexmond, R., Rietkerk, M., Silliman, B. R., Joosten, H., & van der Heide, T. (2022). Recovering biogeomorphic feedbacks to restore wetlands as the world's biotic carbon hotspots. *Science, 376*, 6593.

Tieleman, J. (2020). Huizen verzakken sneller door droogte, schade loopt in de tientallen miljarden. De Volkskrant 8 september 2020, https://www.volkskrant.nl/nieuws-achtergrond/huizen-verzakken-sneller-door-droogte-schade-loopt-in-de-tientallen-miljarden~b9e59bc3/?referrer=https://www.google.com/. Accessed 3 Nov 2023.

Timashev, S. A. (2019). Cyber reliability, resilience, and safety of physical infrastructures. *IOP Conference Series: Materials Science and Engineering, 481*, 012009.

Trepel, M. (2010). Assessing the cost-effectiveness of the water purification function of wetlands for environmental planning. *Ecological Complexity, 7*, 320–326.

Umweltbundesamt. (2020). Methodenkonvention 3.1 zur Ermittlung von Umweltkosten. Umweltbundesamt, Berlin.

UN Global Peatlands Initiative. (2023). *What is the Global Peatlands Initiative?*. https://global-peatlands.org/index.php/about. Accessed 3 Nov 2023.

UNEA. (2019). *United Nations Environment Assembly UNEP/EA.4/Res.16: Conservation and Sustainable Management of Peatlands*. https://www.informea.org/en/decision/conservation-and-sustainable-management-peatlands. Accessed 3 Nov 2023.

UNEP. (2018). *Brazzaville Declaration*. United Nations Environment Programme https://wedocs.unep.org/20.500.11822/25329. Accessed 3 Nov 2023.

UNEP. (2022). *Global Peatlands Assessment – The State of the World's Peatlands: Evidence for action toward the conservation, restoration, and sustainable management of peatlands*. Main Report. Global Peatlands Initiative. United Nations Environment Programme, Nairobi. https://wedocs.unep.org/bitstream/handle/20.500.11822/41222/peatland_assessment.pdf. Accessed 3 Nov 2023.

Van den Born, G. J., Kragt, F., Henkens, D., Rijken, B., van Bemmel, B., & van der Sluis, S. (2016). Dalende bodems, stijgende kosten. Mogelijke maatregelen tegen veenbodemdaling in het landelijk en stedelijk gebied. PBL Planbureau voor de Leefomgeving Den Haag, PBL-publicatienummer 1064.

Wichtmann, W., Schröder, C., & Joosten, H. (Eds.). (2016). *Paludiculture – productive use of wet peatlands. Climate protection – biodiversity – regional economic benefits*. Schweizerbart Science Publishers.

Wilson, D., Blain, D., Couwenberg, J., Evans, C. D., Murdiyarso, D., Page, S. E., Renou-Wilson, F., Rieley, J. O., Sirin, S. A. M., & Tuittila, E.-S. (2016). Greenhouse gas emission factors associated with rewetting of organic soils. *Mires and Peat, 17*, Article 04, 1–28.

World Resources Institute. (2022). *World Forest Review*. Primary forest loss. https://research.wri.org/gfr/forest-extent-indicators/primary-forest-loss. Accessed 3 Nov 2023.

Worldometer. (2023). *Oil left in the world*. https://www.worldometers.info/oil/. Accessed 3 Nov 2023.

Zeitz, J. (2016). Drainage induced peat degradation processes & impact of drainage on productivity. In W. Wichtmann, C. Schröder, & H. Joosten (Eds.), *Paludiculture – productive use of wet peatlands* (pp. 7–13). Schweizerbart Science Publishers.

Ziegler, R., Wichtmann, W., Abel, S., Kemp, R., Simarda, M., & Joosten, H. (2021). Wet peatland utilisation for climate protection – An international survey of paludiculture innovation. *Cleaner Engineering and Technology, 5*, 100305.

Humus Enrichment of Soils

The Many Ways of Regenerative Agriculture

Stefan Schwarzer and Hans Peter Schmidt

Industrial farming systems succeed in producing large quantities of food for global markets. But this leads to various adverse effects, such as significant soil erosion, biodiversity loss, and pollution of water bodies (Mateo-Sagasta et al., 2017; Mekonnen & Hoekstra, 2015; Moss, 2008). They also promote high dependence on agribusiness and its products, lead to high freshwater and nitrogen consumption, and result in up to 34% of all anthropogenic greenhouse gas emissions (IPCC, 2014; UNEP, 2017). Population growth, climate change (with increasing incidence of weather extremes such as droughts and storms), potential shortages of mineral fertilizers (such as phosphorus), soil erosion and decline in soil fertility, high dependence on fossil fuels, decline in pollinators, and other factors collectively pose a major challenge to the current agricultural system.

Might there be alternative approaches, employing a set of diverse tools, able to increase soil fertility and to regenerate soil resources—creating win-win solutions such as sequestering carbon in soil to mitigate climate change? A whole range of new and innovative approaches for such purposes are presented on the following pages.

This chapter is based on United Nations Environment Programme. (2019, May). *Putting carbon back where it belongs—The potential of carbon sequestration in the soil* (Foresight Brief No. 012). Available at: https://wedocs.unep.org/handle/20.500.11822/28453. Accessed 19 September 2023.

S. Schwarzer (✉)
Climate Landscapes, Germany
e-mail: stefan@climate-landscapes.org

H. P. Schmidt
Ithaka Institute for Carbon Strategies, Switzerland
e-mail: www.ithaka-institut.org

© The Author(s) 2024
K. Wiegandt (ed.), *3 Degrees More*,
https://doi.org/10.1007/978-3-031-58144-1_10

Soil and Humus Loss Through Industrial Agriculture

The "modern" or "industrial" agriculture of the early years of the twenty-first century faces many problems and challenges. One of the largest—albeit less noticed in our society—threats to humanity and the planet is the loss of soil and hence of soil fertility through agricultural practices. The fragility of the thin layer of topsoil, which is the basis for almost all that grows and almost all we eat, thus calls into question the "sustainability" of industrialized farming. In many regions of the world, soil fertility has been declining for decades; large quantities of fertile soil have been (and continue to be) washed into rivers, lakes, and oceans—disappearing forever. Soil degradation leads to the production of carbon dioxide (CO_2), which is produced by the oxidation of soil organic matter (SOM, commonly known as "humus") and released into the atmosphere. All this also has significant economic impact.

Each year, we lose around 24 billion metric tons of fertile topsoil on our fields; 10 million hectares of arable land are degraded annually as a result (FAO and ITPS, 2015; Pimentel & Burgess, 2013). The lost topsoil would fill 192 million rail cars; the degraded area is nearly the size of Greece—every year! In the U.S., the loss is 15.7, in Europe 2.5 metric tons per hectare per year (Panagos et al., 2015). "Overall, soil is being lost from agricultural land 10–40 times faster than the rate of soil formation, threatening humanity's food security" (Pimentel & Burgess, 2013). In addition to the loss of topsoil, there are other processes impairing soil quality, with the result that around 25% of the Earth's agricultural land is now considered degraded (Bai et al., 2013).

Around one third of the CO_2 released into the atmosphere by human activities between 1850 and 1998 came from agricultural activities (Houghton & Nassikas, 2017). Estimates[1] of carbon released since the beginning of agriculture due to soil erosion and loss of soil organic matter from forest clearing and burning range from 133 to 379 billion metric tons (Le Quéré et al., 2016; Sanderman et al., 2018). In cultivated soils, about 50–70% of soil carbon has been lost (Machmuller et al., 2015). Agricultural land today often contains less than 2% humus, whereas at the time of conversion from grassland or forests it still contained 8–15% or more.

The key question is: can excess CO_2 be recycled from the air and stored in the soil, thus helping to mitigate climate change? This is a crucial issue because scientists have calculated that extensive terrestrial removal of CO_2 from the atmosphere through controlled biomass and soil carbon sequestration is required to avoid the currently projected temperature increase.

---

[1] Conversion: $C \times 3.67 = CO_2$; $CO_2 - C$: 0.27; proportion of C in humus: 58%; SOC is calculated accordingly as SOM $\times$ 0.58.

How Much Carbon Can Soils Absorb?

The amount of carbon in the atmosphere is about 860 billion metric tons, and there are 450–650 billion metric tons in the biological pool (Lal, 2004). One study put the global stocks of soil organic content (SOC) at about 850, 1800, and 3000 billion metric tons in relation to the top 30, 100, and 200 cm respectively (Sanderman et al., 2018). According to the area classified as arable land by the International Geosphere Biosphere Programme (IGBP), this corresponds, respectively, to an average of 60, 130, and 200 metric tons of carbon per hectare. Soils thus store far more carbon than plants, especially in boreal forests, wetlands (especially peatlands), and grasslands (especially temperate steppes) (Fig. 1). Soils used for agriculture usually contain only small amounts of carbon or humus, and the trend is downward.

The average historical loss is estimated at 20–30 metric tons of carbon per hectare in forests or woodlands and 40–50 metric tons in steppes, savannas, and grassland ecosystems. On average, the conversion of (previously unused) grassland into cropland leads to a loss in carbon content of about 50% (Lal, 2018).

The best-known carbon sequestration initiative "4p1000" (see also section "Policy Implications") has calculated that with a SOC increase of 0.4% per year (or 4p1000,

Fig. 1 Underestimated soils: average amount of carbon stored in vegetation and soils (up to 1 m depth) in different ecosystems—in metric tons per hectare. (Data from IPCC (2021))

hence sometimes called the 4p1000 initiative) in all land uses, including forests, the CO_2 concentration in the atmosphere could be effectively reduced. Based on the baseline values used in its calculation, an additional 2.8 billion metric tons of carbon could be stored each year in the top 30 cm of soil (Soussana et al., 2017). This would lead to a net reduction of CO_2 in the atmosphere because the current annual increase in CO_2 emissions worldwide is "only" about 200 million metric tons of carbon.

Equally important ecologically, however, is that increasing the carbon content of the soil leads to many other benefits that improve crop and pasture yields. It

- Increases available water capacity,
- Improves the nutrient supply of the plants,
- Restores the soil structure, and
- Minimizes the risk of soil erosion.

Estimates for carbon sequestration through improved practices vary considerably, as the understanding of interactions and especially the knowledge of soil response is still limited. Various studies show theoretical potentials of 0.8–8 billion metric tons of carbon per year (NAS and National Academies of Sciences, Engineering, and Medicine, 2018; Griscom et al., 2017; Lal, 2016a, b, 2018; Minasny et al., 2017; Paustian et al., 2016; Smith et al., 2008; Zomer et al., 2017), while more realistic values are probably in the range of 1.5–2.5 billion metric tons. Given global CO_2 emissions from fossil fuels and industry of 9.9 billion metric tons (plus 1.3 billion metric tons from land use changes such as deforestation), the potential for carbon sequestration through regenerative agricultural practices is promising.

But the conditions for implementation are not the same everywhere. Funding and cooperation among scientists, policy makers, practitioners, and various other stakeholders is needed. Global efforts to gradually change land use practices are not easy to implement, which reduces the theoretical mitigation potential. Furthermore, the sink capacity of soils is not infinite—and it is reversible if not managed properly.

Humus Formation in Theory …

Humus is not stable. Build-up and decomposition processes are characteristics of an active and diverse soil life. The various humus compounds have different durations, which can range from weeks to decades. Increasing warming due to climate change tends to lead to faster decomposition due to increased activity of soil organisms and thus makes humus build-up more difficult.

Humus-forming agricultural practices (Fig. 2) include a broad crop rotation, the use of catch crops, of nitrogen-fixing legumes, the use of plant species and varieties with greater root mass and deeper roots, the integration of animals into the cropping system, agroforestry systems, improved grassland management, leaving crop residues and additives such as manure, compost, and biochar (see box below) (Lal, 2010, 2016a; Minasny et al., 2017; Griscom et al., 2017; Paustian et al., 2016; Lugato et al., 2014; Soussana et al., 2017).

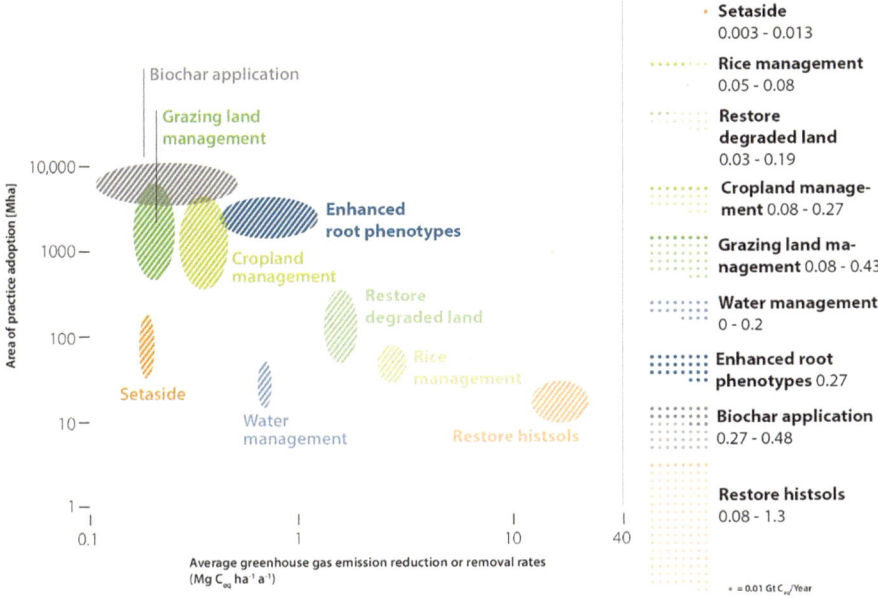

Fig. 2 Global carbon sequestration potentials for various agricultural practices. (Data from Paustian et al. (2016))

Increasing humus levels and adapting agricultural practices accordingly requires a deep understanding of the fundamentally important relationships between plants and soil life. Plants interact intensively with various microorganisms, especially with certain microbes and fungi in the soil. A single gram of healthy soil contains more bacteria, fungi, and other microorganisms than there are people on earth. These influence plant growth and health as well as nutrient and water storage in the soil (Bardgett & van der Putten, 2014; Cao et al., 2011; Eisenhauer et al., 2017; Steinauer et al., 2016). The underground "wood wide web" shares nutrients and water with the plants and receives signals from them, which influence the defense against insect predators and leaf-infesting fungi. Plants in turn transfer and share 20–30%, sometimes even 50% of their photosynthesis products (mainly carbohydrates, but also amino acids) via root exudates[2] with this very diverse life system (Eisenhauer et al., 2017; Jones, 2008; Leigh et al., 2009; Steinauer et al., 2016), and thus form a complex natural symbiosis. Plant diversity and soil microbial diversity interact positively, supporting plant health and plant mineral concentrations. "In fact, roots and their plant health-promoting microbiome may hold the key to the next green revolution" (Pieterse et al., 2014).

---

[2] Exudates are liquids, often rich in carbohydrates, released by a plant through roots and other pores.

Biochar as a Carbon Sink

Biochar,[3] produced by pyrolysis[4] from biomass, is a long-term stable form of carbon. Biochar has abundant advantages, many of which are not yet understood. It is resistant to decomposition (Lehmann et al., 2015; Zimmermann & Gao, 2013) and can stabilize organic matter added to the soil (Weng et al., 2017). Biochar can also create long-term carbon pools in the soil (Griscom et al., 2017; Paustian et al., 2016; Soussana et al., 2017; Woolf et al., 2010). Its use offers a range of benefits for soil fertility and quality, such as the promotion of fungal and bacterial growth, improved water and nutrient retention, reduced impact of pathogens, (Lehmann et al., 2011; Hagemann et al., 2017), and even higher crop yields (Jeffery et al., 2011; Kammann et al., 2015). Plant biomass, after subtracting water, consists of about half of the carbon that was removed from the atmosphere through photosynthesis during the plant's growth. If the plant dies, it begins to biodegrade, with the absorbed carbon returning to the atmosphere in the form of CO_2. To prevent this, biomass can be carbonized. To do this, it is pyrolysed, that is, thermally treated at a temperature of at least 400 °C in the absence of air. This way, much of the biochar is bound into molecular structures that can remain stable in soils for many centuries. The product of this pyrolysis process, biochar, is seen as a way of limiting anthropogenic climate change.

Biochar is used especially as a soil conditioner and carrier matrix for fertilizers and as a feed additive, stable bedding, and manure additive. For some years now, new fields of application have been emerging outside agriculture, for example in the construction and plastics industries, where biochar improves the functional properties of concrete and plastics, or in water and air pollution control, where it can replace conventional activated carbon made from fossil raw materials. Biomass and the biochar produced from it are thus increasingly becoming lucrative agricultural (by-)products that are also used in industry and environmental technology. What all applications have in common is that the biochar is not burned. The carbon (C) content of the biochar, which had been removed from the atmosphere, thus remains stored in the terrestrial system for the long term. The transformation of biomass carbon into biochar as well as pyrolysis oils is the only already available climate technology that has been extensively tested, can be quickly scaled up globally, and can establish and stably maintain relevant carbon sinks for many

---

[3] A long-term stable form of charcoal produced by pyrolysis of biomass; also called *vegetable carbon*.

[4] Pyrolysis is a process that can be applied to any organic (carbon-based) product, exposing the material to high temperatures in the absence of oxygen.

(continued)

centuries. In contrast to all comparable technologies, the biochar industry is already in exponential growth and offers multiple added values that go beyond pure carbon storage. To limit climate change to 2 °C, at least 220 billion metric tons of carbon will need to be stored in additional sinks by 2100—equivalent to around 800 billion metric tons of CO_2. To achieve a third of this sink capacity using biochar and pyrolysis oil, around 400,000 industrial pyrolysis plants would need to be commissioned worldwide by 2050. However, to construct such an enormous number of industrial plants so fast, sustained exponential growth in plant construction over the next 20 years would be needed, from around 100 plants per year presently to some 50,000 plants per year at the peak.

Even more difficult than increasing the capacity for industrial construction of pyrolysis plants is raising the biomass productivity of agriculture in such a way that sufficient feedstock can be provided each year for the production of biochar. This goal can be promoted by using methods such as agroforestry, forest pastures, forest gardens, humus build-up, and mixed cropping, which produce food and biomass simultaneously through multiple uses of the land. With algae farms, additional ocean areas can be developed for biomass production. The essential increase in biomass capacity can also be promoted precisely through the application of biochar, namely by using plant-carbon-based fertilizers. This method mixes biochar with dissolved nutrients so that the nutrient-rich solution is completely absorbed by the pore system of the biochar. So far, nutrient solutions have mainly been of organic origin, such as cattle urine, biogas slurry, press water from tofu production, but also compost extracts, or other commercial liquid fertilizers. In principle, synthetic and mineral fertilizers can also be used. Biochar acts here primarily as a carrier matrix for nutrients, reducing their leaching and improving their uptake by plants as well as the charge equalization between roots and soil.

Affording substantial efficiency gains, the combination of organic fertilizers with biochar offers a promising prospect, especially for organic farming. A 2020 meta-study was the first to evaluate scientific publications that examined exclusively the effects of combining biochar and fertilizers (Ye et al., 2020). The authors could show that biochar is not only an aid for tropical agriculture but, if used correctly, leads to significant yield increases in temperate climates as well. Compared to fertilization with the same amount of nutrients without biochar, the application of biochar fertilizers resulted in an average yield increase of 15%.

In the tropics, average yield increases of 25% can already be achieved with biochar (Jeffrey et al., 2017). Even 50% yield gains were achieved with biochar-based organic fertilizer applied directly to the root zone of vegetable, fruit, and cereal plants (Schmidt et al., 2017).

(continued)

Evaluation of 26 meta-analyses, which included 19,000 data sets from over 1500 scientific publications, show that the use of biochar improved all 66 investigated agronomically relevant parameters by an average of over 20%. It increased yield and plant health, biological soil activity, root growth, water use efficiency, and humus content while also reducing greenhouse gas emissions. Depending on the climate zone and the crop grown, the use of biochar-based fertilizers can not only improve ecosystem services but also increase yields by at least 10–25%. If we were to apply biochar to the entire global agricultural area of 51 million square kilometers, the yield gains would mean that at least 10% less land would be needed to produce the same amount of food and feed. On this freed-up land, a wide range of carbon sequestration measures could be applied, from reforestation to the production of biomass for producing biochar. The latter should ideally be integrated into the remaining crops to achieve structures that are suitable for increasing biodiversity, stabilizing soils, and creating a balanced microclimate.

At a productivity of 10 metric tons of biomass (dry matter) per hectare per year, the above-mentioned 10% of global agricultural land could produce 5.1 billion metric tons of biomass—which could be converted by pyrolysis into 1.7 billion metric tons of carbon.[5] Over a 50 year period, this would amount to 85 billion metric tons of carbon or just under 400 billion metric tons of CO_2e, about half of what would be minimally needed by the end of the century to prevent dangerous climate change. Since significantly higher yield increases and thus land savings in favor of biomass production can be expected in the tropics, the estimated potential of 400 billion metric tons of CO_2e is entirely realistic.

In pointing out all these potentials, it is important to note, to insist, that any food competition is excluded. For any land taken out of food production, there must be a substitute or else yield increases, based on sustainable land management, must be achieved on the remaining land.

If food neutrality is ensured, the use of biochar technology is a promising option that also optimizes various ecosystem services. Specifically, it reduces

- Nutrient leaching and groundwater contamination,
- Emission of greenhouse gases from agriculture, and
- The uptake of pollutants by plants.

---

[5] 5.1 billion metric tons × 48% carbon content × 70% carbon efficiency when converted to a carbon sink.

(continued)

In addition, it improves

- Water storage of soils,
- The resilience of agricultural systems, and
- The buildup of humus, which removes additional carbon from the atmosphere.

In conclusion, it should be admitted that the implementation of the biochar approach is complex. In many places, experts, authorities, advisors, industries, and not least farmers would have to show good will. Still, half of the additional carbon sinks needed to save the climate could be created around the world—with an investment of only 500 billion euros, 50% of the world's annual military spending (Roser et al., n.d.). The emission reductions agreed for the period till 2050 must, however, be achieved in addition, as a necessary precondition for a climate-neutral carbon balance.

Whereas metabolic products of bacteria as well as their cell bodies make up an important part of the soil's carbon pool, fungi that establish symbiotic relationships with plants (mycorrhizae) produce a sticky, carbon-rich glycoprotein known as "glomalin," which is crucial for soil stability and water storage and forms an important carbon reservoir—carbon that has been taken from the atmosphere. Moreover, through their exsudation, roots increase the carbon pool by more than twice what the composting process of dead above-ground biomass can do (Kätterer et al., 2011).

Because a system with a higher carbon return also results in more nutrients being returned to the plant, plant productivity is increased and the need for fertilizer is reduced. In conventional agriculture, chemical fertilizer is one of the main sources of greenhouse gas emissions through both its energy-intensive production and the resulting reaction of microbes. It is important to note that "the gain [of regenerative farming systems] is positively correlated with soil organic matter" (LaCanne & Lundgren, 2018). It is unsurprising therefore that humus losses can lead to an enormous social loss of natural capital.

... and Humus Formation in Practice

The following agricultural practices can help sequester carbon in the soil:

Because ploughing the soil is one of the main drivers of humus mineralization and soil erosion, switching to practices that *reduce or avoid ploughing* can have a positive impact on soil organisms and carbon levels, saving up to 70% of energy and fuel expenditures and of investments in machinery. Under most "no-till" systems, which avoid ploughing before sowing, carbon increases in the upper soil layer (up to about 10 cm depth) and partially decreases below that (Mäder & Berner, 2012; Powlson et al., 2014). Nonetheless, research shows that the activity of bacteria, especially fungi, is increased, and soil structure often improves. No-till helps protect soils but is often combined with the use of herbicides, such as glyphosate, which can have negative effects on soil biology and other living organisms as well as on human health. To benefit from no-till and to store additional carbon, this practice must be integrated into more diverse agroecosystems, for example through the *use of green manure mixtures* that help loosen the soil with deep-reaching roots, transfer carbon to the rhizosphere, stabilize soil aggregation, and suppress weeds and pests.

Management practices that can store additional carbon include *selecting crop species and varieties with greater root mass* and deeper roots, using *crop rotations* with greater carbon inputs, using *catch crops* during fallow periods, *leaving crop residues* on the field, and *additives such as compost and biochar.* Intercropping (the simultaneous production of multiple crops on the same area of land), increased crop rotations and catch crops can improve soil fertility by covering the soil, feeding the microbiome year-round, fixing nitrogen in the soil by nitrogen-fixing plants, and thus increasing soil carbon content (Poeplau & Don, 2015). Such crops also reduce soil erosion and suppress weeds and pests. For example, it has been calculated that through use of catch crops two million metric tons of carbon per year could be stored away in France (Pellerin et al., 2019). *Increasing the diversity of crop species* both within a crop and between successive crops can lead to significant economic gains (higher yields, less pesticide use) through greatly reduced weeds and insect pests, as this positively changes the supply of natural enemies (e.g., aphids) (Lundgren & Fausti, 2015). Plant species with deep roots (particularly useful for catch crops) can play the following key roles: sequester more carbon, break up plough seals, use the subsoil for additional nutrient enrichment, aerate the soil, create favorable conditions for earthworms and other soil life, and positively influence the root diameter of the next crop.

Earthworm abundance is an important indicator of soil activity and soil health. Improving their living conditions is crucial as they dig (bio)pores that help aerate the soil, infiltrate water, and store it quickly. In addition, through their activity and nutrient-rich excretions, they increase humus content by integrating organic matter into the soil and facilitate access to the nutrient-rich subsoil. Leaving crop residues and *mulching with biological material* are important approaches to increasing soil fertility and soil carbon while limiting soil erosion.

Mixed or intercropping, that is, the simultaneous cultivation of several crops on the same area, can increase net plant growth and thus sequester more carbon into the soil, increase yields, and reduce weeds at the same time. This can be explained by a larger leaf surface, increased mycorrhizal activity, increased communication and exchange via root networks, and complementary demands on the soil (plant species use different mineral nutrients in diverse quantities) (Walder et al., 2012; Brooker et al., 2015).

Undersowing helps protect the soil when the main crop does not completely cover the soil. Such "living mulch" helps suppress weeds and can promote the growth of the main crops. The use of legumes in undersowing can provide additional organic nitrogen while increasing soil carbon content.

Another factor is that in the temperate regions the potential photosynthesis rate is highest during the summer months. But, with cereal crops maturing, this energy is not being used for producing carbohydrates. The undersown crop, by contrast, remains photosynthetically active at this time of year, producing carbohydrates among other things, thereby adding carbon to the soil while providing nectar, pollen, and seeds to insects and birds while also enhancing biological pest control.

The *application of compost* to cropland and grassland stimulates both above- and below-ground net primary productivity[6] and, even with only one application, can lead to carbon accumulations of two to five metric tons per hectare in subsequent years (Ryals & Silver, 2013). It increases soil life through the fungi and bacteria in the compost itself. And it stimulates soil life activity while adding extra carbon and nutrients to the soil, which improve the soil's structure and water storage capacity.

Native Pastures: often pastures are regularly cultivated with shallow-rooted species (such as Kentucky bluegrass in the U.S.) and with a low diversity of grasses. But the "natural" prairies of the U.S. (as well as Europe's steppe regions) consisted of a variety of native plants, many of which were deeply rooted into the soil and therefore stored carbon (Teague et al., 2016). While typical seeded grasses reach depths of no more than 50 cm, native plants easily go down several meters, with different root forms occurring and complementing one another.

Combining livestock and cropping, that is, using animals to graze catch crops or stubble, creates synergies between system components, which can improve resilience and sustainability while fulfilling multiple ecosystem functions. It can increase both humus content and economic yield, diversify agricultural production systems, improve drought resilience, and reduce soil erosion (Bonaudo et al., 2014; Franzluebbers & Stuedemann, 2008, 2014). The use of grazing animals not only improves the soil through their bacteria- and nutrient-rich excreta, but can simultaneously replace the use of herbicides (such as glyphosate). "Cereal-pasture mixed cropping systems" (pasture cropping) go one step farther; they combine perennial pastures with annual crops and deliver impressive results in terms of increased soil carbon content (9 metric tons of carbon per hectare per year for the years 2008–2010), biodiversity, and yields (Seis, 2006; Glover et al., 2011).

Improved grassland management, such as lower stocking densities, different types of rotational or short-term grazing, seasonal grazing, inclusion of legumes and a variety of crops, can lead to sequestration of up to 1.8 billion metric tons of carbon annually (Paustian et al., 2016; Teague et al., 2016). Especially effective is adaptive multi-paddock (AMP) grazing (also called holistic grazing management or mob grazing), where herds graze in a rather small plot for a very short period (usually from half a day to 2–3 days) before being led to the next plot, with grazed plots given several weeks or months to regenerate after grazing.

---

[6] Net primary productivity or NPP expresses the net amount of CO_2 taken up by vegetation per unit of area, describing a plant's productivity.

In contrast to a continuous grazing approach, where the net impact of carbon reduction can be offset by N_2O and CH_4 emissions from animals and their excreta, there is new research and an increasing number of practitioners reporting gains in humus content, soil fertility, biomass, and plant diversity. There is a net gain in carbon even when the methane emissions from animals are taken into account (Teague & Barnes, 2017).

Best Practice Examples
Gabe Brown is a prominent conventional farmer in the U.S. who has turned his farm, formerly based on a monocultural model, into a productive operation. With reduced herbicide use, he has managed to increase the water-holding capacity and humus content of his soils. While he had less than 2% humus in his soils in the early 1990s, he now enjoys contents of more than 6%. Brown uses a wide crop rotation with diversified intercropping, has integrated livestock into his cropping system through a holistic grazing management plan, and has stopped ploughing his fields (Brown, 2020). Another well-known North American farmer, *Joel Salatin*, makes intensive use of mob grazing technology, which he has expanded to include a so-called follower system in which different animals such as cows, sheep, chickens, and turkeys follow one another according to their feed needs and take turns. This greatly increased the fertility of his soils and the plant diversity on his meadows (Polyface Farms). In Germany, *Michael Reber*, a farmer from Schwäbisch-Hall, has been able to completely dispense with the use of fungicides and insecticides in recent years by implementing measures such as minimum tillage, diverse catch crops, and mixed crops in the entire crop rotation. He uses herbicides only for maize, when necessary. The use of mineral fertilizer is also being further reduced each year by upgrading the organic fertilizer available on the farm. The aim is to cultivate all crops as mixed crops and, in the medium term, to integrate animal husbandry back into the farm's land use (Reber, 2017).

Agroforestry, that is, the integration of trees and shrubs into cropland and livestock systems, can bring multiple environmental, economic, and social benefits. First of all, it has a positive effect on humus content: between 0.2 and 5.3 billion metric tons of carbon per year can be fixed in soils (in addition to carbon bound in wood), with the best results achieved in the tropics and subtropics (Griscom et al., 2017; Shi et al., 2018). In addition, other positive "side effects" are also at work here, from increased biodiversity to diversified yields. Agroforestry and conservation agriculture approaches in sub-Saharan Africa and tropical countries showed that often significantly larger increases in soil carbon levels are achievable than "only" 0.4%, while delivering higher economic and environmental value (Corbeels et al., 2018). In short, integrating trees into regenerative agricultural practices or holistic pasture management can increase rates of carbon sequestration by a factor of 5–10 and soil carbon stocks by a factor of 3–10 (Toensmeier, 2018).

Finally, it is possible to develop intensive silvopastoral systems (combining trees, animals, and pastures) that not only lead to more humus in the soil but also to a net sequestration of 4–12 metric tons of carbon per hectare per year—offsetting the methane production of the animals. In addition, the production of meat and milk can be increased (Montagnini et al., 2013).

Policy Implications

There is some progress around the world. The Australian Coalition Government has in the years 2018–23 invested around \$450 million in a *Regional Land Partnership Program* plus \$134 million in a Smart Farms Program to improve soil health. The Andhra Pradesh government has launched a *Scale-Out-Plan* to convert six million farms/farmers to 100% chemical-free agriculture by 2024. This program is a contribution to the United Nations Sustainable Development Goals. Late in 2020, the EU has presented a soil strategy and announced a "Soil Health Law" for 2023.

Putting the above methods into practice is a challenge, of course, because it requires much knowledge and must be adapted to local conditions. Some of these efforts will require several years of persistent implementation to achieve reliable results and to overcome concerns about financial risks and other criticisms from the more conservative farming community. There is already a small but growing number of farmers successfully using these techniques. It is increasingly likely that others will follow. Interest in field days by these innovative farmers is steadily increasing around the world.

An important conclusion is that only a combination of approaches can help mitigate climate change. But it is even more important to show how agricultural practices that increase soil organic matter also support improved food production, greater biodiversity, increased water storage and drought resilience, and other important ecosystem services, thereby creating a win-win solution for farmers and society at large. The current structures underpinning the "industrialized agricultural system" are complex and well-established, involving farmers, machinery, and chemical manufacturers, markets and trade, taxes and subsidies, not least resulting in low consumer prices. A broad implementation of the approaches described above can only be achieved with the active support of governments, while the development of the regenerative agricultural movement is currently mainly bottom-up.

Although many of the practices described come with costs, some of them will generate revenue and cost savings. The costs we are willing to bear for them determines the amount of carbon removed from the atmosphere. Price tags vary but suggest that at \$20–\$100 per metric ton of carbon much of the technical carbon sequestration potential could be realized (UNEP, 2017; McKinsey & Company, 2009).

The Five Principles

… of carbon storage in soil and regenerative agriculture are based on the motto "do as nature does":

1. Protect the soil surface,
2. Minimize soil disturbance,
3. Use a high diversity of plants and animals,
4. Preserve living plant-root networks,
5. Integrate animals into arable farming

Taking this into account, the following cross-cutting measures should be prioritized by policy makers whenever the aim is to increase humus content and thus to transfer carbon back into the soil:

- *Combat soil degradation and support land regeneration.* Agricultural practices have reduced soil fertility and degraded large parts of the land surface. Given the regenerative powers of nature, such areas can, with suitable expertise, be restored.
- *Promote agroecological practices that increase the amount of humus and pay farmers for storing carbon in the soil.* A small but growing number of farmers are using various innovative methods that use nature as a model for increasing humus levels and thus many other "ecosystem services." These best practices should be supported, communicated, and, if successful, widely disseminated, at both the national and international levels.
- *Popularize agroecology and holistic food system approaches in politics, education, and research.* Holistic thinking in the above-mentioned methods can be seen as a paradigm shift in the agricultural sector which, however, impedes an instant breakthrough. Knowledge about these agroecological approaches should be promoted through politics, education, and research to enable a faster and more efficient transition.
- *Improve knowledge, communication, training, and networking of/for practitioners to increase humus content, sustainable soil management, and agroecological practices and approaches.* The dissemination of this knowledge currently happens through local initiatives and small regional to international networks. Governments and other institutions should support these efforts toward a new future for agriculture.
- *Focus not only on yield, but also on other "ecosystem services" that farmers can contribute to* (carbon sequestration, climate regulation, water storage and filtration, erosion control, biodiversity, nutrient-rich food, and others). Our current system mainly looks at the parameter "yield per hectare" as an indicator of success, neglecting other important factors of sustainable practice. These should be made more prominent through education.
- *Successively restructure fossil-energy and agrochemical subsidies to encourage diversification of agroecological practices.* The current practice of industrial agriculture is heavily dependent on inputs and threatens the underlying basis of its own production system—soil, biodiversity, water, and climate. Shifting the

focus to diversified agroecological practices can help promote the very resources we depend on to produce diverse and healthy food.

- *Support agriculture and forestry as sectors that can potentially contribute to climate change mitigation.* Agriculture and forestry can be important sectors for climate mitigation as they have the potential to store large quantities of carbon in the biophysical realm, while providing important benefits to our society.
- *Support campaigns to conserve and revitalize soils, such as SaveOurSoils and 4p1000.* There are several international initiatives working to promote this issue as part of the political agenda.

The 4P1000 Initiative
… is the most prominent and politically active movement to advance the issue of carbon sequestration in combination with agroecological practices.

Launched by France at COP-21 in December 2015, this initiative brings together public and private sector stakeholders (local, regional, and national governments, businesses, trade organizations, NGOs, research institutions, etc.) under the Lima-Paris Action Plan (LPAP). Over 40 countries and over 1000 institutions and organizations worldwide have joined this movement. The 4p1000 initiative provides a space for collaborative interaction among scientists, policy makers, and practitioners to ensure that actions are scientifically sound. The initiative is very active at the policy level and promotes science, as it has also proposed a research program to support the initiative's goals. In addition, Regeneration International, a cooperation of more than 350 companies, farmers, and institutions, is working to raise awareness and scientific knowledge in this area and on the application side.

- *Help initiate emissions trading and/or expanding it to new sectors such as agriculture and agroforestry.* Although the success of existing emissions trading is limited, a prominent concern on our political agenda should be to integrate agriculture and forestry into existing systems and to adapt them to promote regenerative practices that support carbon sequestration.
- *Develop strategies for the provision of agricultural products that promote sustainable land management through public procurement where appropriate.* The transition to sustainable land management practices may increase costs and/or reduce returns to farmers in the early years. As the current economic model does not usually factor land degradation into the cost of production, farmers should receive support from governments, markets, and consumers to develop appropriate farming practices.
- *Improve research for soil carbon sequestration methods to generate knowledge to support action.* Best practices must be identified, monitored, verified, publicized, and promoted with science-based harmonized protocols and standards to increase reliable knowledge of successful approaches.

The potential for carbon sequestration in soils through agriculture can play an important role in mitigating climate change. Although the calculated values represent important contributions, the hope of putting all these techniques into practice quickly on a global scale is not realistic. But because the benefits of regenerative agriculture are so rich, as outlined above, there should be an overarching interest in investing in regenerative agricultural methods.

References

Bai, Z., Dent, D., Wu, Y., & de Jong, R. (2013). Land degradation and ecosystem services. In R. Lal, K. Lorenz, R. F. Hüttl, B. U. Schneider, & J. von Braun (Eds.), *Ecosystem services and carbon sequestration in the biosphere* (pp. 357–381). Springer. https://doi.org/10.1007/978-94-007-6455-2_15

Bardgett, R. D., & van der Putten, W. H. (2014). Belowground biodiversity and ecosystem functioning. *Nature, 515*(7528), 505–511. https://doi.org/10.1038/nature13855

Bonaudo, T., Bendahan, A. B., Sabatier, R., Ryschawy, J., Bellon, S., Leger, F., Magda, D., & Tichit, M. (2014). Agroecological principles for the redesign of integrated crop–livestock systems. *European Journal of Agronomy, 57*, 43–51. https://doi.org/10.1016/j.eja.2013.09.010

Brooker, R. W., Bennett, A. E., Cong, W.-F., Daniell, T. J., George, T. S., Hallett, P. D., Hawes, C., et al. (2015). Improving intercropping: A synthesis of research in agronomy, plant physiology and ecology. *New Phytologist, 206*(1), 107–117. https://doi.org/10.1111/nph.13132

Brown, G. (2020). *Aus toten Böden wird fruchtbare Erde: Eine Familie entdeckt die regenerative Landwirtschaft.* Kopp Verlag.

Bundesinformationszentrum Landwirtschaft. (n.d.). *Bodenzustandserhebung: So viel Humus steckt unter deutschen Äckern und Wiesen.* https://praxis-agrar.de/pflanze/ackerbau/bodenzustandserhebung-humus

Cao, Z., Li, D., & Han, X. (2011). The fungal to bacterial ratio in soil food webs, and its measurement. *Shengtai Xuebao/Acta Ecologica Sinica, 31*(16), 4741–4748.

Corbeels, M., Cardinael, R., Naudin, K., Guibert, H., & Torquebiau, E. (2018). The 4 per 1000 goal and soil carbon storage under agroforestry and conservation agriculture systems in sub-Saharan Africa. *Soil and Tillage Research, 188*, 16–26. https://doi.org/10.1016/j.still.2018.02.015

Eisenhauer, N., Lanoue, A., Strecker, T., Scheu, S., Steinauer, K., Thakur, M. P., & Mommer, L. (2017). Root biomass and exudates link plant diversity with soil bacterial and fungal biomass. *Scientific Reports, 7*, 44641. https://doi.org/10.1038/srep44641

FAO and ITPS. (2015). *Status of the world's soil resources* (pp. 1–94). FAO.

Franzluebbers, A., & Stuedemann, J. (2008). Soil physical responses to cattle grazing cover crops under conventional and no tillage in the Southern Piedmont USA. *Soil and Tillage Research, 100*(1–2), 141–153. https://doi.org/10.1016/j.still.2008.05.011

Franzluebbers, A. J., & Stuedemann, J. A. (2014). Crop and cattle production responses to tillage and cover crop management in an integrated crop–livestock system in the southeastern USA. *European Journal of Agronomy, 57*, 62–70. https://doi.org/10.1016/j.eja.2013.05.009

Glover, J., Duggan, J., & Jackson, L. (2011). *A novel perennial pasture and winter wheat conservation agriculture intercrop system for central USA.* Science and Technology Policy Fellow, U. 4.

Griscom, B. W., Adams, J., Ellis, P. W., Houghton, R. A., Lomax, G., Miteva, D. A., Schlesinger, W. H., et al. (2017). Natural climate solutions. *Proceedings of the National Academy of Sciences, 114*(44), 11645–11650. https://doi.org/10.1073/pnas.1710465114

Hagemann, N., Kammann, C. I., Schmidt, H.-P., Kappler, A., & Behrens, S. (2017). Nitrate capture and slow release in biochar amended compost and soil. Edited by Jorge Paz-Ferreiro. *PLoS One, 12*(2), e0171214. https://doi.org/10.1371/journal.pone.0171214

Houghton, R. A., & Nassikas, A. A. (2017). Global and regional fluxes of carbon from land use and land cover change 1850-2015: Carbon emissions from land use. *Global Biogeochemical Cycles, 31*(3), 456–472. https://doi.org/10.1002/2016GB005546

Intergovernmental Panel on Climate Change [IPCC]. (2014). *Climate change 2013: The physical science basis; Working Group I contribution to the fifth assessment report of the Intergovernmental Panel on Climate Change.* Cambridge University Press.

IPCC. (2021). *Climate change 2021: The physical science basis* (ipcc.ch). Cambridge University Press. Accessed 11 August 2023.

Jeffery, S., Verheijen, F. G. A., van der Velde, M., & Bastos, A. C. (2011). A quantitative review of the effects of biochar application to soils on crop productivity using meta-analysis. *Agriculture, Ecosystems & Environment, 144*(1), 175–187. https://doi.org/10.1016/j.agee.2011.08.015

Jeffery, S., Abalos, D., Prodana, M., Bastos, A.C., van Groenigen, J.W., Hungate, B.A., Verheijen, F. (2017). Biochar boosts tropical but not temperate crop yields. Environmental Research Letters 12, 053001. https://doi.org/10.1088/1748-9326/aa67bd

Jones, C. E. (2008). Liquid carbon pathway unrecognised. *Australian Farm Journal, 8*(5), 15–17.

Kammann, C. I., Schmidt, H.-P., Messerschmidt, N., Linsel, S., Steffens, D., Müller, C., Koyro, H.-W., Conte, P., & Joseph, S. (2015). Plant growth improvement mediated by nitrate capture in co-composted biochar. *Scientific Reports, 5*(1), 11080. https://doi.org/10.1038/srep11080

Kätterer, T., Bolinder, M. A., Andrén, O., Kirchmann, H., & Menichetti, L. (2011). Roots contribute more to refractory soil organic matter than above-ground crop residues, as revealed by a long-term field experiment. *Agriculture, Ecosystems & Environment, 141*(1–2), 184–192. https://doi.org/10.1016/j.agee.2011.02.029

LaCanne, C. E., & Lundgren, J. G. (2018). Regenerative agriculture: Merging farming and natural resource conservation profitably. *PeerJ, 6*, e4428. https://doi.org/10.7717/peerj.4428

Lal, R. (2004). Agricultural activities and the global carbon cycle. *Nutrient Cycling in Agroecosystems, 70*(2), 103–116. https://doi.org/10.1023/B:FRES.0000048480.24274.0f

Lal, R. (2010). Managing soils and ecosystems for mitigating anthropogenic carbon emissions and advancing global food security. *Bioscience, 60*(9), 708–721. https://doi.org/10.1525/bio.2010.60.9.8

Lal, R. (2016a). Beyond COP 21: Potential and challenges of the "4 per thousand" initiative. *Journal of Soil and Water Conservation, 71*(1), 20A–25A. https://doi.org/10.2489/jswc.71.1.20A

Lal, R. (2016b). Soil health and carbon management. *Food and Energy Security, 5*(4), 212–222. https://doi.org/10.1002/fes3.96

Lal, R. (2018). Digging deeper: A holistic perspective of factors affecting soil organic carbon sequestration in agroecosystems. *Global Change Biology, 24*(8), 3285–3301. https://doi.org/10.1111/gcb.14054

Le Quéré, C., Andrew, R. M., Canadell, J. G., Sitch, S., Korsbakken, J. I., Peters, G. P., Manning, A. C., et al. (2016). Global carbon budget 2016. *Earth System Science Data, 8*(2), 605–649. https://doi.org/10.5194/essd-8-605-2016

Lehmann, J., Rillig, M. C., Thies, J., Masiello, C. A., Hockaday, W. C., & Crowley, D. (2011). Biochar effects on soil biota—A review. *Soil Biology and Biochemistry, 43*(9), 1812–1836. https://doi.org/10.1016/j.soilbio.2011.04.022

Lehmann, J., Czimczik, C., Laird, D., & Sohi, S. (2015). Stability of biochar in soil. In *Biochar for environmental management: Science, technology and implementation* (pp. 235–282). Taylor & Francis Ltd.

Leigh, J., Hodge, A., & Fitter, A. H. (2009). Arbuscular mycorrhizal fungi can transfer substantial amounts of nitrogen to their host plant from organic material. *New Phytologist, 181*(1), 199–207. https://doi.org/10.1111/j.1469-8137.2008.02630.x

Lugato, E., Bampa, F., Panagos, P., Montanarella, L., & Jones, A. (2014). Potential carbon sequestration of European arable soils estimated by modelling a comprehensive set of management practices. *Global Change Biology, 20*(11), 3557–3567. https://doi.org/10.1111/gcb.12551

Lundgren, J. G., & Fausti, S. W. (2015). Trading biodiversity for pest problems. *Science Advances, 1*(6), e1500558. https://doi.org/10.1126/sciadv.1500558

Machmuller, M. B., Kramer, M. G., Cyle, T. K., Hill, N., Hancock, D., & Thompson, A. (2015). Emerging land use practices rapidly increase soil organic matter. *Nature Communications, 6*, 6995. https://doi.org/10.1038/ncomms7995

Mäder, P., & Berner, A. (2012). Development of reduced tillage systems in organic farming in Europe. *Renewable Agriculture and Food Systems, 27*(01), 7–11. https://doi.org/10.1017/S1742170511000470

Mateo-Sagasta, J., Zadeh, S. M., Turral, H., & Burke, J. (2017). *Water pollution from agriculture: A global review – Executive summary*. FAO.

McKinsey & Company. (2009). *Pathways to a low carbon economy*. McKinsey & Company.

Mekonnen, M. M., & Hoekstra, A. Y. (2015). Global gray water footprint and water pollution levels related to anthropogenic nitrogen loads to fresh water. *Environmental Science & Technology, 49*(21), 12860–12868. https://doi.org/10.1021/acs.est.5b03191

Minasny, B., Malone, B. P., McBratney, A. B., Angers, D. A., Arrouays, D., Chambers, A., Chaplot, V., et al. (2017). Soil carbon 4 per mille. *Geoderma, 292*, 59–86. https://doi.org/10.1016/j.geoderma.2017.01.002

Montagnini, F., Ibrahim, M., & Murgueitio Restrepo, E. (2013). Silvopastoral systems and climate change mitigation in Latin America. *Bois & Forets Des Tropiques, 67*(316), 3–16.

Moss, B. (2008). Water pollution by agriculture. *Philosophical Transactions of the Royal Society B: Biological Sciences, 363*(1491), 659–666. https://doi.org/10.1098/rstb.2007.2176

NAS and National Academies of Sciences, Engineering, and Medicine. (2018). *Negative emissions technologies and reliable sequestration: A research agenda*. National Academies Press. https://doi.org/10.17226/25259

Panagos, P., Borrelli, P., Poesen, J., Ballabio, C., Lugato, E., Meusburger, K., Montanarella, L., & Alewell, C. (2015). The new assessment of soil loss by water erosion in Europe. *Environmental Science & Policy, 54*, 438–447. https://doi.org/10.1016/j.envsci.2015.08.012

Paustian, K., Lehmann, J., Ogle, S., David Reay, G., Robertson, P., & Smith, P. (2016). Climate-smart soils. *Nature, 532*(7597), 49–57. https://doi.org/10.1038/nature17174

Pellerin, S., Bamière, L., Launay, C., Martin, R., Schiavo, M., Angers, D., Augusto, L., et al. (2019). *A model-based assessment of the soil C storage potential at the national scale: A case study from France*. https://doi.org/10.15454/1.5433098269609653E12

Pieterse, C. M. J., Zamioudis, C., Berendsen, R. L., Weller, D. M., Van Wees, S. C. M., & Bakker, P. A. H. M. (2014). Induced systemic resistance by beneficial microbes. *Annual Review of Phytopathology, 52*(1), 347–375. https://doi.org/10.1146/annurev-phyto-082712-102340

Pimentel, D., & Burgess, M. (2013). Soil erosion threatens food production. *Agriculture, 3*(3), 443–463. https://doi.org/10.3390/agriculture3030443

Poeplau, C., & Don, A. (2015). Carbon sequestration in agricultural soils via cultivation of cover crops—A meta-analysis. *Agriculture, Ecosystems & Environment, 200*, 33–41. https://doi.org/10.1016/j.agee.2014.10.024

Polyface Farms. www.polyfacefarms.com

Powlson, D. S., Stirling, C. M., Jat, M. L., Gerard, B. G., Palm, C. A., Sanchez, P. A., & Cassman, K. G. (2014). Limited potential of no-till agriculture for climate change mitigation. *Nature Climate Change, 4*(8), 678–683. https://doi.org/10.1038/nclimate2292

Reber, M. (2017). *Aufbauende Landwirtschaft (2017)*. www.youtube.com/watch?v=WP3NEyNgYq4

Roser, M., et al. (n.d.). *Military personnel and spending*. Published online at OurWorldInData.org. https://ourworldindata.org/military-spending

Ryals, R., & Silver, W. L. (2013). Effects of organic matter amendments on net primary productivity and greenhouse gas emissions in annual grasslands. *Ecological Applications, 23*(1), 46–59. https://doi.org/10.1890/12-0620.1

Sanderman, J., Hengl, T., & Fiske, G. J. (2018). Soil carbon debt of 12,000 years of human land use. *Proceedings of the National Academy of Sciences, 114*(36), 9575–9580.

Schmidt, H. P., et al. (2017). Biochar-based fertilization with liquid nutrient enrichment: 21 field trials covering 13 crop species in Nepal. *Land Degradation and Development, 28*(8), 2324–2342.

Seis, C. (2006). Pasture cropping as a means to managing land. *Australian Organic Journal, 66*, 42–43.

Shi, L., Feng, W., Jianchu, X., & Kuzyakov, Y. (2018). Agroforestry systems: Meta-analysis of soil carbon stocks, sequestration processes, and future potentials. *Land Degradation & Development, 29*(11), 3886–3897. https://doi.org/10.1002/ldr.3136

Smith, P., Martino, D., Cai, Z., Gwary, D., Janzen, H., Kumar, P., McCarl, B., et al. (2008). Greenhouse gas mitigation in agriculture. *Philosophical Transactions of the Royal Society B: Biological Sciences, 363*(1492), 789–813. https://doi.org/10.1098/rstb.2007.2184

Soussana, J.-F., Lutfalla, S., Ehrhardt, F., Rosenstock, T., Lamanna, C., Havlík, P., Richards, M., et al. (2017). Matching policy and science: Rationale for the "4 per 1000 – Soils for food security and climate" initiative. *Soil and Tillage Research, 188*, 3–15. https://doi.org/10.1016/j.still.2017.12.002

Steinauer, K., Chatzinotas, A., & Eisenhauer, N. (2016). Root exudate cocktails: The link between plant diversity and soil microorganisms? *Ecology and Evolution, 6*(20), 7387–7396. https://doi.org/10.1002/ece3.2454

Teague, R., & Barnes, M. (2017). Grazing management that regenerates ecosystem function and grazingland livelihoods. *African Journal of Range & Forage Science, 34*(2), 77–86. https://doi.org/10.2989/10220119.2017.1334706

Teague, W. R., Apfelbaum, S., Lal, R., Kreuter, U. P., Rowntree, J., Davies, C. A., Conser, R., et al. (2016). The role of ruminants in reducing agriculture's carbon footprint in North America. *Journal of Soil and Water Conservation, 71*(2), 156–164. https://doi.org/10.2489/jswc.71.2.156

Toensmeier, E. (2018). Perennial staple crops and agroforestry for climate change mitigation. In *Integrating landscapes: Agroforestry for biodiversity conservation and food sovereignty* (pp. 439–451). Springer.

UNEP. (2017). *The emissions gap report 2017.*

Walder, F., Niemann, H., Natarajan, M., Lehmann, M. F., Boller, T., & Wiemken, A. (2012). Mycorrhizal networks: Common goods of plants shared under unequal terms of trade. *Plant Physiology, 159*(2), 789–797. https://doi.org/10.1104/pp.112.195727

Weng, Z. H., Van Zwieten, L., Singh, B. P., Tavakkoli, E., Joseph, S., Macdonald, L. M., Rose, T. J., et al. (2017). Biochar built soil carbon over a decade by stabilizing rhizodeposits. *Nature Climate Change, 7*(5), 371–376. https://doi.org/10.1038/nclimate3276

Woolf, D., Amonette, J. E., Alayne Street-Perrott, F., Lehmann, J., & Joseph, S. (2010). Sustainable biochar to mitigate global climate change. *Nature Communications, 1*(5), 1–9. https://doi.org/10.1038/ncomms1053

Ye, L., Camps-Arbestain, M., Shen, Q., Lehmann, J., Singh, B., & Sabir, M. (2020). Biochar effects on crop yields with and without fertilizer: A meta-analysis of field studies using separate controls. Edited by Leo M. Condron. *Soil Use and Management, 36*(1), 2–18. https://doi.org/10.1111/sum.12546

Zimmermann, A. R., & Gao, B. (2013). The stability of biochar in the environment. In *Biochar and soil biota.* Taylor & Francis Group. https://www.taylorfrancis.com/books/e/9781466576513/chapters/10.1201%2Fb14585-3

Zomer, R. J., Bossio, D. A., Sommer, R., & Verchot, L. V. (2017). Global sequestration potential of increased organic carbon in cropland soils. *Scientific Reports, 7*(1), 15554. https://doi.org/10.1038/s41598-017-15794-8

Strengthen Terrestrial Water Cycles

Evaporative Cooling as a Forgotten Climate Opportunity

Stefan Schwarzer

Vegetation plays an important, often neglected role in regulating the climate. Imagine the difference you feel when, on a hot summer day, you stand either in an open field or in a dense forest. It is self-evident that the major change brought about by the conversion of forests into farmland or urban areas greatly affected the climate.

Of the solar radiation that hits an area densely covered with vegetation, only 1% is used for photosynthesis (Fig. 1), between 5% and 10% heats the air ("sensible heat"). More than 70% of the radiation is used by plants for transpiration, converting liquid water into water vapor, which is a very energy-intensive process ("latent heat"). Including unplanted areas and water surfaces, about 50% of the solar energy reaching the Earth's surface is used for evaporation and transpiration of water ("evapotranspiration"[1]) (Pokorny et al., 2010; Jasechko et al., 2013). As air masses rise, the water vapor condenses and releases the same amount of energy that was consumed at ground level, with some escaping into space. The resulting clouds reflect incident solar radiation and are the source of new precipitation.

On land surfaces, annual precipitation amounts to about 120,000 km^3, of which around 50% originate from the oceans and 50% from land (Fig. 2) (Ellison et al., 2019). About 60–80% of this land-sourced atmospheric moisture derives from plant transpiration (Wei et al., 2017), which shows the important role of vegetation in the

---

[1] Sum of direct evaporation and release mainly by plants (transpiration).

This chapter is based on United Nations Environment Program. (2021, July). *Working with plants, soils and water to cool the climate and rehydrate Earth's landscapes* (Foresight Brief No. 025). Available at: https://wedocs.unep.org/20.500.11822/36619. Accessed: 19 September 2023.

S. Schwarzer (✉)
Climate Landscapes, Germany
e-mail: stefan@climate-landscapes.org

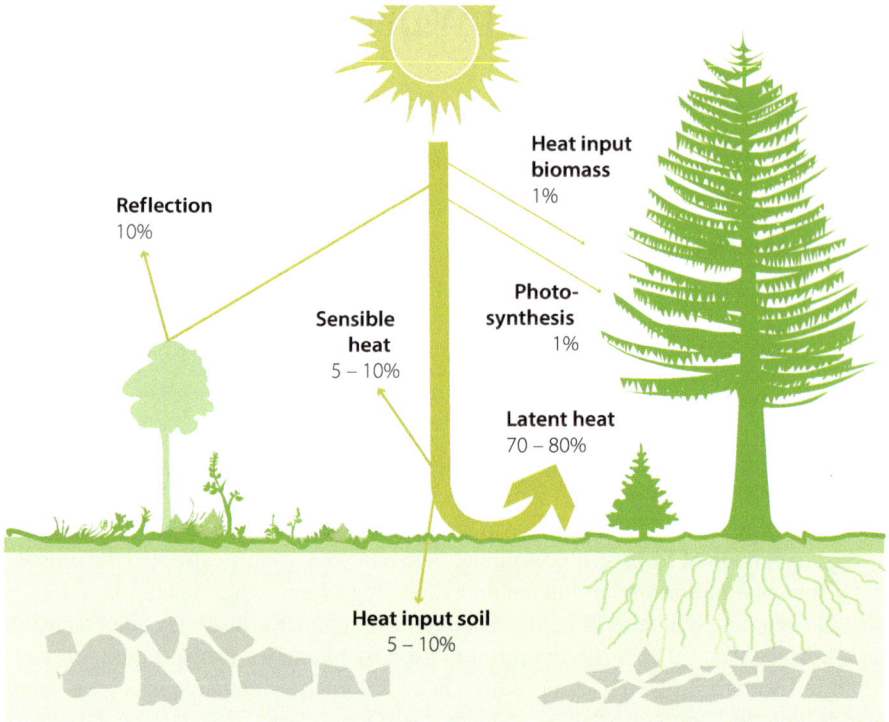

Fig. 1 Distribution of solar energy incident on vegetation (Latent and sensible heat are types of energy that are released or absorbed in the atmosphere. Latent heat refers to phase changes among liquid, gaseous, and solid. Sensible heat refers to temperature changes of a gas or object without a phase change) (Pokorny et al., 2010)

precipitation cycle as well as in the transfer of energy from the soil to the upper atmosphere.

It was assumed until recently that human influence on atmospheric water vapor is negligible, as the focus was mainly on industrial processes. It is now known that anthropogenic changes in land cover make this influence substantial (Kravčík et al., 2007; van der Ent et al., 2010; Mahmood et al., 2014)—with the main factor being deforestation which, since the beginning of agriculture, has wiped out nearly half of the world's forests.

Trees Provide Cooling and Generate Water Vapor

Every tree in the forest is a fountain that with its roots sucks water from the ground, pumps it through its trunk, branches, and leaves, and releases it through its leaves into the atmosphere as water vapor. On a normal sunny day, a single tree can evaporate several hundred liters of water, providing its surroundings with 70 kWh of

Fig. 2 Global water flows. Of the 120,000 km³ of precipitation over land, 50% comes from the oceans and 50 % from land areas. Of the latter amount, about 70% comes from plant transpiration and 30% from water bodies and soils. 32,000 km³ of evapotranspiration on land returns to the ocean via atmospheric moisture; 40,000 km³ are discharged into the oceans via rivers. (Data from van der Ent et al. (2010))

cooling per 100 liters, which is equivalent to the cooling effect of two air-conditioners running for 24 h (Pokorny, 2012; Ellison et al., 2017). Billions of trees together generate huge rivers of water in the air ("flying rivers")—rivers that form clouds and generate precipitation hundreds or even thousands of kilometers away (Nobre, 2014; Weng et al., 2018).

Globally, 50% of precipitation falling over land comes from moisture produced by evapotranspiration over land, mainly from transpiring trees (Eltahir & Bras, 1994; van der Ent et al., 2010; Keys et al., 2016; Ellison et al., 2017; Staal et al., 2018). In some regions of the world, the share is 70% (or more) of precipitation (van der Ent et al., 2010), with higher shares inland (Fig. 3).

Tropical evergreen deciduous forests occupy only about 10% of the Earth's land surface but contribute 22% of global evapotranspiration, (Wang-Erlandsson et al., 2014) which underscores their importance for the trans-regional hydrological cycle. The typical distances that moisture evaporated from land travels in the atmosphere before falling back onto land are in the order of 500–5000 km; the typical time scale ranges from 8 to 10 days (van der Ent & Savenije, 2011; van der Ent & Tuinenburg, 2017). For example, moisture evaporating from the Eurasian continent is responsible for 80% of China's water resources (van der Ent et al., 2010). The main source

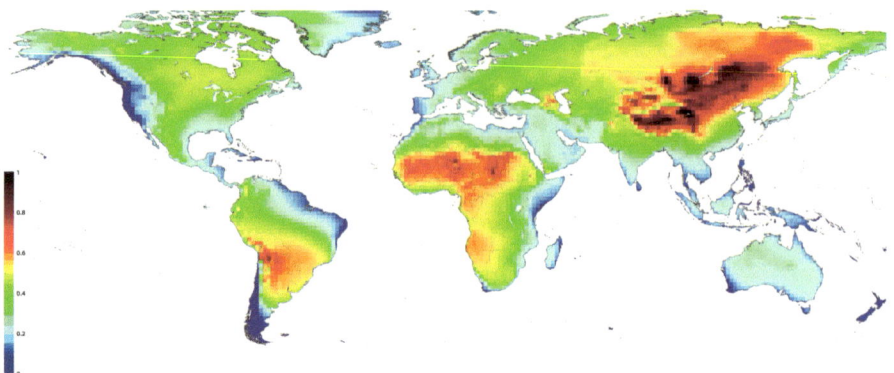

Fig. 3 Average continental precipitation recycling ratio (1999–2008). The higher the number, the more precipitation comes from land evaporation (van der Ent et al., 2010; van der Ent, 2014)

of rainfall in the Congo Basin is moisture evaporated over East Africa, while the Congo Basin is in turn an important source of rainfall in the Sahel (van der Ent et al., 2010). The condition of the West African rainforest is especially important for precipitation in the Ethiopian highlands and thus in turn for the runoff of the Nile (Gebrehiwot et al., 2019). This explains why even in large river basins such as those of the Amazon, the Congo, and the Yangtze, rainfall is more influenced by land use changes outside than inside the basins. Even in river basins that do not span several countries, runoff has been significantly affected by land use in other countries (Wang-Erlandsson et al., 2018).

Altered Heat Flows, Altered Atmospheric Patterns

Models show that local conversion of forests or grasslands to cropland can reduce annual terrestrial evapotranspiration by 30–40% (Sterling et al., 2013). On a global scale, land cover change between 1950 and 2000 resulted in a reduction in terrestrial evapotranspiration by 4–5%, or 3000–3500 km^3, and a 6.8% increase in surface water runoff (Gordon et al., 2005; Sterling et al., 2013). On the other hand, scientists have found that vegetation has a cooling effect due to increased efficiency in the vertical movement of heat and water vapor between the land surface and the atmosphere (Chen et al., 2020).

Satellite observations indicate that forests have a major influence on cloud formation, not only in the tropics but also in temperate zones: disappearance of forests can lead to a substantial decrease in local cloud cover and thus in precipitation (Teuling et al., 2017). Model calculations have shown that large-scale global deforestation between 1700 and 1850 has led to a decline in monsoon rainfall over the Indian subcontinent and southeastern China, and a consequent weakening of the

Asian summer monsoon circulation (Takata et al., 2009). In the tropics, deep cumulus convection has changed significantly due to land use changes (especially the conversion of forest to cropland). This affects not only local precipitation, but also has long-distance impacts through processes known as remote effects (or "teleconnections"). These can affect higher latitudes, significantly altering the weather in these regions (Sheil & Murdiyarso, 2009; Gebrehiwot et al., 2019). Even relatively small disturbances in land cover in the tropics can lead to effects at higher latitudes (Chase et al., 2000); for example, changes in Amazonia can lead to effects in the north-west of the United States (Medvigy et al., 2013). Forest disappearance can also lead to lower rainfall and longer dry seasons, as reported from Rondônia in Brazil (Coe et al., 2017) and from Borneo, where water catchment areas with the greatest forest loss were found to have suffered a 15% decline in rainfall (McAlpine et al., 2018). In India, patterns of decreasing rainfall during the monsoon were accompanied by changes in forest cover, due to reduced evapotranspiration with subsequent decline in the recycled rainfall component (Paul et al., 2016).

Back Radiation from Bare Ground

Normally, more than 50% of the solar radiation hitting the Earth's surface is converted into latent heat by evapotranspiration, which in turn enters the atmosphere, feeds the precipitation cycle, and partially radiates back into space.

On bare surfaces, such as fallow fields, dry meadows (in summer and after the hay harvest) as well as on concrete or asphalt surfaces, the ground absorbs more incident solar radiation, thus heats up more, generates sensible heat, and releases thermal energy to the atmosphere which, according to the Stefan-Boltzman law, increases in proportion to the fourth power of its absolute temperature (Fig. 4). The differences in surface temperature between these bare areas and the forested areas can be up to 20 °C on Central European summer afternoons (Fig. 5) (Hesslerová et al., 2013). On the Indonesian island of Sumatra, temperature differences between forested and bare areas of up to 10 °C have been observed, which in turn can be explained by the evaporative cooling effect of the forests, which outweighs the albedo heating effect resulting from forest areas being darker (Sabajo et al., 2017). Local biophysical processes triggered by forest loss can therefore cause a net increase in summer temperatures in all regions of the world (Alkama & Cescatti, 2016).

Historical deforestation has actually reduced latent heat flow on land and increased sensible heat on the ground (Bounoua et al., 2002; Brovkin et al., 2006). Deforestation has caused significant warming from 2003 to 2013, averaging up to 0.28 °C in tropical areas and up to 0.32 °C in temperate regions of the southern hemisphere (Li et al., 2016). At the current rate of deforestation, the loss of tropical forests could increase global temperatures by 1.5 °C by 2100, not taking into account other human-induced temperature increases (Mahowald et al., 2017).

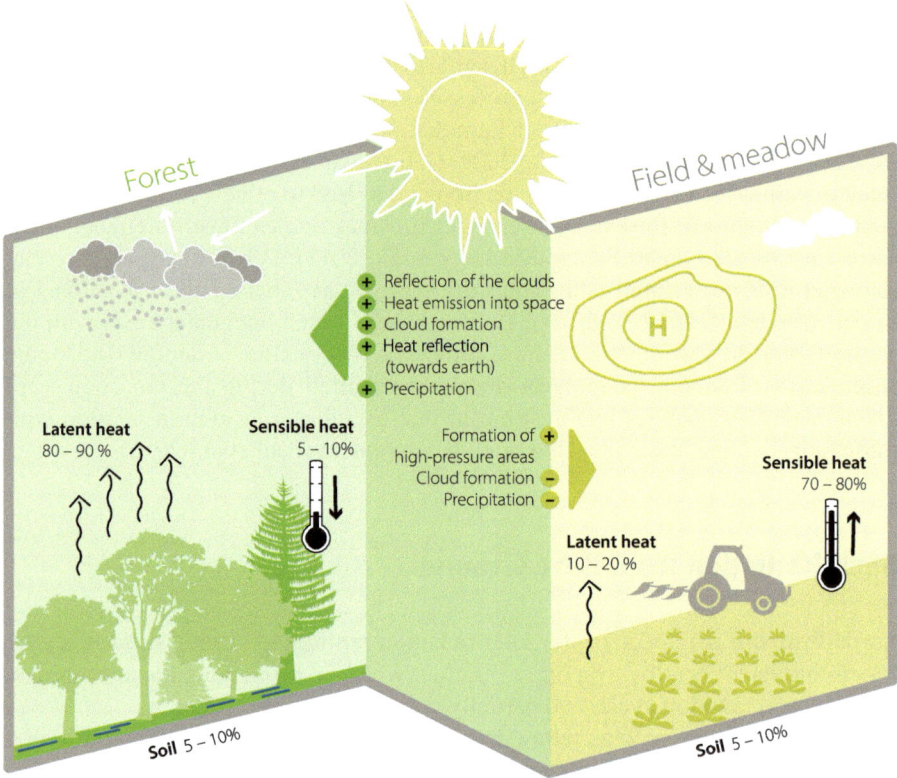

Fig. 4 Evapotranspiration lowers ground temperature and it increases cloud albedo, radiation to space during the condensation process, cloud formation, and thus precipitation. Removal of vegetation increases ground temperature, radiates exponentially increasing amounts of heat energy as ground temperature increases, creates high-pressure areas that impede the passage of low pressure areas (and thus potential precipitation bringers), reduces cloud formation potential and thus precipitation. (Data from: various textbooks on climatology, own design)

Between 1950 and 2000, surface temperature increased by 0.3 °C worldwide due to changes in land cover (Sterling et al., 2013). Disturbances in the surface energy balance caused by vegetation changes between 2000 and 2015 have led to an average 0.23 °C increase in local surface temperatures (Duveiller et al., 2018). The average warming due to land cover changes could explain 18–40% of current global warming trends through the reduction in evapotranspiration that overcompensates for the increase in surface albedo (Ban-Weiss et al., 2011; Alkama & Cescatti, 2016; Wolosin & Harris, 2018).

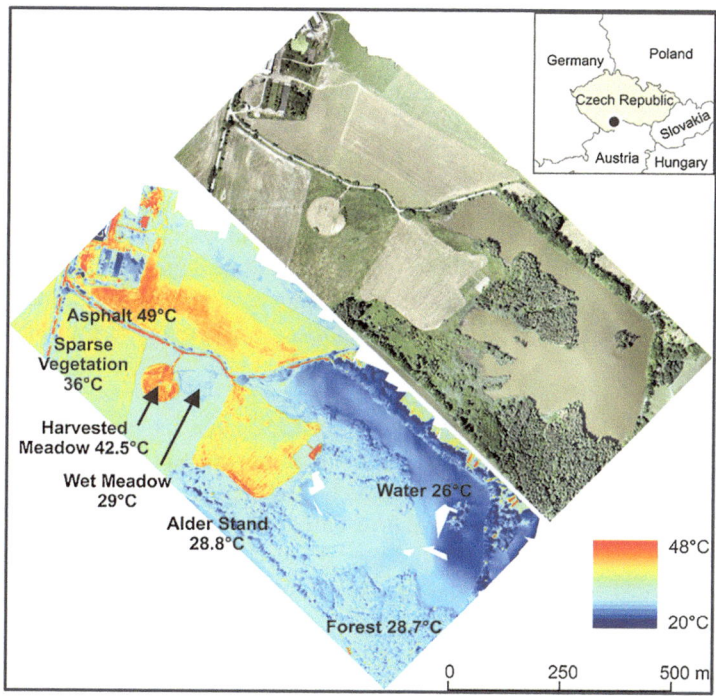

Fig. 5 Surface temperature distribution in a mixed landscape (Hesslerová et al., 2013; Ellison et al., 2017)

Biogenic Aerosols for Cloud Formation

In addition to the importance of forests for energy flows and as sources of precipitation, large forests also appear to be biogeochemical reactors in which the biosphere and atmospheric photochemistry generate nuclei for cloud and precipitation formation, thus maintaining the hydrological cycle. Trees produce volatile organic compounds and release microorganisms, bacteria, fungal spores, pollen, and other biological particles, which get into the air after rainfall from leaf surfaces, especially of trees. Once in the atmosphere, they contribute to the formation of condensation nuclei, which in turn affect cloud formation and precipitation. Biogenic aerosols can also help increase the freezing temperature by forming ice nuclei, without which freezing would occur only at cloud temperatures of −15 °C or below. Facilitated by such ice nuclei, freezing can occur at temperatures close to 0 °C, which enables efficient cloud formation and promotes (local) rain events.

Policy Implications

Vegetation, fertile soils, and water retention must be recognized as key regulators of the water, energy, and carbon cycles. Some of the policy implications are listed below and should be implemented if possible.

- We should be aware of the positive feedback loops. As explained earlier, deforestation makes land areas and the climate drier and warmer. This leads to conditions that increase the risk of forest and vegetation fires, which in turn release CO_2 and lead to further deforestation, creating a vicious cycle. Climate change, deforestation, drought, and forest fires form a triple cycle of reinforcing feedbacks (Fig. 6).
- In view of the long-distance effects of large forest ecosystems, these should be seen as a global goods. For example, the REDD+ mechanism developed under the UNFCCC could provide a model for recognizing and financing the international water and energy services that such forests provide.
- Especially important and sensitive forest areas should be protected and managed accordingly.
- It is of utmost importance to stop deforestation and to increase reforestation worldwide.
- Agricultural practices should focus on soil building, year-round plant cover, and use of agroforestry methods.

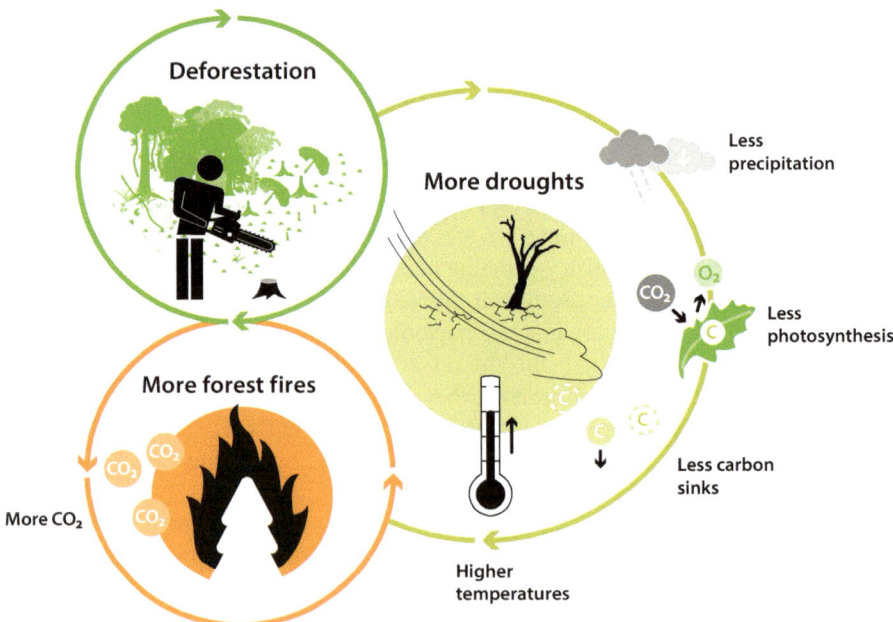

Fig. 6 Triple positive feedbacks between deforestation, drought, forest fires, and climate change. (Modified from Peduzzi (2012))

Conclusion

It is important to understand that the carbon, water, and energy cycles on land are closely interlinked. Restoring the atmospheric and terrestrial moisture cycles in vegetation, soils, and the atmosphere is paramount to cooling the planet and safeguarding precipitation patterns globally. The penalty for failure is the drying-out of terrestrial landscapes.

Stopping deforestation, increasing reforestation, and introducing agroforestry practices are essential if we are to avoid climate catastrophe. A systemic approach is needed to understand and harness the underlying patterns of rain formation. To restore rainfall to areas like the Sahel, trees must be planted not only in the region, but also on the coast to draw moist air from the ocean onto land (Ellison & Speranza, 2020).

Another important approach to supplying water and energy cycles involves increasing soil fertility, water retention, and soil protection through the practice of the regenerative organic movement, such as year-round vegetation cover through intercropping and undersowing or introduction of agroforestry. Finding ways to build up additional soil organic matter is one of the keys to success for large areas of the world which are currently under cultivation.

In general, we need a paradigm shift that values the hydrological and climate-cooling effects of vegetation, and particularly of forests, in addition to their carbon sequestration potentials. We must pay more attention to the beneficial effects that cover with vegetation, especially trees, has on the local, regional, and continental climate.

References

Alkama, R., & Cescatti, A. (2016). Biophysical climate impacts of recent changes in global forest cover. *Science, 351*, 600–604. https://doi.org/10.1126/science.aac8083

Ban-Weiss, G. A., Bala, G., Cao, L., et al. (2011). Climate forcing and response to idealized changes in surface latent and sensible heat. *Environmental Research Letters, 6*, 034032. https://doi.org/10.1088/1748-9326/6/3/034032

Bounoua, L., Defries, R., Collatz, G. J., et al. (2002). Effects of land cover conversion on surface climate. *Climatic Change, 52*, 29–64.

Brovkin, V., Claussen, M., Driesschaert, E., et al. (2006). Biogeophysical effects of historical land cover changes simulated by six Earth system models of intermediate complexity. *Climate Dynamics, 26*, 587–600. https://doi.org/10.1007/s00382-005-0092-6

Chase, T. N., Pielke, R. A., Sr., Kittel, T. G. F., et al. (2000). Simulated impacts of historical land cover changes on global climate in northern winter. *Climate Dynamics, 16*, 93–105. https://doi.org/10.1007/s003820050007

Chen, C., Li, D., Li, Y., et al. (2020). Biophysical impacts of Earth greening largely controlled by aerodynamic resistance. *Science Advances, 6*, eabb1981. https://doi.org/10.1126/sciadv.abb1981

Coe, M. T., Brando, P. M., Deegan, L. A., et al. (2017). The forests of the Amazon and Cerrado moderate regional climate and are the key to the future. *Tropical Conservation Science, 10,* 194008291772067. https://doi.org/10.1177/1940082917720671

Duveiller, G., Hooker, J., & Cescatti, A. (2018). The mark of vegetation change on Earth's surface energy balance. *Nature Communications, 9,* 679. https://doi.org/10.1038/s41467-017-02810-8

Ellison, D., & Speranza, C. I. (2020). From blue to green water and back again: Promoting tree, shrub and forest-based landscape resilience in the Sahel. *Science of the Total Environment, 739,* 140002. https://doi.org/10.1016/j.scitotenv.2020.140002

Ellison, D., Morris, C. E., Locatelli, B., et al. (2017). Trees, forests and water: Cool insights for a hot world. *Global Environmental Change, 43,* 51–61. https://doi.org/10.1016/j.gloenvcha.2017.01.002

Ellison, D., Wang-Erlandsson, L., van der Ent, R., & van Noordwijk, M. (2019). Upwind forests: Managing moisture recycling for nature-based resilience. *Unasylva, 70,* 13.

Eltahir, E. A. B., & Bras, R. L. (1994). Precipitation recycling in the Amazon basin. *Quarterly Journal of the Royal Meteorological Society, 120,* 861–880. https://doi.org/10.1002/qj.49712051806

Gebrehiwot, S. G., Ellison, D., Bewket, W., et al. (2019). The Nile Basin waters and the West African rainforest: Rethinking the boundaries. *Wiley Interdisciplinary Reviews: Water, 6,* e1317. https://doi.org/10.1002/wat2.1317

Gordon, L. J., Steffen, W., Jonsson, B. F., et al. (2005). Human modification of global water vapor flows from the land surface. *Proceedings of the National Academy of Sciences, 102,* 7612–7617. https://doi.org/10.1073/pnas.0500208102

Hesslerová, P., Pokorný, J., Brom, J., & Rejšková-Procházková, A. (2013). Daily dynamics of radiation surface temperature of different land cover types in a temperate cultural landscape: Consequences for the local climate. *Ecological Engineering, 54,* 145–154. https://doi.org/10.1016/j.ecoleng.2013.01.036

Jasechko, S., Sharp, Z. D., Gibson, J. J., et al. (2013). Terrestrial water fluxes dominated by transpiration. *Nature, 496,* 347–350. https://doi.org/10.1038/nature11983

Keys, P. W., Wang-Erlandsson, L., & Gordon, L. J. (2016). Revealing invisible water: Moisture recycling as an ecosystem service. *PLoS One, 11,* e0151993. https://doi.org/10.1371/journal.pone.0151993

Kravčík, M., Pokorný, J., Kohutiar, J., et al. (2007). *Water for the recovery of the climate – A new water paradigm* (pp. 1–94). People and Water NGO.

Li, Y., Zhao, M., Mildrexler, D. J., et al. (2016). Potential and actual impacts of deforestation and afforestation on land surface temperature: Impacts of forest change on temperature. *Journal of Geophysical Research: Atmospheres, 121,* 14,372–14,386. https://doi.org/10.1002/2016JD024969

Mahmood, R., Pielke, R. A., Hubbard, K. G., et al. (2014). Land cover changes and their biogeophysical effects on climate. *International Journal of Climatology, 34,* 929–953. https://doi.org/10.1002/joc.3736

Mahowald, N. M., Ward, D. S., Doney, S. C., et al. (2017). Are the impacts of land use on warming underestimated in climate policy? *Environmental Research Letters, 12,* 094016. https://doi.org/10.1088/1748-9326/aa836d

McAlpine, C. A., Johnson, A., Salazar, A., et al. (2018). Forest loss and Borneo's climate. *Environmental Research Letters, 13,* 044009. https://doi.org/10.1088/1748-9326/aaa4ff

Medvigy, D., Walko, R. L., Otte, M. J., & Avissar, R. (2013). Simulated changes in northwest U.S. climate in response to Amazon deforestation*. *Journal of Climate, 26,* 9115–9136. https://doi.org/10.1175/JCLI-D-12-00775.1

Nobre, A. D. (2014). *The future climate of Amazonia: Scientific assessment report* (pp. 1–42). CCST-INPE, INPA and ARA.

Paul, S., Ghosh, S., Oglesby, R., et al. (2016). Weakening of Indian summer monsoon rainfall due to changes in land use land cover. *Scientific Reports, 6,* 32177. https://doi.org/10.1038/srep32177

Peduzzi, P. (2012). *Risk and global change: Developing scientific methods for advocacy and awareness raising* (PhD thesis) (p. 33). Université de Lausanne.

Pokorny, J. (2012). *What can a tree do?*

Pokorny, J., Brom, J., Cermak, J., et al. (2010). Solar energy dissipation and temperature control by water and plants. *International Journal of Water, 5,* 311. https://doi.org/10.1504/IJW.2010.038726

Sabajo, C. R., le Maire, G., June, T., et al. (2017). Expansion of oil palm and other cash crops causes an increase of the land surface temperature in the Jambi province in Indonesia. *Biogeosciences, 14,* 4619–4635. https://doi.org/10.5194/bg-14-4619-2017

Sheil, D., & Murdiyarso, D. (2009). How forests attract rain: An examination of a new hypothesis. *Bioscience, 59,* 341–347. https://doi.org/10.1525/bio.2009.59.4.12

Staal, A., Tuinenburg, O. A., Bosmans, J. H. C., et al. (2018). Forest-rainfall cascades buffer against drought across the Amazon. *Nature Climate Change, 8,* 539–543. https://doi.org/10.1038/s41558-018-0177-y

Sterling, S. M., Ducharne, A., & Polcher, J. (2013). The impact of global land-cover change on the terrestrial water cycle. *Nature Climate Change, 3,* 385–390. https://doi.org/10.1038/nclimate1690

Takata, K., Saito, K., & Yasunari, T. (2009). Changes in the Asian monsoon climate during 1700-1850 induced by preindustrial cultivation. *Proceedings of the National Academy of Sciences, 106,* 9586–9589. https://doi.org/10.1073/pnas.0807346106

Teuling, A. J., Taylor, C. M., Meirink, J. F., et al. (2017). Observational evidence for cloud cover enhancement over western European forests. *Nature Communications, 8,* 14065. https://doi.org/10.1038/ncomms14065

van der Ent, R. J. (2014). *A new view on the hydrological cycle over continents.* https://doi.org/10.4233/uuid:0ab824ee-6956-4cc3-b530-3245ab4f32be

van der Ent, R. J., & Savenije, H. H. G. (2011). Length and time scales of atmospheric moisture recycling. *Atmospheric Chemistry and Physics, 11,* 1853–1863. https://doi.org/10.5194/acp-11-1853-2011

van der Ent, R. J., & Tuinenburg, O. A. (2017). The residence time of water in the atmosphere revisited. *Hydrology and Earth System Sciences, 21,* 779–790. https://doi.org/10.5194/hess-21-779-2017

van der Ent, R. J., Savenije, H. H. G., Schaefli, B., & Steele-Dunne, S. C. (2010). Origin and fate of atmospheric moisture over continents. *Water Resources Research, 46,* W09525. https://doi.org/10.1029/2010WR009127

Wang-Erlandsson, L., van der Ent, R. J., Gordon, L. J., & Savenije, H. H. G. (2014). Contrasting roles of interception and transpiration in the hydrological cycle – Part 1: Temporal characteristics over land. *Earth System Dynamics, 5,* 441–469. https://doi.org/10.5194/esd-5-441-2014

Wang-Erlandsson, L., Fetzer, I., Keys, P. W., et al. (2018). Remote land use impacts on river flows through atmospheric teleconnections. *Hydrology and Earth System Sciences, 22,* 4311–4328. https://doi.org/10.5194/hess-22-4311-2018

Wei, Z., Yoshimura, K., Wang, L., et al. (2017). Revisiting the contribution of transpiration to global terrestrial evapotranspiration: Revisiting global ET partitioning. *Geophysical Research Letters, 44,* 2792–2801. https://doi.org/10.1002/2016GL072235

Weng, W., Luedeke, M. K. B., Zemp, D. C., et al. (2018). Aerial and surface rivers: Downwind impacts on water availability from land use changes in Amazonia. *Hydrology and Earth System Sciences, 22,* 911–927. https://doi.org/10.5194/hess-22-911-2018

Wolosin, M., & Harris, N. (2018). *Tropical forests and climate change: The latest science* (pp. 1–14). World Resources Institute.

Part III
Call to Action: On the Power of Informed Citizens in a Democracy

Germany Under Climate Stress

Consequences for Our Social Coexistence

Jutta Allmendinger and Wolfgang Schroeder

In these long months and years of the pandemic, the population is asked, implored, and reminded again and again to please stick together, to show solidarity, to have trust, to take responsibility. Cohesion, solidarity, trust, and responsibility are all concepts that move away from the "I" toward a "we" that places people in the context of others. They are relational concepts. One can neither hold together nor be in solidarity with oneself. Goodbye to singletons, individualists, distinctiveness—what counts now is the big picture.

Aligning one's own actions with the common good probably does not come naturally. This is shown by a vaccination rate of only 74% at the end of 2021 and the consequent fierce struggle over compulsory vaccination. But the effort is not entirely hopeless, as shown by vaccination rates of over 90% in Bremen and Hamburg, where politics has convinced, educated, promoted, and created structures and cultures that connect one's own actions more easily and more naturally with the conduct of others.

As we have known for a long time, Corona is a local, national, and global tour de force, posing a challenge to all. It requires interdisciplinary and international research to explore the SARS virus, to develop vaccines, to understand the social

---

We thank Jannik Zindel for support.

J. Allmendinger (✉)
Social Science Research Center Berlin (WZB), Humboldt University of Berlin, Berlin, Germany
e-mail: jutta.allmendinger@wzb.eu

W. Schroeder
Social Science Research Center Berlin (WZB), University of Kassel, Kassel, Germany
e-mail: wolfgang.schroeder@uni-kassel.de

239

effects and side effects of the Corona measures. It requires cross-sectoral politics that takes advice—from science, business, and civil society. It requires a robust and participatory economy. And it preeminently requires a civil society that displays mutual understanding, builds trust, and follows rules.

Yet, compared to climate change, the virus is nothing. Corona can be contained and probably soon be cured with pharmaceuticals. In a few months, or years at most, the virus will be endemic, with herd immunity and a return to our old way of life. Climate change, by contrast, cannot be vaccinated away and will not be stopped in the foreseeable future. Despite, or precisely because of this difference, the virus is a challenging test case for the much larger tasks of tomorrow.

Climate Change as a Political Challenge

At present, climate change continues to advance, threatening the very basis of life for us all. A technological solution that would allow us to continue living as we do now is not in sight. But technologies that help us on the road to climate neutrality do exist, and more such technologies must be developed. We urgently need them. Equally urgent is state coordination that invites participation and co-ownership, fosters civil society solidarity, and lays the foundation for universal conduct change. At present, many domains feature high social selectivity: certain groups cause greater emissions than others, are less affected by climate change, and financially contribute less to measures against it. Climate change thus affects lower-income households much more severely than the better-off. Measures taken against climate change further exacerbate these differences. This can be clearly seen in the domains of work, mobility, housing, food, and health, which we will examine. In these and many other domains, climate change raises fundamental questions of socio-economic distribution.

All climate policies should therefore be examined for possible adverse distributive effects, which should be corrected through redistributive or compensatory governmental or societal measures. Quite apart from egalitarian principles being fundamental to our democracy, a huge challenge such as climate change cannot be met against resistance by large segments of the population.

Conflicts on the way to a climate-neutral world are not only about distribution but also about aspects of mutual recognition. Only when diverse situations and lifestyles are respected and accepted as fundamental for a joint way forward can the discourse about an effective climate policy succeed under liberal democratic conditions. It is therefore important persistently to question the fundamental orientation of our value system and behavior patterns. Is our conduct oriented toward our own well-being or that of the whole population? Are we considering only our own lifespan or also the lives of future generations? Are we in Europe acknowledging the situation and anxieties of people in the global South, whom climate change often deprives of their livelihoods and forces to migrate? Are we prepared to meet them on an equal footing, to integrate them or to compensate them? These questions add a socio-cultural dimension that mainly concerns matters of recognition.

Issues of socioeconomic distribution and socio-cultural recognition are easier to approach and to resolve the less divided and fractured a society is and the more deeply its members accept social solidarity and care about the common good. Without social cohesion, the climate goals cannot be achieved; it is the starting and end point of successful climate politics.

This chapter begins with reflections on the relationship between social cohesion and climate change. We then discuss, as outlined, socio-cultural recognition and socio-economic distribution as a dual challenge to climate policy. An assessment of the current climate-political projects and their open questions concludes.

Social Cohesion and Climate Change

In the discourses on social fragmentation, segmentation, and polarization, the topic of social cohesion has regained importance since at least the 1990s. Social cohesion is based on respect and trust. These in turn can solidify when there is a minimum of cohesion among people. Georg Simmel describes trust as "a hypothesis about future conduct firm enough to base practical action upon it." Such hypotheses can only arise when people have knowledge about one another and share values and norms that are subject of frequent reassurance and joint development. This requires shared spaces and social encounters, an overlapping of social circles (Simmel). Rainer Forst (2020) organizes all these aspects into an appropriate system. He speaks of five dimensions of social cohesion. First, social cohesion is politically constituted, the result of social and political processes. Second, it concerns individual and collective attitudes and patterns of conduct. Third, it is about relationships of mutuality, which are mutually binding and socially reciprocal. Fourth, these attitudes and conduct patterns must be secured through a structural institutionalization as a precondition for establishing and maintaining enduring, reliable, and resilient social coexistence. Fifth, the foundations and rules of these institutional structures must be regularly adjusted through an open social discourse and political decisions based on it.

What does all this mean for climate change? First of all, it should be noted that the majority of the population takes climate change seriously and is willing to contribute to combating it. People understand that this will require profound social change. In a recent survey, 72% of respondents shared this view (El-Menouar & Unzicker, 2021). There is awareness of the problem and readiness to change conduct. Politics can build on this.

But doing so is not easy, the headwinds are strong. Many people have high expectations and demands on democratic and state institutions but are also skeptical about the capacity of politics for action especially in response to major and usually global challenges. Politics is, as it were, disenchanted (Forst, 2020). People's high expectations, perceived state overload, and lengthy democratic-procedural processes of balancing competing interests can easily lead to a loss of trust, which would contribute to further disillusionment with liberal narratives of progress (Reckwitz, 2020). The future would then no longer be shapeable and thus lose its optimistic connotations. On the contrary, one would expect a problematic development of our society, a dystopia, discouraging citizen engagement.

Equally adverse to a successful climate politics is the increasing longing by many for a nostalgically charged normality. Many critics of environmental and climate policies are strongly anchored in right-wing populist and anti-science milieus and attitudes. The more they fear being disadvantaged by climate policies, the more susceptible people are to (right-wing) populist views (Humpert et al., 2021). Civil protest might turn into civil disobedience. The result would be an increase in polarized conflicts and further disintegration processes.

These attitudes can only be countered by resolutely keeping an eye on social cohesion, strengthening it, and not endangering it. This requires, first of all, a strong politics that, in communication with science, business, and civil society, defines and communicates climate goals and policies. Special care must be taken to ensure that socio-economic distribution and socio-cultural recognition do not fall by the wayside. If a society is socially divided, jointly upheld norms and values erode, polarization takes hold and subverts trust. But if climate politics resolutely ends the social selectivity of current climate policies, relieves the poor, and places more burdens on higher-income groups, then the transformation can succeed with those contributing more to climate change bearing higher costs.

Attention to socio-cultural recognition exemplifies a deeper approach that is not about allocating financial burdens but about patterns of conduct that are mindful of social cohesion across time and space. The recognition that certain effects of personal and collective behavior will manifest themselves only in the future leads to sustainable conduct informed by contracts and trust across generation. This approach must also be an international one: our conduct has monumental effects for the global South. Climate migration is one of them. What follows delves into this in more detail, showing where socio-cultural conflicts over recognition and socio-economic conflicts over distribution exist, and how climate politics might address them.

Climate Change and Socio-cultural Conflicts Over Recognition

Climate change requires conduct changes whose material effects some population groups can cope with better than others. Social frictions can arise when today's climate policy decisions adversely affect future quality of life or when climate impacts trigger major refugee and migration flows. We call these socio-cultural conflict situations.

Generational Conflict

Climate change has been gathering pace for decades and is raging unabated, even if initially its effects were not immediately visible and tangible for all. The success of climate policy decisions is likewise not immediately apparent. Measures to protect the environment will in specific domains take effect with delay. Individual costs and benefits can be far apart. Costs, such as consumption foregone, are incurred in the present, the resulting benefits are enjoyed only in the future—perhaps even by future generations. To accept such costly conduct, people require some assurance of effectiveness and correctness, as conduct motivated by abstract altruism is exceedingly rare. Such assurance of correctness must ultimately come through political decisions and their persuasive communication.

The effects of decades of climate-damaging personal and commercial activities are by now evident, especially among the younger generations. For example, people born in 2020 face seven times greater exposure to weather extremes than those born in 1960 (Ryan et al., 2021).[1] Accordingly, many in the younger generation demand enhanced representation in the political system as well as heightened intergenerational solidarity. At present, older people enjoy greater representation in the social, economic, and political system and are better able than the young to articulate their interests broadly and audibly and to prevail politically. Due to a lack of opportunities for participation, younger generations, by contrast, are increasingly governed by others.

The Federal Constitutional Court (Bundesverfassungsgericht, BVerfG) is aware of this imbalance. In its judgement of 24 March 2021, it largely accepts a constitutional complaint against the Federal Government's Climate Protection Act of 2019. The Court found that the Climate Protection Act requires too modest emission reductions from the present generation and thereby fails to specify, for the post-2030 period, a freedom-preserving transition to climate neutrality. By shifting the burden onto future generations, the Climate Protection Act endangers the constitutionally protected freedom of young and future generations and is therefore inconsistent with the fundamental rights that safeguard freedom across time. The BVerfG thus concludes that the federal government has disregarded the interests of young and future generations.

In this context, one should highlight the founding of Fridays For Future (FFF), whose aim is compliance with the 1.5-degree target of the Paris Climate Agreement. Protest movements and civil society associations for environmental and climate protection have existed for a long time, but only this self-organization resulted in a broad protest mobilization, which is supported and sustained by previously established movements. On this basis, FFF rose to become a high-profile mouthpiece for the previously latent interests and needs of parts of the younger generations. Here FFF was helped also by its peaceful protest culture in contrast to other protest

---

[1] Children from the USA and Germany are not exempt from these events, although they are less likely to be affected by crop failures.

movements, such as Extinction Rebellion, whose more reckless actions of civil disobedience exacerbate polarization (BMFSFJ, 2020). FFF is supported by various subsidiary movements from all parts of society (e.g., Scientists for Future) that were founded in its wake.

The reach of the movement is overestimated in media reporting. Only a minority of the young generation takes part in the protests (BMU/UBA, 2020). Moreover, FFF displays high social selectivity: participants are mostly left-wing (78%), have or pursue high educational qualifications (92%[2]) and hail from the middle class (70%) (Institut für Protest- und Bewegungsforschung, 2019). Large parts of the younger generation with other socio-demographic backgrounds are underrepresented. The analysis of the 2021 federal election results shows that most eligible voters across all age groups have not aligned themselves with the climate and nature conservation interests of the younger generation. It also reveals an intergenerational (young vs. old) as well as an intragenerational conflict structure, with the latter showing that, even within the younger age cohorts, the commitment to climate and nature conservation interests varies (cf. Fig. 1).

The reasons for this surprisingly low commitment to the climate goals are probably some of the characteristics of the FFF movement. Though distinguished by a strong affinity with science, it is often accused of not being open enough to the plurality of perspectives and the fallibility of science, of excluding contrary opinions and thereby promoting the moralization and polarization of public discourse.

In conclusion, climate conflicts have an intergenerational and an intragenerational conflict dimension. Due to advancing climate change and a heretofore inadequate climate politics, it is to be expected that these conflicts will further intensify on the side of the FFF movement and its opponents.

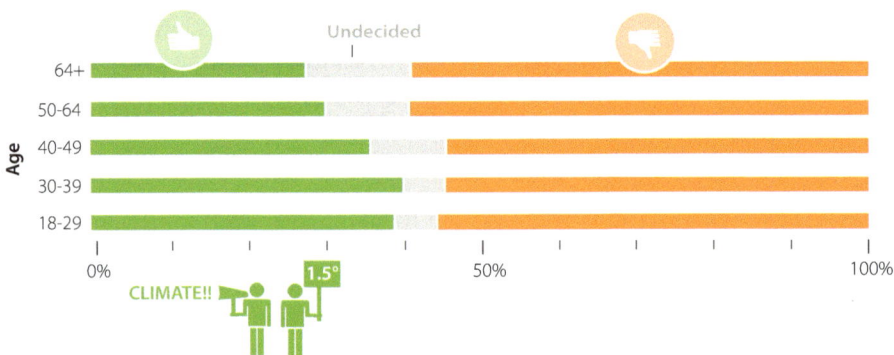

Fig. 1 What role did the climate and nature conservation interests of young generations play in the 2021 federal election decisions in different age groups? (NABU, 2021)

---

[2] Advanced technical college entrance qualification/Abitur: 56%, completed studies: 31% and doctorate: 5%.

Conflict Over Migration

In 2019, extreme weather events alone displaced 23.9 million people from their homes (BMZ, n.d.). In addition, creeping effects of climate change, such as coastal erosion and changing rainy seasons, have increasingly adverse effects on food production, economic activities, and lifestyles, and thereby on the food security of many people. Heavily dependent on agriculture, poorer population groups in the global South are most affected.

The topic of migration and climate change has attracted increasing attention since the 1990s.[3] Since then, two basic positions have stood in opposition. On one side there is the "alarmist" position, which assumes that climate change alone drives migration decisions. On the other side there is the "skeptical" counter-position which assumes that there are many reasons for migration and formulates a multi-causal model (Schraven, 2019). In academia, a "skeptical" position predominates due to the 2011 "Foresight Report on Migration and Global Environmental Change." The media and the political discourse, by contrast, are "alarmist," (Tangermann & Kreienbrink, 2019) drawing renewed public attention to the migration debate.

In view of the polarized social attitudes resulting from the so-called "refugee crisis" of 2015, the migration issue is once again fertile ground for right-wing populist arguments. This is interesting in connection with the fact that in 2015 it was precisely the initial media disregard and one-sidedly positive reporting that led to an alienation of population groups who saw their fears unrepresented. Trust in the media eroded, echo chambers formed in social networks, frustration and hatred arose (Haller, 2017), leading ultimately to social movements such as Pegida. The mixing of legitimate concerns and xenophobic positions helped the AfD and created an irreconcilable debating climate which continues into the present.

The propagated threat scenario of a new, now climate-induced "refugee crisis" is further damaging social cohesion. The fear of (felt) social decline and of social devaluation and marginalization processes leads to intensified struggles for recognition. Groups that already feel marginalized may experience these fears more strongly.

In conclusion, that what stirs up fears and thus increases the dynamics of social polarization and disintegration is not so much the real threat situation as the climate change-related migration debate orchestrated by media and politics (see also chapter "Escape from Heat, Drought, and Extreme Weather").

---

[3] With its prediction that by 2050 there would be 200 million climate refugees, Norman Myer's 1997 work still enjoys much media resonance today (Schraven, 2019).

Climate Change and Socio-economic Conflicts Over Distribution

Climate change affects us all, worldwide. But some people harm the climate more than others, and measures to combat climate change also affect people differentially. Low-income households contribute less to climate change, for example, but are disproportionately affected by climate policies. A few figures suffice to show this. Between 1990 and 2015, the richest 10% of the German population caused about 26% of total emissions in Germany, nearly as much as the entire poorer half. Over time, this imbalance has increased worldwide; in 2015, the emissions of the top 10% were higher than those of the poorer 50% of the population (Oxfam, 2020). In addition, those who contribute the least to climate change are hit especially hard by the effects of climate change and climate policies.

Figure 2 shows selected policy domains that have a high relevance for distribution policy and, due to their high emission shares and steep reduction paths, are crucial for the green transformation. The emission trajectory in the transport sector is especially discouraging.

There is consensus now in the political discourse that social hardships must be avoided in the transition to a climate-neutral society. But the coalition agreement of the new federal government is still quite vague about the real effects on low-income households of worsening climate change and contemplated climate policies.

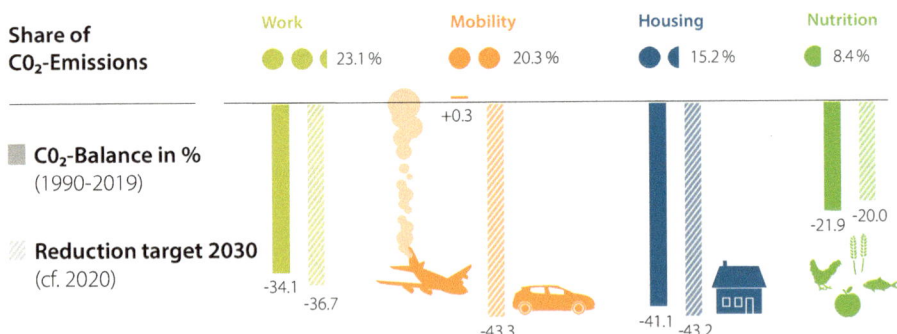

Fig. 2 Climate ledger showing the policy domains with their emission shares and reduction targets for 2019. Especially sobering are the data from the transport sector where CO_2 emissions increased between 1990 and 2019. It is to be feared that the now legally fixed reduction targets for 2020 to 2030 will give rise to substantial distributional conflicts (UBA, 2021a; BMU, 2021)

Gainful Employment

Gainful employment stands for the satisfaction of material and social needs. It is crucial for individual personality formation and shapes the distribution of social recognition and social participation. It thereby shapes the relationship network between individual and society and functions as a central mechanism for social (dis) integration.

The world of work in the twenty-first century is still strongly shaped by the historically grown, emission-intensive economic structures of the industrial age. Climate protection was thus initially regarded exclusively as a "job destroyer," especially in the heavily industrial-fossil sectors of the economy. Ecology and economy were seen as incompatible. Today, people have jettisoned this black-and-white thinking. This is due to forecasts always being difficult. Changes in labor demand can be attributed not only to climate protection, but also to automation processes and the outmigration of energy-intensive industries. This is why prognoses diverge. According to calculations by Prognos AG, unemployment is not expected to rise until 2050 (Hoch et al., 2019). There is even talk of an increase in employment: "The results show that climate protection is associated with positive economic effects overall. The ex-post analysis clearly shows that the provision of climate protection technologies and services generates substantial employment" (ibid., p. 42).

Be that as it may, aggregate effects conceal quite diverse sectoral and activity-specific developments. In certain sectors, especially in the automotive industry, jobs will continue to be lost due to climate protection measures. According to a study by the Fraunhofer Institute, such job losses till 2030 will however be much smaller than heretofore assumed (Herrmann et al., 2020). In addition to the job losses, there will also be a restructuring of jobs. Here, too, the automotive industry can be cited as an example. The switch from combustion engines to electric motors makes early retraining and qualification of employees important. In the medium term, measures against climate change can also stimulate economic growth. New technologies emerge, are deployed, and create new jobs. Good examples are job growth in the wind and solar energy sectors as well as in the construction industry.

Mobility

Spatial mobility is the second-largest contributor to greenhouse gas emissions in Germany. Between 2000 and 2018, distance traveled by air grew 64.9%, by rail 30.1%, and by car 7.5% (Fig. 3). Despite slower growth, individual car transport is the most important means of mobility in absolute terms. Traffic volume, number of cars, and distances travelled, have continued to increase. The car is an integral part of the material needs of life, of social participation, and of social interaction. But not everyone can afford a car: only 77% of all households even own one (UBA, 2021b).

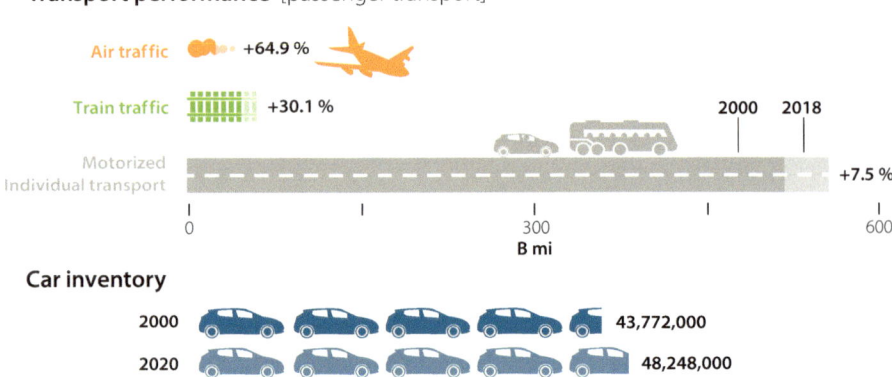

Fig. 3 Key figures in the transport sector (ibid.)

Cars increase environmental pollution. The more frequently cars are used, the greater the environmental impact; and the farther the number of cars increases, the more of the gains from technological emission reductions are erased. Both trends have been steeply rising in the last 18 years. Environmentally damaging air transport has also seen a sharp increase, while rail travel, though increasing, is lagging far behind private motorized transport (Fig. 3). Transport is the sole sector that did not achieve any emission reductions between 1990 and 2019 (Fig. 2). The ecological transformation dubbed "transport turnaround" has failed thus far.

Various instruments are used to reduce traffic volume and to replace high-emission private transport, among them the CO2 tax which increases the gasoline price and thus has a highly selective social effect.

Lower-income households are disproportionately burdened by this levy (Held et al., 2021). Income poverty becomes "mobility poverty." Lower-income households, if they own a car at all, are "forced (…) to spend more money on transport costs than they can really afford" (Daubitz, 2016, p. 440). They have limited opportunities to reallocate expenses, by spending less in vital areas such as housing and food. The subsidy strategy meant to replace "combustion engines" with e-cars is not a viable option for them, nor for almost three quarters of the population (72%), despite the subsidy (acatech, 2021).[4] Local public transport is therefore extremely important for these households, but not up to the task.

The federal government is aware of the social imbalance. To cushion the social impact, it was decided to increase the commuting allowance and to introduce a mobility subsidy, both valid till the end of 2026 and for commutes of 21 km or more. The mobility subsidy[5] is intended as cost compensation for people on low incomes. As a result, however, higher-income households are again the beneficiaries, as they

---

[4] In the countryside (29%), this option is desired slightly more than in the city (21%).

[5] The mobility subsidy is for low-income earners whose taxable income is below the basic allowance threshold and who therefore do not benefit from the increase in the commuting allowance.

are barely affected by the price increases and benefit disproportionately from the new instruments. Despite the mobility flat rate, the largest net burden still falls on the 40% with the lowest incomes (Held et al., 2021).

Housing

Housing is a human right, included in the Declaration of Human Rights. A home offers space to live and is the basis for our social coexistence. When a suitable accommodation is lacking, many other human rights are threatened, such as the rights to health and life or the right to participation. Although the German Basic Law (constitution) does not explicitly recognize a right to housing, no one questions that housing is an existential good that should be provided to all.

The housing issue is one of the most pressing problems of our time. The gap between owning and renting is widening, especially between those who can afford expensive housing and those who depend on an affordable rent. This is because households at risk of poverty spend almost 60% of their disposable income on housing, while the rest of the population on average spend less than 25%. The housing situation is getting worse, especially in the cities, in part because of people moving in from rural areas (urbanization). At the same time, the stock of subsidized housing is declining, rental costs in conurbations are rising and in inner-city neighborhoods wealthy households are displacing those with lower incomes (gentrification). Social dislocations reach into the middle class (Fig. 4). Social cohesion is threatened.

The building sector's decreasing harm to the climate cannot hide the fact that this sector still accounts for almost one sixth of German greenhouse gas emissions, which must be reduced by almost half by 2030 (Fig. 2). Two policies directly pursue this goal: CO2 pricing in the building sector and energy-saving renovation. A third measure concerns the construction industry which, due to its use of raw materials,

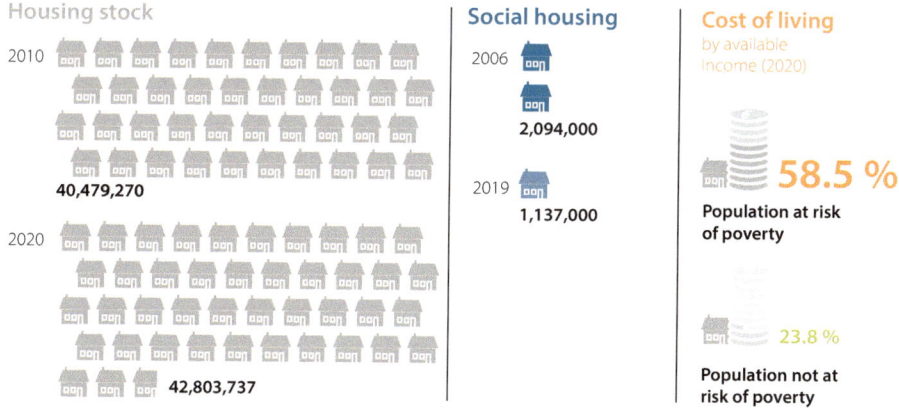

Fig. 4 Key figures on housing (Destatis, 2020, 2021; Statista, 2021)

is responsible for substantial CO2 emissions. Emission reductions in the housing sector can succeed only if housing construction is not continuously impeded by new building regulations and shortages of building materials.

Since January 2021, the CO2 price applies also to the building sector. This triggers additional costs for private households due to the increased cost of electricity, heating oil, and natural gas. The price increases for electricity are accompanied by an expansion of renewable energies. The green power surcharge (EEG) and its impact on the electricity markets have led to large excess costs for private households. The CO2 levy primarily affects low-income households and those in the lower middle class, as the levy forces them to spend a higher proportion of their total budget on covering their basic needs. Moreover, the costs are imposed entirely on tenants; landlords can fully pass on the costs of the CO2 levy. The CO2 price also disproportionately affects low-income tenants because they often live in poorly insulated buildings in which energy costs tend to be much higher than in renovated houses. To bring relief, it was decided to use the revenue from the CO2 levy to increase the housing allowance, which will primarily benefit lower-income households.

The second pillar of climate politics is based on the massive expansion of energy-focused renovation. The originally intended effect: to stimulate energy-focused renovations through a price on CO2, is not achievable with the current design. Why should landlords invest in energy-efficient renovation when the resulting cost savings go exclusively to their tenants? The crucial instrument for energy-efficient renovation of existing buildings is therefore the modernization levy pursuant to §559 of the German Civil Law (BGB), which can lead to an increase in the basic rent, even though the levy has been limited to 8% since 2019 (previously 11%) by the Rent Adjustment Act. For landlords, modernization is especially worthwhile in conurbations with a substantial housing shortage, as rents can there be raised more easily. Climate policy thus leads to rent increases and to processes of displacement and social segregation. In the entire policy domain of housing, climate policies increase social inequality, which shows the urgent need for policy revisions.

An important field is therefore the construction of subsidized housing units that are both ecologically sound and affordable to rent. Since the number of subsidized housing units has been cut by nearly one-half since the turn of the century, substantial efforts will be necessary.

Food

Food and food security are a human right which, though not explicitly laid down in the German Constitution, is implied by the welfare state. Agriculture is therefore the basis of our coexistence and social reproduction.

Agriculture has become much more efficient in recent decades. Today, one farmer feeds twice as many people as in 1990. This efficiency gain is due to massive consolidation in agriculture accompanied by reduction in employment. Since the mid-1990s, the number of farms has halved. In terms of total area farmed, organic farming is still insignificant (Fig. 5).

Germany produces more meat and milk than its own population consumes but is otherwise dependent on imports. Foodstuffs such as fruits and vegetables are sourced mainly from abroad (DBV, 2020). In the agricultural producer countries of Asia and Africa, global warming of perhaps 3 degrees and more is leading to disastrous conditions, with floods, storms, and droughts becoming the norm. Due to this development, the price of wheat might rise 55% by 2050, that of rice by 37%, and that of corn/maize by 11%. Large and volatile price fluctuations are also expected (Cameron, 2015). Higher consumer prices would make staple foods from abroad unaffordable to lower-income households and would also burden middle-class household budgets. Conversion to organic agriculture might further increase dependence on imports (cf. Fig. 5). Increasing demands on sustainably produced

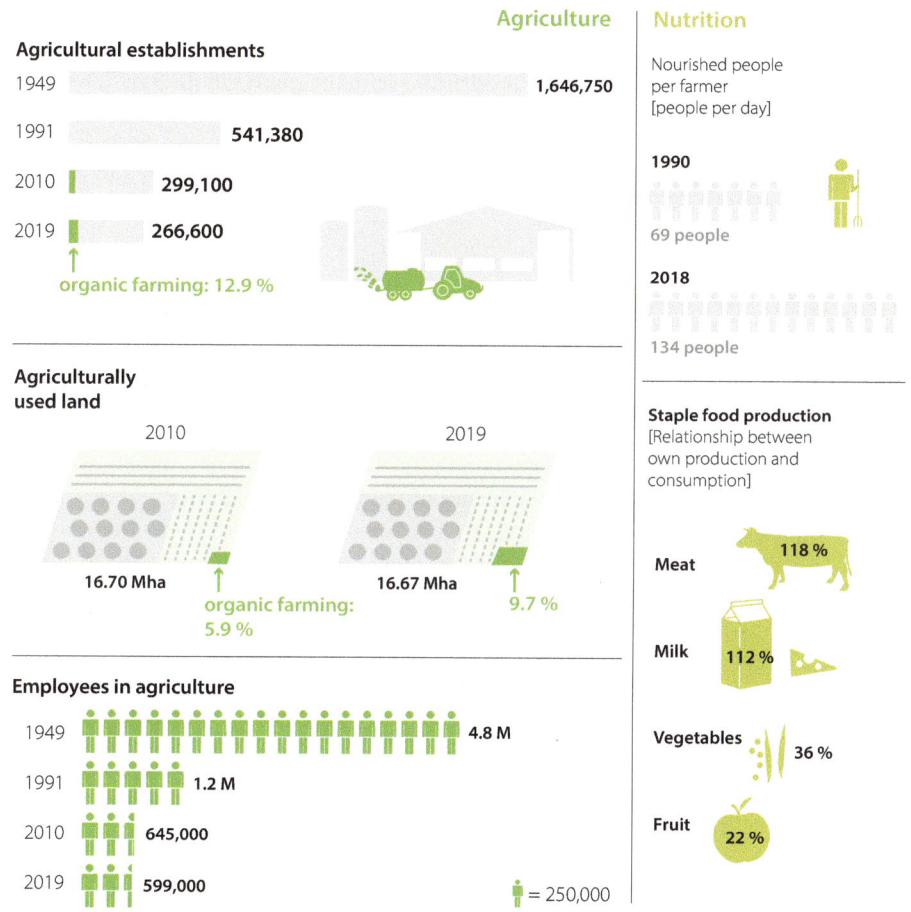

Fig. 5 Key figures on food/agriculture (DBV, 2020; Forum Moderne Landwirtschaft, 2021)

foodstuffs might result in international competitive disadvantages and even in out-sourcing effects.

In Germany, climate policy is primarily concerned with the ecological restructur-ing of agricultural production, which is especially affected by the effects of climate change: extreme weather events and permanent climatic changes can lead to crop failures or declining yields (as well as to losses in livestock production)—though these losses will be less dramatic than in the global South. Individual farms might suffer enormous income losses, which would necessitate investments in precaution-ary and adaptive measures. Such investments would be especially difficult to afford for small, financially weaker farms. Already today, more than one sixth of farmers (17%) consider ecological readjustment unaffordable, and almost one third (29.5%) fear a negative financial cost-benefit balance. However, a large majority of farms (80%) are willing to reduce greenhouse gas emissions with appropriate cost com-pensation (Stumpenhorst, 2020).

Consumers today are more conscious of their diet than in earlier times, even if only 37% buy organic food frequently or exclusively. Almost two-thirds (64%) find it too expensive (Forum Moderne Landwirtschaft, 2021). By causing crop failures and rising world market prices for foodstuffs, climate change will drive up retail food prices, burdening all income groups but especially low-income households whose food expenditures account for a larger share of their overall budget. This development, too, might provide fertile soil for polarization and disintegration processes.

Health

Health is much more than physical integrity. Over time, proper nutrition and exer-cise have become integral parts of social interaction. Health is also more than the health of individual persons. The pandemic has clearly shown how much our eco-nomic and social life depends on the health of the whole population.

Climate change is one of the greatest challenges for health. Nonetheless, the health effects of climate change are not yet much discussed in our country. Large parts of the medical profession, associations, and scientific societies do not even participate in the debate (Lehmkuhl, 2019). It was only recently, in 2017, that the "German Alliance for Climate Change and Health" (KLUG) was founded. This is surprising because low-income households have long been especially affected by environmentally harmful conditions: "people with a low social status and socially disadvantaged urban neighborhoods tend to be affected more frequently by often multiple environmental burdens relevant to health, such as noise and/or air pollut-ants and a lack of green spaces" (BMAS, 2021, p. 342).

As a result of climate change, temperatures are rising, there are more "hot days." With especially high exposure to urban climate effects, urban populations suffer most from this. With much of the ground sealed or covered with buildings, heat builds up and maximum temperatures rise. "Heat stress" in indoor areas will continue to increase, resulting in a higher mortality rate. Germany already ranks third in heat-related deaths; in 2018, 22,000 people over 65 died from heat-related causes (The Lancet, 2020). The increasing frequency of extreme weather events such as storms, heavy rainfall, floods, avalanches, or landslides has a direct impact on people's health, causing diseases and injuries, stress, mental disorders, anxiety, and depression. These climate-related health risks especially affect children and older, weakened, or health-impaired people. Socio-economic status is another risk factor. Especially people of "lower social status, who are less educated, who live in neighborhoods that have a high proportion of low-income population, who are unmarried, or who have lost their spouse" (Eis et al., 2010) are at heightened risk of mortality.

If one links the various dimensions of social inequality, it becomes apparent that people with low incomes are more affected by climate change than others. They are more vulnerable to its impacts, and their low income also reduces their ability to preemptively adapt to climate change and other difficulties. People face a dilemma: if they do not adapt to the new conditions, their health will suffer. If they try to protect themselves, they bump up against their financial limits due to increased rents, among other things. Protection against climate change once again becomes a social issue.

Conclusion

Climate change poses major socio-political challenges for political, economic, and social actors. Climate policy can further divide society, but it can also strengthen social cohesion by bringing social groups closer together. In view of the unequal social impact, there is much to suggest that the pessimistic scenario, a deepening division, is the more likely prospect. This must be prevented by political decisions and their engaged communication.

Entailing rising costs in the areas of work, mobility, housing, food, and health, climate policy measures have a drastic impact on the lives of lower-income and less-educated households and on parts of the lower middle class. Higher-income households contribute more than others to CO_2 emissions while being financially less affected by them. They are better able to cope with the effects of climate change because they are less affected, more frequently take precautions, and are more likely to make use of alternative options. Climate change will thus further widen the gap between rich and poor in key areas of society.

The new federal government has endorsed compensating lower-income households through tax-financed covering of the EEG levy, for example, or through additional climate payments (Die Bundesregierung, 2021). Subsidized housing is to be expanded as well with an additional one billion euros in federal funds for 2022. The evidence suggests, however, that this will not be enough to adequately compensate for the expected increase in burdens.

If the climate goals are still to be met, rapid and strong control and framing of the complex and simultaneous processes of dismantling, renovating, and constructing is required. Here, too, the new German government is setting new accents in climate policy. The state as framework provider and the economy as innovation and technology driver are to cooperate more closely:

> We want more innovation, more competitiveness, more efficiency, good jobs, and climate-neutral prosperity. For this, we need a decade of investment in the future and more speed. Our goal is a social-ecological market economy. (ibid., p. 25)

Climate protection policy also attempts to change individual conduct. Politically, this falls under diverse headings. Due to the restrictive, prohibition-oriented measures (bans on New Year's Eve fireworks or diesel engines, etc.), climate protection has a reputation for being the antagonist of freedom. Veggieday is a good example of this. To avoid the political consequences of restrictive policies, more and more parties are relying on technological progress and positive incentives, including the Green Party: "The use of modern technologies enables climate neutrality. It is the task of politics, therefore, to activate people's ingenuity toward developing suitable technologies and deploying them intelligently" (Die Grünen, 2021, p. 22).

Which innovations should be introduced is controversial as some prioritize the market while others prefer the state to take a leading role. The preference for market-based and technological solutions is strongest in the Liberal Party (FDP), which wants climate protection to be run primarily by companies and citizens. Opposed to this is the view that a consistent climate policy cannot be implemented without (radical) intervention in the private and economic spheres. The federal government is trying to establish an intermediate path.

Climate policy can strengthen societal coherence if a joint effort in solidarity were to succeed, and an overburdening of weaker groups could be avoided. This would require broad inclusion of all, new patterns of participation, as well as procedures and reforms of the existing forums.

Climate policy must then face the dual socio-political challenge and offer proactive answers to socio-economic and socio-cultural challenges. These answers will pose a major challenge to the state, to markets, and to society's own initiatives, but will ultimately co-determine the success or failure of climate policy.

References

acatech. (2021). *Mobilitätsmonitor 2021*. www.acatech.de/mobilitaetsmonitor-2021-alle-ergebnisse/. Accessed 30 Sept 2021.

BMAS. (2021). *Lebenslagen in Deutschland: Der Sechste Armuts- und Reichtumsbericht der Bundesregierung*. Berlin.

BMFSFJ. (2020). *Bericht über die Lage junger Menschen und die Bestrebungen und Leistungen der Kinder- und Jugendhilfe. 16. Kinder- und Jugendbericht*. Berlin.

BMU. (2021). *Lesefassung des Bundes-Klimaschutzgesetzes 2021 mit markierten Änderungen zur Fassung von 2019*.

BMU/UBA. (2020). *Zukunft? Jugend fragen! – Was junge Menschen bewegt*. Berlin.

BMZ. (n.d.). *Migration und Klima*. www.bmz.de/de/entwicklungspolitik/klimawandelund-entwicklung/migration-und-klima. Accessed 17 Nov 2021.

Cameron, E. (2015). *Klimawandel: Was er für die Landwirtschaft bedeutet, Kernergebnisse aus dem Fünften Sachstandsbericht des IPCC*. Cambridge, Hamburg.

Daubitz, S. (2016). Mobilitätsarmut: Die Bedeutung der sozialen Frage im Forschungs- und Politikfeld Verkehr. In O. Schwedes, W. Canzler, & A. Knie (Eds.), *Handbuch Verkehrspolitik*. Springer.

DBV. (2020). *Trends und Fakten zur Landwirtschaft. Situationsbericht 2020/21*. Berlin.

Destatis. (2020). *Anteil der Wohnkosten am verfügbaren Haushaltseinkommen*.

Destatis. (2021). *Wohnungsbestand Ende 2020: 42,8 Millionen Wohnungen*.

Die Bundesregierung. (2021). *Koalitionsvertrag zwischen SPD, Bündnis 90/Die Grünen und FDP*. www.bundesregierung.de/breg-de/aktuelles/koalitionsvertrag-2021-1990800. Accessed 3 Jan 2022.

Die Grünen. (2021). *Grundsatzprogramm*. https://cms.gruene.de/uploads/documents/20200125_Grundsatzprogramm.pdf. Accessed 3 Jan 2022.

Eis, D., et al. (2010). *Klimawandel und Gesundheit: Ein Sachstandsbericht*. Berlin.

El-Menouar, Y., & Unzicker, K. (2021). *Klimawandel, Vielfalt, Gerechtigkeit. Wie Werthaltungen unsere Einstellungen zu gesellschaftlichen Zukunftsfragen bestimmen: Programm "Lebendige Werte"*. Bertelsmann Stiftung.

Forst, R. (2020). *Gesellschaftlicher Zusammenhalt: Zur Analyse eines sperrigen Begriffs*. Leibniz-Informationszentrum Wirtschaft. http://hdl.handle.net/10419/233764

Forum Moderne Landwirtschaft. (2021). *Stadt. Land. Fakten. Moderne Landwirtschaft in Zahlen: Die moderne Landwirtschaft und ihre Bedeutung für die Gesellschaft*. www.moderne-landwirtschaft.de/wp-content/uploads/2021/06/FML_Stadt_Land_Fakten.pdf. Accessed 17 Nov 2021.

Haller, M. (2017). Die "Flüchtlingskrise" in den Medien: Tagesaktueller Jismus zwischen Meinung und Information. *OBS Arbeitsheft, 93*, 1–182.

Held, B., Leisinger, C., & Runkel, M. (2021). *Sozialverträgliche Kompensation der CO2-Bepreisung im Verkehr*. Studie im Auftrag des vzbv.

Herrmann, F., et al. (2020). *Beschäftigung 2030: Auswirkung von Elektromobilität und Digitalisierung auf die Qualität und Quantität der Beschäftigung bei Volkswagen*. Fraunhofer-Institut für Arbeitswirtschaft und Organisation.

Hoch, M., et al. (2019). *Jobwende: Effekte der Energiewende auf Arbeit und Beschäftigung*. Mainz.

Humpert, F., et al. (2021). *Auf Kosten des Volkes: Rechtspopulistische Positionen zu Klima und Umwelt*. Europa-Universität Flensburg/TU Dortmund.

Institut für Protest- und Bewegungsforschung. (2019). *Fridays for future: Eine neue Protestgeneration?* Ergebnisse einer Befragung von Demonstrierenden am 15. März 2019 in Berlin und Bremen. https://protestinstitut.eu/wp-content/uploads/2019/03/Befragung_Fridays-for-Future_online.pdf. Accessed 23 Oct 2021.

Lehmkuhl, D. (2019). Das Thema Klimawandel und seine Bedeutung im Gesundheitssektor: Entwicklung, Akteure, Meilensteine. *Bundesgesundheitsblatt – Gesundheitsforschung – Gesundheitsschutz, 62*, 546–555.

NABU. (2021). *NABU-Umfrage zum Klimaschutz. Interessen der jungen Generation werden bei der Wahl ignoriert*. NABU-Pressedienst.

Oxfam. (2020). *Das reichste 1 Prozent schädigt das Klima doppelt so stark wie die ärmere Hälfte der Welt*.

Reckwitz, A. (2020). *Das Ende der Illusionen: Politik, Ökonomie und Kultur in der Spätmoderne*. Berlin.

Ryan, E., Luthe, S., & Wakefield, J. (2021). *Born into the climate crisis: Why we must act now to secure children's rights*. Save the Children International.

Schraven, B. (2019). *Der Zusammenhang zwischen Klimawandel und Migration*. Bundeszentrale für politische Bildung.

Statista. (2021). *Bestand der Sozialmietwohnungen in Deutschland in den Jahren von 2006 bis 2019.*

Stumpenhorst, C. S. (2020). *Umfrage: Landwirte wollen klimafreundlich arbeiten.*

Tangermann, J., & Kreienbrink, A. (2019). *Zur Prognose des Umfangs klimabedingter Migrationen.* Bundeszentrale für politische Bildung.

The Lancet. (2020). *The 2020 report of The Lancet Countdown on health and climate change: Responding to converging crises.*

UBA. (2021a). *Treibhausgasemissionen sinken 2020 um 8,7 Prozent.*

UBA. (2021b). *Mobilität privater Haushalte.* www.umweltbundesamt.de/daten/privatehaushalte-konsum/mobilitaet-privaterhaushalte#-hoher-motorisierungsgrad. Accessed 3 Jan 2022.

People Must Know What They Are in For!

Solutions, Financing, and the Power of Civil Society

Klaus Wiegandt

The description of a 3-degree warmer world, as presented in the various scientific contributions to this volume, depicts a devastating future for humanity. In such a future world, we will have to deal with a radicalization of weather patterns and with temperatures that will be as much as 6 degrees higher on average over land areas. Such a transformation will have grave effects on global agriculture, massively damage global infrastructure, and significantly impair or even destroy large ecosystems.

The majority of human beings will be affected by unprecedented restrictions on their living and survival conditions, and many will be killed. For example, regions south of the Sahara will become uninhabitable, forcing millions of people to migrate. As "climate refugees," their main destination will be Europe. Unlike today, the Mediterranean region will not offer them a new home—increasing dryness and droughts will lead to desertification there as well.

Furthermore, the economic damage in the 3-degree scenario outlined here will exceed 10% of the gross world product annually. In previous global crises, whose economic damage was incomparably smaller, extended recovery phases for the world economy finally set in. This will be dramatically different in the case of future climate-related crises. The intervals between crises will become ever shorter—until the burdens on state budgets and companies grow to the point where entire national economies will gradually implode and the global economy will be in danger of collapse. We can already foresee the violent conflicts and wars that will result from the growing competition for ever scarcer resources. What will happen when such a global collapse occurs?

We have not even mentioned non-linear effects and the occurrence of tipping points in the climate system. They can lead to a much stronger and faster rise in sea levels or to a collapse of the Antarctic Circumpolar Current or the Atlantic Meridional Overturning Circulation. No one can accurately predict what this will mean for the

K. Wiegandt (✉)
Forum für Verantwortung, Seeheim-Jugenheim, Germany
e-mail: info@forum-fuer-verantwortung.de

© The Author(s) 2024
K. Wiegandt (ed.), *3 Degrees More*,
https://doi.org/10.1007/978-3-031-58144-1_13

survival of humanity, but it is highly likely that billions of people will starve, die of thirst, and lose their lives to warfare.

Against this background, it becomes clear that the whole climate discussion of the last decades has failed adequately to discuss and to research the limits of societies' ability to adapt to increasing global warming.

Anders Levermann of the Potsdam Institute for Climate Impact Research warned in a Frankfurter Allgemeine guest article on 14 April 2011 that civil societies—in relation to rising global warming—also have a tipping point at which all essential social systems gradually collapse (Levermann, 2011). Science cannot predict at what degree of warming this collapse will set in. It is highly likely to be between +3 and +6 °C. According to Levermann, warming will stop at that point because societies will collapse and human CO_2 emissions will collapse with them.

In 2024, 13 years later, it can be assumed that the tipping point will probably be reached already at +3 °C. In this context, an article by climatologist Mojib Latif in Focus, 27 March 2023, should be cited, in which he argues that it is a huge mistake to believe that Germany's society and economy will be able to adapt to a world that is 2–3 degrees warmer (Latif, 2023).

The world community can still avert such a scenario. But decision-makers in politics, business, and civil society must form a realistic picture of both the consequences of this scenario and the set of measures now necessary to avert it, including their affordability.

In democracies, such a gigantic undertaking will only become politically feasible if the vast majority of the population becomes aware of the consequences of a +3 °C scenario and thus accepts the necessary measures to avert it. However, there will be such acceptance only if the rich upper class of our societies bears most or all of the cost of these future measures.

However, educational campaigns have so far been lacking in all democracies and must therefore be set in motion by governments as a matter of priority. Against the background of the current narrative, which essentially focuses on the consequences of global warming of +1.5 °C, no successful climate policy can be implemented in any democracy. Since the remaining window of opportunity in climate change is narrow, time is of the essence.

In this context, I would like to recall that governments mobilized trillions of U.S. dollars, overnight, as it were, both in 2008 to deal with the world financial crisis and in 2020 in the wake of the rampant Corona pandemic! We must realize that global warming poses an incomparably greater challenge.

Why the Paris Agreement Needs to Be Supplemented

From today's perspective, the resolutions of the 2015 Paris Agreement are insufficient and must urgently be improved and supplemented so that it corrects the failed climate protection policies of the last two decades while also addressing new challenges.

These challenges include a further global population increase by about two billion by 2050 and, above all, the economic rise of another two billion people into the so-called middle class of consumers in the emerging and developing countries. This new middle class will strive for a lifestyle and consumption style just as wasteful as those of the rich industrialized nations. This importantly includes an energy-rich diet with substantially increased consumption of meat.

Efficiency Improvements and Switch to Renewable Energies

Taken together, these changes will overwhelm the Paris Agreement in its current form. In my opinion, its two main pillars, efficiency improvements and the switch to renewable energies, cannot be implemented fast and comprehensively enough to achieve the treaty's goal.

In the area of efficiency, stricter requirements than previously envisioned would mean deep structural changes in key sectors of the global economy. The massive increase in unemployment to be expected would be a heavy burden, and the measures would therefore be politically risky and hard to enforce.

In addition, we must remember that, although relatively great progress has been made over the last 20 years in improving energy and resource efficiency, much of this progress has been cancelled out by so-called rebound effects. Future efficiency improvements will suffer the same fate.

In this context, I would like to make a few brief comments on basic research. According to the International Energy Agency, a little more than $30 billion is currently spent worldwide on publicly subsidized basic research. The economic historian Adam Tooze commented on this in Die ZEIT of 6 July 2023 as follows: "We do not even know the possibilities of progress yet. Why don't we increase the research tasks in the energy sector tenfold and see what happens?" (Tooze, 2023).

More intensive investment is needed in research into the concept of carbon capture and storage (CCS). Its large-scale use is not yet possible at economically feasible prices. But if the concept is politically desired and socially accepted, there are ways to accelerate this process (Schellnhuber, 2015). The consulting firm McKinsey, for example, even estimates that, to achieve climate neutrality for Europe by 2050, carbon capture and storage (CCS) will have to account for 25% of the reduction in CO_2 emissions (McKinsey & Company, 2020).

The German public is mostly skeptical about the use of CCS because it poses a new problem for us: the safe storage of CO_2. However, in my opinion, all options should be explored against the background of the danger of unchecked climate change. In this context, we must take into account that many poorer countries, in particular, have large coal reserves and want to and will use them for as long as possible. In the long term, CCS is certainly the far lesser danger to humanity.

The subject of nuclear fusion is also viewed very controversially in our society. Around 70 years have been invested in basic research toward developing the commercial application of nuclear fusion. Calls for an end to such funding are growing louder. However, according to a large majority of physicists, nuclear fusion has the potential to make a significant contribution to the future energy mix because, like nuclear fission, it is low in CO_2 emissions, has a high energy density, and can be used in base-load operation. In addition, nuclear fusion has the advantages over nuclear fission that it basically cannot lead to a meltdown and only produces radioactive waste with a comparatively short half-life. Therefore, as with electricity storage, we should persist, covering a period of up to 100 years in basic research.

There is, in any case, an urgent need for a massive increase in spending on basic research, as this is the only way to achieve fundamental breakthroughs for technological innovations in future-oriented fields.

When it comes to the switch to renewable energies, the difficulty is that even the targets adopted to-date will only be achieved with great difficulty. For this reason, a substantial acceleration in the pace of expansion must be envisioned, both in renewable energy production and in electricity grids. Developing countries, in particular, urgently need financial aid to build up their renewable energy systems.

Today, solar and wind power account for around 12% of global electricity generation. To reach 90% by 2050, only 26 years remain. This underscores what enormous efforts must be made to restructure the global energy industry.

At the same time, we must not ignore the fact that renewable energy production also consumes substantial resources such as raw steel, copper, aluminum, chrome, and cement. For example, a single wind turbine (WTG) uses up to 200 tons of material in total, including steel, copper, and other industrial metals (Misereor, 2018). These materials are energy-intensive in their production and, in some cases, also scarce, which is why experts are already talking about our approaching an inevitable peak metal point.

This is not an objection to the energy transition because, compared to fossil fuel plants, the overall raw material balance of renewables plants is still better. However, it is to be feared that, in the course of the necessary massive expansion of renewable energies, the price of copper, for example, will rise so sharply that only the rich industrialized countries will be able to afford this transition.

Sufficiency, or Why Less Is More

In the developed countries, the relatively easy efficiency improvements have been largely exhausted, especially in business. Efficiency improvements beyond the level achieved today require much higher capital input and take much more time, so the question of sufficiency inevitably belongs here. Politically, a call for "less" is a difficult issue worldwide because politicians—especially in the industrialized nations—do not have the courage to show citizens how wasteful their lifestyles and consumption patterns are (as are most of the production processes that accompany them).

When balancing efficiency and sufficiency, we must be clear that every climate-harming action we avoid constitutes active climate protection; every flight we do without (= sufficiency) immediately improves the climate balance. We will therefore not be able to avoid a political and social discourse on sufficiency. It is long overdue, because we live on a planet with limited resources and energies as well as limited sinks for (harmful) substances. If we do not want to seriously endanger our existence, our activities must not exceed the biophysical limits of the Earth system and, where they have already done so, we must quickly reverse such transgressions.

In the field of energy, we in the industrialized countries are engaged in breathtaking waste. In his book The Uninhabitable Earth (Wallace-Wells, 2019), British journalist David Wallace-Wells refers to a study according to which two thirds of the energy produced in the USA is wasted (Stark, 2016). We are similarly wasteful and thoughtless with regard to many resources. In the industrialized countries, we have had saturated consumer markets for decades. As a result, efforts are made daily in a great many companies to develop new consumption needs or to slightly modify or improve existing products in order to push them into the saturated markets with great advertising pressure. A vivid example of this is the cell phone. Although economically usable for many years, consumers are induced to replace their current device with a new one every time the model changes. It is therefore not surprising that for years now more than 1 billion new cell phones were sold annually; in 2021, the figure was 1.36 billion. At the same time, hundreds of millions of devices lie discarded in drawers without their valuable components being recycled.

For more than two decades, material scientists have been calling for the dematerialization of our product world. The incessant rise in material consumption goes largely unnoticed due to the fixation on CO_2 emissions during operation. Yet, 40 years ago, for example, the VW Golf weighed 800 kg, while a comparable car today weighs 1200–1500 kg and carries an "ecological backpack" that, according to the Wuppertal Institute for Climate, Environment and Energy, is 80–100 tons. This concept of the ecological backpack describes the chain from raw material extraction, production, packaging, transport, and use to the disposal of a product.

These few examples illustrate that far too little attention is paid to resource consumption, even though every raw material consumed has caused significant energy consumption in the production chain. Every kilowatt hour not consumed in the process helps avoid investments in renewable energies and thus in valuable non-renewable resources.

The identified challenges to the Paris Agreement require this treaty to be amended and supplemented. This centrally involves a commitment by countries to rapidly invest 2% of their gross domestic products in climate protection policies and in comprehensively launching the nature-based solutions that the treaty envisages.

Why We Need Nature-Based Solutions

In addition to the two main pillars of the Paris Agreement, efficiency improvements and a comprehensive energy transition, we must rapidly implement also its third pillar: nature-based solutions. These help set milestones in climate protection policy in a timely, relatively cost-effective, and socially acceptable manner and help demonstrate to the global population that major progress in climate policy is finally being made.

The most important measures are an immediate halt to deforestation of rainforests, the reforestation of 350 billion trees in the tropics and subtropics, the global rewetting of drained peatlands, and the highest possible humus enrichment on agricultural land. These nature-based solutions are largely envisaged in the climate agreement and the nationally determined contributions (NDCs), but so far without binding funding commitments, which is why they require rapid political support for timely implementation.

While the threatening destruction of rainforests as well as the demand for regeneration of degraded rainforests has been increasingly noted in the public debate for two decades, the possibilities of strengthening nature-based solutions via climate protection policy have remained under researched thus far.

The rewetting of peatlands is one of these solutions, because draining peatlands for gaining peat and agricultural land turned out to be an ecological disaster and one of the worst CO_2 sources in agriculture. The resulting global CO_2 emissions are now close to two billion metric tons per year. Nonetheless, it has only been in the last few years that the immensely urgent rewetting of degraded peatlands has attracted public attention as a climate protection measure.

Another nature-based solution is humus enrichment in soils as a climate-relevant carbon sink. This will be achieved through various humus-building agricultural practices such as agroforestry systems, use of crops with greater root mass, improved grassland management, etc. At the Paris climate summit, France launched the "4-Promille Initiative" and 39 nations and more than 190 NGOs and other institutions signed a voluntary agreement. The initiative aims to foster collaboration among scientists, policy makers, and practitioners to ensure science-based actions. However, implementation in agricultural practice is very demanding.

Far too little attention was paid also to another potential carbon sink. To stabilize the climate, climate researcher Hans-Joachim Schellnhuber calls for a new direction in the building sector, with a return to sustainable timber construction. This would offer an opportunity for long-term CO_2 storage and would measurably reduce two energy-intensive sources of CO_2 emissions in the construction sector: steel and cement. Around 40% of greenhouse gases are emitted during the construction and use of buildings. Here wood must become the globally dominant raw material of the future.

Rainforest Rescue

The absolute priority, however, is to put a stop to deforestation of the rainforests as quickly as possible. This stop would reduce annual human-induced CO_2 emissions by around 4.7 billion metric tons. All public and private funds have failed in the last 20 years to significantly slow the process of deforestation. Even "The New York Declaration on Forests," a UN initiative established in 2014 to end rainforest deforestation in two stages by 2030, says it has failed for lack of funding. The destruction of rainforests continues unabated. It is undeniable, however, that without these many initiatives the rate of deforestation would be even higher today.

What are the main reasons for this failure? The newly industrializing and developing countries can, without burdensome formalities, receive a lot of export revenue, directly and indirectly, from palm and soy plantations established, or large-scale cattle breeding practiced, on deforested land. In addition, trade in precious tropical woods is still a highly lucrative business. Governments justify their support for such activities by appeal to the poverty of their countries—ignoring both the longer-term disastrous effects and the unsustainability of this approach.

Deforestation can be successfully stopped only if the global community is prepared to compensate these countries for the opportunity cost in lost income. An important step is to create a better legal framework for forest conservation. The ongoing legislative process on the EU law against deforestation and forest degradation will be a model here for trade in raw materials that may involve forest destruction. Supply chains to and within Europe must then be designed to ensure that only products whose raw materials come from land already used for agriculture before 2022 are used. The current process, REDD+ (Reducing Emissions from Deforestation, and Forest Degradation and the role of conservation, sustainable management of forests, and enhancement of forest carbon stocks in developing countries), developed under the umbrella of the UN, will hardly achieve this goal in the future. It is too bureaucratic, too slow, and also lacks sufficient funding. In addition, funds only flow on a performance basis, based on clearly demonstrated CO_2 reductions.

To appreciate the urgent need to stop deforestation of the rainforests, we must realize that this destruction is ongoing and requires immediate action. Brazil's former President Bolsonaro viewed Brazil's rainforest as nothing more than untapped economically potential and promoted its deforestation with slash-and-burn. This brought the tipping point for the desertification of the Amazon rainforest ever closer. In Indonesia, there is a real danger that its 21 million hectares of peat swamp rainforests will be cleared in the next 10–15 years. Since they store 20 times as much carbon as normal rainforests, their destruction will also emit 20 times as much CO_2.

There are no reliable figures on the revenue losses emerging and developing countries would incur if deforestation were to stop. Older estimates were in the range of \$40–\$45 billion annually. The demand of the former Brazilian president, who was prepared to preserve the Amazon rainforest for \$10 billion annually, is of the same order of magnitude.

REDD+ has so far offered the countries of the global South $5 per metric ton of CO_2 averted. In the meantime, there are initial projects in which $10 have been agreed, which would amount to $47 billion if the deforestation process were to be completely stopped globally. If, in the spirit of the political scientist and philosopher Thomas Pogge, this money flow from the industrialized nations could be dedicated toward a basic income paid out in the emerging and developing countries, such an arrangement would certainly be widely approved by their populations. In the West, the argument that these funds would be misused by corrupt governments would no longer apply.

What does the global community gain from a treaty-based freeze and simultaneous compensation for lost revenue? Global CO_2 emissions would be reduced by 4.7 billion metric tons per year (or about 11% of global emissions), which is more than all of Europe emits annually. At the same time, the largest possible milestone would be set in the short term to preserve the unique tropical biodiversity.

In my view, a corresponding agreement with the countries of the tropics is possible in 4–5 years, but only on condition that the industrialized nations are prepared to provide around US$47 billion annually for this purpose. Under the new Brazilian president, Lula da Silva, it should be possible to end the deforestation and burning of the Amazon rainforest within 2–3 years.

There is no alternative way of achieving CO_2 savings of this magnitude in such a short time span and also in a socially acceptable way. Nor is there one that can be implemented so easily, because we basically have to do nothing to achieve this, we have to refrain from doing something by stopping logging immediately!

In the medium and long term, global CO_2 emissions can be reduced by up to 25% through further nature-based solutions such as reforestation, rewetting of the world's drained peatlands, humus enrichment, and long-term storage of CO_2 through widespread use of wood in the construction sector. Estimates suggest that the annual investment required to achieve this reduction will be between $200 and $300 billion by 2050 (see also chapter "Stop Rainforest Deforestation").

Reservations About Climate-Based Solutions

How can it be explained that politicians have over the last two decades paid very little attention to nature-based solutions? This is especially remarkable because such solutions can be promptly implemented, are reasonably cheap (certainly relative to the damages climate change will cause), and threaten no jobs but, on the contrary, would have highly positive employment effects for decades, especially in the emerging and developing countries.

In my opinion, a large proportion of politicians remain convinced that the Paris Agreement, limiting global warming to well below 2 °C, can only be fulfilled with technical innovation. Ideas for investing in nature-based solutions on the scale required are therefore very slow to get off the ground at this time. For the most part, nature-based solutions have so far been viewed as a simple, unhelpful offset mechanism for economy-based emissions. This view does not do justice to their great potential importance—also in regard to the preservation of biodiversity.

The prevailing policy stance is supported by various other arguments against nature-based solutions. With regard to afforestation, it has been repeatedly argued that, to have meaningful climate effects, the areas converted into forests would have to be so large that valuable agricultural land would be lost. This would indeed be unjustifiable, but it is not necessary. There are enough degraded rainforest areas that lie fallow and are still suitable for reforestation. Under the UN's Bonn Challenge initiative, countries in Africa, South America, and Asia, in particular, have identified some 200 million hectares (equivalent to about 200 billion trees) of degraded rainforest and pledged to restore tree-rich landscapes. Implementation failed, again due to lack of funding.

From time to time, the term "neocolonialism" appears in connection with afforestation in the tropics and subtropics. Without doubt, it must be ensured through the UN that the afforested areas remain the property of the countries of the global South or the relevant indigenous populations. The planting of about 350 billion trees, financed by the industrialized countries, would build substantial national wealth in the developing countries. A seedling costing at most $3 or $4 becomes a tropical tree worth $400–$500 in 20 years. If the rainforests are managed sustainably, this can create the basis for a functioning and profitable bioeconomy in the long term. Central to this, of course, is the preservation and protection of the rights of indigenous peoples. With its two platforms "The New York Declaration on Forests" and the "Bonn Challenge," the UN has built up enough expertise to address these legitimate concerns.

In addition to the accusation of neocolonialism, many climate activists are accused of "selling indulgences" on the motto: if you plant trees, you can continue business as usual. If, on top of that, compensation payments in the billions would have to be made to corrupt countries in order to protect the rainforests, any willingness to engage in a constructive discussion about such solutions disappears.

Such rejection fails to recognize the crucial role that investments in nature-based solutions can play in achieving the goals of the Paris Agreement and for preserving biodiversity. None of these measures should impede the social-ecological transformation of the economy, the reduction of greenhouse gas emissions, and the achievement of climate neutrality. But through them we can gain a time buffer that would make the socio-ecological transformation more socially acceptable and its political implementation much easier.

Last but not least, I believe that the industrialized nations are reluctant to take part in implementing nature-based solutions. As already mentioned, their role would consist in providing financial support to the poorer countries of the global South. This is not a high priority for them because they get more credit for their domestic emission reductions. Their reluctance then spreads to the countries of the South which—lacking financial resources from, and also role models in the richer North—prioritize other goals, such as achieving basic prosperity for their population, over "saving the world."

Such mutual demotivation fails to recognize, however, that nature-based solutions are a global asset in the fight against global warming. If investments in these areas are not financially secured through a climate treaty, they will remain irresponsibly absent, and the destruction will continue. Under no circumstances must this continue if we are to avoid climate collapse. Nature-based solutions such as those described are timely, socially acceptable, and extremely cost-effective measures compared to the threat of enormous damage.

How We Can Finance the Measures

Since the beginning of global climate protection policy, it has been impossible, year after year, to mobilize the political will to provide the financial resources needed to limit global warming to +1.5 °C. For politicians, investments in climate protection policy were and are primarily costs that must be minimized to please taxpayers.

As early as 2006, the economist Nicholas Stern, commissioned by the British government, and the management consultancy McKinsey presented, independently of each other, expert opinions showing that limiting global warming to the then target of +2 °C could only be achieved if by 2050 1% of the gross world product (at that time around $500 billion annually) were invested in climate protection. Failure to do so would mean having to cope with many times this amount in material damage later on. This fact was ignored by politicians around the world.

In 2021, Stern estimated that investments of 2% of gross world product were needed to limit global warming to +1.5 °C. He explicitly emphasized that he is not talking about costs, but about profitable investments: "It is not about adding costs to things we are already doing today. It's about investing in new things and in a radically different way." The same view is now held by other well-known institutions such as Morgan Stanley and McKinsey. No one can reliably estimate today the total volume of investment needed to secure the 1.5 degree target. It is likely, however, that the quoted 2% of gross world product will turn out to be rather optimistic.

In the coming decades, trillions of U.S. dollars will have to be invested in the transformation of today's fossil fuel energy systems in industrialized nations and in the development of climate-neutral energy systems in emerging and developing countries. Investments in the infrastructures of both the industrialized nations and the emerging and developing countries will be similarly high. At the same time, the now unbearable gap between rich and poor must be reduced both between and within countries. In addition, debt relief for developing countries in the order of $800 billion is needed if we are to kick-start private investment in climate protection in those countries.

Let the Polluter-Pays Principle Prevail

In addition to the unmanageable challenges outlined above, poor countries face the additional problem that, if global warming is to be limited to +1.5 °C, a large part of their known coal, oil, and gas reserves must remain in the ground. But how can this be achieved when fossil fuels often represent the overwhelming majority of the nation's wealth?

The answer: we will have to provide fair financial compensation to the countries of the South. And not merely because we benefit from limiting greenhouse gas emissions, but because the pressing issue of achieving climate justice obliges us to do so. After all, since industrialization, the material prosperity of rich countries has been based primarily on the unrestrained use of fossil fuels for energy production and on the exploitation of other non-renewable raw materials in today's emerging and developing countries. Taking these long-standing inequities into account, it becomes obvious that the industrialized countries must bear the cost of implementing climate-protective measures.

In energy production, in particular, a large part of the environmental costs were externalized (while the profits benefited the companies). No one should expect these costs to be borne by countries and populations that had no role in the relevant decisions and received no share of the created wealth.

Climate fairness requires that the richest 1% of the world's population, in particular, bear the "lion's share" of future costs or investments to cap global warming at +2 °C. They have built up their current fortunes through massive use of fossil fuels without paying for the true ecological costs of their conduct.

How Are These Gigantic Investments to Be Financed?

In the interests of intergenerational justice, consumer spending should not be financed by an expansion of government debt. In 2022, according to IMF, global government debt has reached 92% of gross world product. Total global debt rose to 238% of gross world product in the same period.

Future investments in climate protection as well as climate adaptation measures should be financed from government budgets, private businesses as well as donations from philanthropists. If these are insufficient to jointly finance the necessary climate protection policies, then additional investment measures for climate protection are justified, also through further indebtedness.

Governments should first look at possible shifts in their budgets. This should include eliminating direct and indirect subsidies of fossil fuels. According to the IEA in Paris, direct subsidies have averaged $500 billion annually over the past 10 years. The International Monetary Fund (IMF) estimates that in 2022 direct subsidies came to $1.3 trillion (up from $0.4 trillion in 2015) while indirect subsidies amounted to $5.7 trillion (up from $4.1 trillion in 2015). Indirect subsidies include, among other things, the externalized costs associated with burning fossil fuels (Statista, 2023). Budget shifts might also involve closing tax loopholes for high-income earners and reducing the $1500–$1800 billion in military budgets around the world. The latter option, however, is likely to be unfeasible for the next 5–10 years because the current geopolitical realignment is leading to massive resistance to any reductions in military budgets and, indeed, to strong pressures to increase military spending.

Since we have only a narrow window of opportunity to set the course for tolerable climate change with a certain degree of reliability, the focus in the short and medium term must be on taxation.

The initial focus here is on a CO_2 tax. Such a CO_2 tax really is equivalent to a reduction in fossil fuel subsidies because it internalizes the cost that fossil-fuel burning imposes on third parties and the planet. As a steering instrument, it is to be gradually increased in the medium and long term. Since even in Germany it probably burdens more than a third of all people disproportionately and thus not in a socially acceptable way, the complete reimbursement for this part of the population is planned. In this respect, this source of revenue will also be reduced by this portion and, due to its gradual introduction, will only come into full effect years later anyway.

Let's therefore take a look at two types of tax that need to come into focus worldwide because they are significant both in terms of their potential yield and allow for an urgently needed social correction: the financial transaction tax and the inheritance tax. Both taxes should flow into new sovereign wealth funds to be established, which may be used exclusively for climate and infrastructure investments.

To reiterate: The 2008 global financial crisis and the Corona pandemic have driven all nations of the world, almost without exception, deeper into debt. Now, trillions of U.S. dollars in investments in climate protection (prevention as well as adaptation) are due over the next three decades. It would be socially irresponsible and politically unfeasible to try to finance these investments through general tax increases. In the interest of future generations, however, countries must not continue to run up debts. Let us now look at how such a sovereign wealth fund can be built.

Financial Transaction Tax

A financial transaction tax is a minimum-turnover or value added tax (VAT) in the financial sector. Anyone who makes everyday goods or other purchases today pays between 7% and 22% VAT, depending on the country. Anyone who buys shares or bonds, invests, or speculates in high-frequency trading does not pay a cent of VAT.

A 2015 report by the German Institute for Economic Planning concluded that a tax of 0.1% on trades in stocks and bonds, and of 0.01% on high-frequency trading would generate around 40 billion euros annually in additional tax revenue for the German government. Extrapolated to the global financial system, governments could collect a high three-digit billion amount (US$) annually with this tax.

I find it regrettable that politicians around the world are still shying away from this potentially profitable and, above all, socially acceptable source of revenue. In democracies, it would meet with great approval if voters were properly informed about the elimination of a tax injustice that has existed for decades. In the meantime, Great Britain and France have introduced this tax on shares and bonds even in the amount of 0.5%, although the entire derivatives sector has been left out.

The world financial crisis of 2008 showed us how far the financial world had distanced itself from the real economy and how, through the extent of its speculative transactions, it had brought not only itself but, above all, the real economy to the brink of global collapse. Trillions of U.S. dollars of taxpayers' money had to be used to save the global financial system then. It is therefore time to finally include this area comprehensively in the sales tax system. All countries are highly indebted today, so it should be possible to convince politicians worldwide to introduce this tax.

Inheritance Tax

According to the CS Global Wealth Report, 1.1% of the world's population own around 45% of its private wealth (UBS, 2023).

This concentration of wealth is due to two main causes: income taxation favoring the super-rich, and extremely low prices of fossil fuels with, more generally, the ruthless exploitation of finite raw materials at sometimes dirt-cheap prices in emerging and developing countries. A timely introduction of "ecologically true prices" would have significantly reduced the concentration of wealth and today's environmental problems. Even if introduced now, far too late, such true pricing can still make a substantial contribution to slowing climate change by reducing emissions. And some of the past mispricing can be corrected through a comprehensive inheritance tax which would recoup some of the undue benefits derived from the past underpricing of fossil fuels, without which the accumulation of today's extreme fortunes would not have been possible. Such an inheritance tax would also reduce the existing extreme wealth concentration, which in my view represents a ticking time bomb in terms of the stability of societies.

Conclusion

Without incurring new debt, the democracies can afford climate protection policies as well as the necessary infrastructure investments if the political course is quickly set for the introduction of a financial transaction tax and a reform of the inheritance tax.

To limit global warming to +1.5 °C, the global community must become climate neutral by 2050. There are only 26 years left to meet this challenge. In the short term, the greatest obstacle will be the provision of sufficient financial resources, in addition to public acceptance.

In democracies, legislative processes to introduce a financial transaction tax or to reform the inheritance tax certainly take a period of at least 4–5 years. In view of the urgency of combating climate change, bridge financing should therefore be considered. On the world's financial markets, there is investment-seeking capital in the trillions that would be available in the short term and would be satisfied with a minimum interest rate. Essential condition: investment security. Governments could set up sovereign wealth funds that, with government-backing, would borrow on world markets. Future debt service—interest and repayment—would be financed from the two types of taxes once these will have been introduced. In this way, years could be gained to immediately initiate nature-based measures and the urgently needed restructuring or development of the energy sector in the global South, for example. Time is of the essence when it comes to achieving these goals.

In order to involve the private sector in the financing of future investments to a much greater extent than in the past, governments must set appropriate framework conditions. These include planning security, openness to technology, reduction of bureaucracy, debt relief for developing countries or government protection of investments in these countries, and acceptance of an appropriate return on these future investments.

Why Mobilization of Civil Society Is Needed

The scenarios and consequences of +3 °C global warming highlighted in the contributions to this volume suggest urgently needed additions to the Paris climate protection measures. So far, politicians have seen little reason to show the citizens of their countries the seriousness of the situation and to appeal for understanding for more comprehensive climate protection measures.

Therefore, a majority of the world's population is still not aware of the dramatic consequences of +3 °C global warming. For far too long the dangers of climate change have been downplayed in the media or displaced to distant regions of the world. The melting of glaciers or the extinction of the polar bear may have caused feelings of consternation, but this did not allow for a realistic assessment of the real economic and biophysical threats that people are facing today.

In July 2022, leading climatologists around lead author Luke Kemp of Cambridge University published an article in the journal *PNAS* entitled "Climate Endgame." In this report, they accuse the Intergovernmental Panel on Climate Change of having downplayed for years the impending consequences of a temperature increase of "3 degrees and more." Furthermore, they call on the Intergovernmental Panel to submit a special report and to address the crucial question: "Can human-induced climate change lead to a global collapse of societies or even the extinction of humanity?"

A public discourse about the true consequences of a global warming of +3 °C, which will threaten the existence of all humankind, has never been conducted with the necessary clarity. This has allowed governments to pursue half-hearted climate policies with impunity for a good 20 years, leading to a dangerous increase in global CO_2 emissions from 22 billion metric tons per year in 1990 to just under 40 billion today (and still rising!). In my opinion, comprehensive education of the world's population about the devastating consequences that decades of completely inadequate climate protection policies will cause would have led to the development of massive resistance against the climate ignorance of politics, at least in the democracies.

In 2018, the great commitment of Swedish climate activist Greta Thunberg and the global Fridays for Future movement initiated by her school strikes had managed to mobilize parts of civil society for the first time. Despite Thunberg's media omnipresence, her active participation in environmental summits, UN meetings, and exclusive talks with governments, however, this has so far not been enough to put politicians under decisive pressure to act. To make matters worse, the Corona pandemic took the debate on climate protection measures out of the focus of the world's media—despite an increase in extreme weather events and catastrophes.

Nonetheless, I remain convinced that the transformation to climate neutrality will not succeed without broad popular support, despite all the declarations of intent by politicians and industry. I find it incomprehensible, even negligent, that politicians around the world claim to believe they can push through the suite of measures that are indispensable for achieving the goals, some of which are drastic, under enormous time pressure. And this without a comprehensive social discourse with and in civil society about the devastating consequences of unchecked climate change.

If policymakers stick to their guns, foundations, in cooperation with companies and wealthy individuals, should initiate science-based dialogues with the public about the consequences of global warming of +3 °C. Without such a discourse, there is a high probability that we will lose further valuable years in the fight against climate change.

There have been few highlights in climate policy over the past 25 years. One highlight was undoubtedly the Paris Agreement in which, for the first time, 196 countries acknowledged the existence of human-made climate change and promised to do everything in their power to limit global warming to well below +2 °C. Since then, however, once again far too little has been done.

Climate policy needs a new global proof of commitment, showing that meaningful progress is really being made. One such milestone would be a binding agreement under international law with the emerging and developing countries to stop the deforestation of tropical and subtropical rainforests within the next 3–4 years. This would achieve global CO_2 emissions equivalent to Europe becoming climate-neutral by 2026 at the latest. At the same time, it would be the greatest possible step toward preserving biodiversity. The estimated annual cost would be "only" around $45 billion.

Realizing an equivalent step—a reduction of five billion metric tons of CO_2 emissions in addition to the planned measures of the climate treaty—via the global economy requires much more time, is much more expensive, and would be politically difficult to implement.

The financial resources required to realize this project can only be raised if the G20 countries make a binding commitment to do it. At the 2020 World Economic Forum in Davos, major global corporations declared their intention to commit financially to the preservation of rainforests without compromising their companies' own efforts toward climate neutrality. Were these corporations to keep their promise, this would substantially lighten the burden of such a commitment on the G20 countries.

From my perspective, stopping the deforestation of the rainforests would have the greatest chance of being realized quickly if a country such as Germany took the lead, as it has done with renewable energies. In line with its economic strength, Germany would have to provide around $2 billion annually and mandate the contributors to the UN platform "The New York Declaration on Forests" to negotiate treaties with countries such as Indonesia, Ecuador, Peru, or Honduras to stop the deforestation of their rainforests. At the same time, this initiative should be put on the agenda of the G7 and eventually the G20 to start negotiations, especially with Brazil, on the protection of the Amazon rainforest. Success in such an endeavor would set a milestones in climate protection and biodiversity policy and would demonstrate to the world that significant progress in climate protection can finally be achieved.

It is therefore up to governments to quickly seek dialogue with their own populations about the dangers posed by climate change and by a +3 °C warmer world. The experience of the last 10 years with the consequences of increasing global warming leads to fears that the tipping point for many societies could occur well before +3 °C is reached. Together we still have the power to prevent this from happening!

References

Latif, M. (2023). *Es ist ausgeschlossen, die Erderwärmung auf 1,5 Grad zu begrenzen*. Focus online https://www.focus.de/earth/experten/klimaforscher-mojib-latif-mojib-latif-schliesst-1-5-grad-grenze-der-erderwaermung-aus_id_189378460.html. Accessed 7 Aug 2023.

Levermann, A. (2011). *Die Gefahr der Kippelemente*. Frankfurter Allgemeine Zeitung via FAZ.net https://www.faz.net/aktuell/politik/energiepolitik/atomunfall-und-klimawandel-die-gefahr-der-kippelemente-1627212.html. Accessed 7 Aug 2023.

McKinsey & Company. (2020). *Net-zero Europe*. https://www.mckinsey.com/capabilities/sustainability/our-insights/how-the-european-union-could-achieve-net-zero-emissions-at-net-zero-cost. Accessed 30 Mar 2021.

Misereor. (2018). *Rohstoffe für die Energiewende – Menschenrechtliche und ökologische Verantwortung in einem Zukunftsmarkt*. https://www.misereor.de/fileadmin/publikationen/studie-rohstoffe-fuer-die-energiewende.pdf. Accessed 3 Feb 2022.

Schellnhuber, H.-J. (2015). *Selbstverbrennung. Die fatale Beziehung zwischen Klima, Mensch und Kohlenstoff*. Bertelsmann.

Stark, A. (2016). *Americans used more clean energy in 2016*. Released by Lawrence Livermore National Laboratory https://www.llnl.gov/news/americans-used-more-clean-energy-2016. Accessed 14 Sep 2023.

Statista. (2023). https://de.statista.com/infografik/31006/volumen-der-weltweiten-subventionen-fuer-fossile-brennstoffe/. Accessed 19 Oct 2023.

Tooze, A. (2023). *Manchmal muss man etwas wagen.* ZEIT online https://www.zeit.de/2023/29/adam-tooze-usa-weltwirtschaft-klimaschutz-investitionen/seite-2. Accessed 7 Aug 2023.

UBS. (2023). *Global wealth report 2023. The global wealth pyramid 2022.* https://www.ubs.com/global/en/family-office-uhnw/reports/global-wealth-report-2023/exploring.html. Accessed 19 Oct 2023.

Wallace-Wells, D. (2019). *The uninhabitable Earth: Life after global warming.* Ludwig Publishing House.

Index